Progress in Inflammation Research

Forthcoming titles:

NPY Family of Peptides in Immune Disorders, Inflammation, Angiogenesis and Cancer, G.Z. Feuerstein, Z. Zukowska (Editors), 2005

Complement and Kidney Disease, P. F. Zipfel (Editor), 2005

Chemokine Biology: Basic Research and Clinical Application, Volume I: Immunobiology of Chemokines, K. Neote, L. G. Letts, B. Moser (Editors), 2005

Chemokine Biology: Basic Research and Clinical Application, Volume II: Pathophysiology of Chemokines, K. Neote, L. G. Letts, B. Moser (Editors), 2005

The Hereditary Basis of Rheumatic Diseases, R. Holmdahl (Editor), 2005

(Already published titles see last page.)

Turning up the Heat on Pain: TRPV1 Receptors in Pain and Inflammation

Annika B. Malmberg
Keith R. Bley

Editors

Birkhäuser Verlag
Basel · Boston · Berlin

Editors

Annika B. Malmberg
Elan Pharmaceuticals
800 Gateway Boulevard
South San Francisco, CA 94080
USA

Keith R. Bley
NeurogesX, Inc.
981F Industrial Road
San Carlos, CA 94070
USA

Library of Congress Cataloging-in-Publication Data

Turning up the heat on pain : TRPV1 receptors in pain and inflammation / Annika B. Malmberg, Keith R. Bley, editors.
p. ; cm. -- (Progress in inflammation research)
Includes bibliographical references and index.
ISBN 978-3-7643-7080-0 (alk. paper)
ISBN 3-7643-7080-7 (alk. paper)
1. Nociceptors. 2. Pain. 3. Inflammation. 4. Inflammation--Mediators. I. Malmberg, Annika B., 1966–
II. Bley, Keith R. III. PIR (Series)

QP451.4.T87 2005
612.8'8--dc22

2005048122

Bibliographic information published by Die Deutsche Bibliothek
Die Deutsche Bibliothek lists this publication in the Deutsche Nationalbibliografie; detailed bibliographic data is available in the internet at http://dnb.ddb.de

ISBN 3-7643-7080-7 Birkhäuser Verlag, Basel – Boston – Berlin
ISBN 978-3-7643-7080-0 Birkhäuser Verlag, Basel – Boston – Berlin

Printed on acid-free paper produced from chlorine-free pulp. TCF ∞
Cover design: Markus Etterich, Basel
Cover illustration: Schematic diagram of TRPV1 activation and ionic permeation

ISBN 3-7643-7080-7
ISBN 978-3-7643-7080-0

ISBN 3-7643-7379-2 (eBook)

9 8 7 6 5 4 3 2 1

www.birkhauser.ch

Contents

Part IV: Vanilloid receptor involvement in disease states

Part V: Therapeutic potential of vanilloid agonists and antagonists

List of contributors

Ole K. Andersen, Center for Sensory-Motor Interaction, Laboratory for Experimental Pain Research, Aalborg University, Fredrik Bajers Vej 7, D3, DK-9220 Aalborg, Denmark

Lars Arendt-Nielsen, Center for Sensory-Motor Interaction, Laboratory for Experimental Pain Research, Aalborg University, Fredrik Bajers Vej 7, D3, DK-9220 Aalborg, Denmark; E-Mail: LAN@smi.auc.dk

Peter J. Barnes, Respiratory Pharmacology Group and Airway Disease Section, National Heart & Lung Institute, Imperial College, Dovehouse Street, London SW3 6LY, UK; E-Mail: p.j.barnes@imperial.ac.uk

Maria G. Belvisi, Respiratory Pharmacology Group, Airway Disease Section, National Heart & Lung Institute, Faculty of Medicine, Imperial College, Dovehouse Street, London SW3 6LY, UK; E-Mail: m.belvisi@imperial.ac.uk

Keith R. Bley, NeurogesX, Inc., 981F Industrial Road, San Carlos, CA 94070, USA; E-Mail: kbley@neurogesx.com

Peter M. Blumberg, Laboratory of Cellular Carcinogenesis and Tumor Promotion, National Cancer Institute, Building 37, Room 4048, 37 Convent Drive MSC 4255, Bethesda, MD 20892-4255, USA; E-Mail: blumberp@dc37a.nci.nih.gov

Derek C. Braun, Laboratory of Cellular Carcinogenesis and Tumor Promotion, National Cancer Institute, Bethesda, MD 20892, USA;
E-Mail: braund@mail.nih.gov

Francisco Cruz, Department of Urology, Hospital S. João and Faculty of Medicine of Porto, Alameda Prof Hernâni Monteiro, 4200–076 Porto, Portugal;
E-Mail: cruzfjmr@med.up.pt

Paulo Dinis, Department of Urology, Hospital S. João and Faculty of Medicine of Porto, Alameda Prof Hernâni Monteiro, 4200–076 Porto, Portugal;
E-Mail: padioli@mail.telepac.pt

Peter Holzer, Department of Experimental and Clinical Pharmacology, Medical University of Graz, Universitätsplatz 4, A–8010 Graz, Austria;
E-Mail: peter.holzer@meduni-graz.at

Noemi Kedei, Laboratory of Cellular Carcinogenesis and Tumor Promotion, National Cancer Institute, Bethesda, MD 20892, USA;
E-Mail: kedein@mail.nih.gov

Jozsef Lazar, Laboratory of Cellular Carcinogenesis and Tumor Promotion, National Cancer Institute, Bethesda, MD 20892, USA;
E-Mail: lazarjo@mail.nih.gov

Annika B. Malmberg, Elan Pharmaceuticals, 800 Gateway Boulevard, South San Francisco, CA 94080, USA; E-Mail: Annika.Malmberg@elan.com

Vladimir Pavlyukovets, Laboratory of Cellular Carcinogenesis and Tumor Promotion, National Cancer Institute, Bethesda, MD 20892, USA;
E-Mail: pavlyukv@mail.nih.gov

Larry V. Pearce, Laboratory of Cellular Carcinogenesis and Tumor Promotion, National Cancer Institute, Bethesda, MD 20892, USA;
E-Mail: pearcel@mail.nih.gov

James D. Pomonis, Algos Therapeutics, 1246 University Ave W, Suite 205, St. Paul, MN 55104, USA; E-Mail: jpomonis@algosinc.com

Ruth A. Ross, School of Medical Sciences, Institute of Medical Sciences, University of Aberdeen, Foresterhill, Aberdeen, AB25 2ZD, Scotland, United Kingdom;
E-mail: r.ross@abdn.ac.uk

Zoltán Sándor, Department of Pharmacology and Pharmacotherapy, Faculty of Medicine, University of Pécs, H–7624 Pécs, Hungary

Carlos Silva, Department of Urology, Hospital S. João and Faculty of Medicine of Porto, Alameda Prof Hernâni Monteiro, 4200–076 Porto, Portugal;
E-Mail: carsil@mail.telepac.pt

Arpad Szallasi, Department of Pathology, Monmouth Medical Center, 300 Second Avenue, Long Branch, NJ 07740, USA; E-Mail: aszallasi@sbhcs.com

János Szolcsányi, Department of Pharmacology and Pharmacotherapy, University Medical School of Pécs, Neuropharmacological Research Group of the Hungarian Academy of Sciences, Szigeti u. 12, H–7624 Pécs, Hungary;
E-Mail: janos.szolcsanyi@aok.pte.hu

Makoto Tominaga, Section of Cell Signaling, Okazaki Institute for Integrative Bioscience, National Institutes of Natural Sciences, Okazaki 444–8787, Japan;
E-Mail: tominaga@nips.ac.jp

Kenneth J. Valenzano, Amicus Therapeutics, 6 Cedarbrook Drive, Cranbury, NJ 08512, USA; E-Mail: valenzano@amicustherapeutics.com

Katharine Walker, Beatson Institute for Cancer Research, Garscube Estate, Switchback Road, Glasgow G61 1BD, Scotland, UK; E-Mail: k.walker@beatson.gla.ac.uk

Janet Winter, Novartis Institute for Medical Sciences, 5 Gower Place, London WC1E 6BN, United Kingdom;
E-Mail: Janet.Winter@Novartis.com, WinterJa@tiscali.co.uk

Preface

Despite tremendous advances in the understanding of the sensory nervous system which have accompanied the recent explosive growth of the neurosciences, remarkably few innovative medicines directed towards pain and inflammation are available. Indeed, many patients are still prescribed analgesic and anti-inflammatory medications that were identified long ago as components of herbal remedies. Similarly, potential new medicines in clinical evaluation based on capsaicin and the capsaicin receptor are both grounded firmly on folk traditions and yet rely upon the most contemporary techniques of drug discovery and delivery.

The first formal report of the pain-relieving properties of capsaicin appeared in 1850 [1]. However, for centuries before this, capsaicin-containing extracts had been used as folk medicines in cultures with access to pepper plants, much in the same way as poppy or willow-bark extracts were. Despite widespread use, it was not until 1878 that the selective action of capsaicin on the sensory nervous system was recognized [2]. In Chapter 1 of this volume, Janos Szolcsányi reviews this early research, which culminated with the seminal studies of Nicholas Jansco and his colleagues in Hungary in the 1940s. Since then, capsaicin and related vanilloid compounds have played a prominent role in analgesia and inflammation investigations because of their ability to selectively activate a subpopulation of sensory neurons and produce sensations of pain and localized erythema. The widespread production of pungent molecules such as capsaicin by plants has recently been explained in terms of advantages with respect to seed dispersal and deterrence of ambulatory seed eaters [3].

Since 1997 there has been profound interest in capsaicin and pungent vanilloids because of the cloning of a specific ion channel that mediates the effect of this class of compounds [4]. Specifically, it has been found that pungent vanilloids mediate their effects by selective agonism of an ion channel, the transient receptor potential vanilloid receptor 1 (TRPV1; or, according to older nomenclature, VR1). TRPV1 is a ligand-gated, non-selective cation channel expressed preferentially in small-diameter, primary afferent neurons, including nociceptive sensory nerves. In addition to being activated by capsaicin and related vanilloids, TRPV1 responds to heat and extracellular acidification, and will integrate simultaneous exposures to these stimuli.

The aim of this volume is to summarize recent insights into the role of TRPV1 in pain and inflammation, and how discuss how modulation of this receptor may lead to important advances in analgesic and anti-inflammatory drug development. The book contains chapters relating to five themes: (1) historical perspectives on capsaicin and its receptor; (2) the molecular and cellular properties of TRPV1; (3) the pharmacology and physiology of TRPV1; (4) evidence for the involvement of TRPV1 in diseases and syndromes; and (5) the therapeutic potential of TRPV1 agonists and antagonists. It is our hope that this book will provide an integrated overview of the actions of classes of compounds that have been used extensively – for a wide range of reasons – by cultures around the world. Possibly more importantly, we hope that this book will also provide insights into the prospects for the therapeutic potential of TRPV1 activation or inhibition, particularly in various painful or inflammatory conditions.

References

1 Turnbull A (1850) Tincture of capsaicin as a remedy for chilblains and toothache. *Dublin Free Press* 1: 95–96

2 Hőgyes A (1878) Beiträge zur physiologischen Wirkung der Bestandteile des *Capsicum annuum*. *Arch Exp Pathol Pharmakol* 9: 117–130

3 Tewksbury JJ, Nabhan GP (2001) Seed dispersal. Directed deterrence by capsaicin in chilies. *Nature* 412: 403–404

4 Caterina MJ, Schumacher MA, Tominaga M, Rosen TA, Levine JD, Julius D (1997) The capsaicin receptor: a heat-activated ion channel in the pain pathway. *Nature* 389: 816–824

Part I
Historical perspective on capsaicin and its receptor

Hot peppers, pain and analgesics

János Szolcsányi

Department of Pharmacology and Pharmacotherapy, University Medical School of Pécs, Neuropharmacological Research Group of the Hungarian Academy of Sciences, Szigeti u. 12, H-7624 Pécs, Hungary

Introduction

The hot pepper, or more precicely capsaicin, the hot component of chilli peppers of the Capsicum family, was the key that opened new horizons in the field of analgesics and the pharmacology of nociceptors. It is surprising that during the last century real breakthroughs in drug development in the field of analgesics were not achieved. Remarkably, medical doctors at the end of the 19th century used analgesics that act on opioid receptors and cyclooxygenase (COX) enzymes, just as contemporary physicians do. Until quite recently, no chapters on drugs acting on nociceptive sensory neurons could be found in modern pharmacology textbooks, in striking contrast to several chapters on the sympathetic or parasympathetic nervous system, for which a continuously growing number of drug targets has been revealed. The roots of this flourishing pharmacology of the autonomic nervous system can be traced back to the analysis of the mode of action of naturally occurring drugs such as nicotine (Langley and Dickinson, 1889[1]), ergotoxine (Dale, 1906[2]) and atropine (Loewi, 1921[3]). These brilliant studies on alkaloids provided the first evidence for the very existence of mediators of neurohumoral transmission and raised clues for drug development (for references, see [1]). An experimental study by Andreas Hőgyes [2], published during this pioneering period of pharmacology, was the first attempt to describe the actions of capsaicin under experimental conditions. These observations led Hőgyes to conclude that this compound acts mainly on sensory nerves ('Capsicol hauptsächlich auf die sensiblen Nerven wirkte'). In traditional treatments of arthritic diseases, capsaicin was applied topically as a 'counterirritant', but its pharmacology remained neglected for seven decades [3, 4].

Early discoveries by Nicholas Jancsó

Capsaicin desensitization

During and soon after the Second World War, Nicholas (Miklós) Jancsó in Hungary was engaged in studying the storage of macromolecules in the reticuloendothelial

Turning up the Heat on Pain: TRPV1 Receptors in Pain and Inflammation, edited by Annika B. Malmberg and Keith R. Bley

system in inflammation. In this context he started to use capsaicin and made a serendipitous observation, which indicated that this pain-producing irritant produces a completely new type of analgesic effect.

Jancsó routinely pretreated mice with high doses of histamine to achieve histamine desensitization in order to reveal the role of histamine in inflammatory processes. By replacing histamine with capsaicin – which he assumed to be a histamine-releasing compound [5] – Jancsó observed a unique type of behavioural unresponsiveness of these animals whereby 'pain sensory receptors' or 'pain receptors' became 'completely insensitive to the strongest known chemical stimuli or caustics' without any obvious change in their responsiveness to physical stimuli (electrical, mechanical and noxious heat) [6]. The effects of all tested irritants were abolished, including those to formaldehyde, KCl, veratridine, choracetophenone, allylamine and nicotine [5, 6]. Using nicotine instead of capsaicin he described another type of desensitization that was chemospecific and suggested to be mediated by a pharmacological receptor: "nicotinic stimulants attach themselves to a specific receptor structure for reaching similarity with the so-called acetylcholine receptor of the ganglionic synapse" [7].

In the effect of capsaicin-induced desensitization – which he always used as a descriptive term – Jancsó emphasized the importance of the acylamide linkage of the tested molecule, since similar desensitization was achieved with piperine, the non-vanilloid acylamide ingredient of black pepper, but not with the vanilloid ketone of zingerone or mustard oil. Nevertheless, he never proposed a 'specific receptor structure' for actions of capsaicin. It is important to emphasize that these sensory effects of capsaicin desensitization were revealed well before the discovery of the distinction between mechanonociceptors and polymodal nociceptors (i.e. that the signalling mechanism for chemically evoked pain differs from that of mechanically evoked pain). Thus mechanisms underlying the phenomenon of capsaicin-induced desensitization, however, remained obscure. Furthermore considerable space remained for doubts and challenges ('...invalidates the claim of Jancsó...') [8], since – although being a classical pharmacologist who invented brilliant histological techniques in the fields of the reticuloendothelial system and kidneys [5] – Jancsó was reluctant to use statistical analyses (e.g. means with standard errors) and usually documented his results with photos, data from individual experiments and qualitative descriptions.

Neurogenic inflammation

After capsaicin pretreatment, not only was the nociceptive effect of irritants abolished, but their inflammatory actions as well. This observation led Jancsó to the reasonable conclusion that pain-producing chemical agents evoke neurogenic inflammation by releasing a putative mediator from the pain-sensory nerve endings [5–7]. He thought that this 'neurohumor' was bradykinin [1].

The first evidence for neurogenic inflammation was described by Ninian Bruce in 1910[4] using mustard oil, which did not evoke plasma extravasation after sensory denervation. He explained this response as being a part of the phenomenon of antidromic vasodilatation discovered by Stricker (1876)[5] and characterized by Bayliss (1901, 1923)[6], and suggested that both of these responses are mediated through axon reflexes (for references, see [9–12]). This field, however, remained abandoned, except the axon reflex flare, which was thoroughly analysed by Thomas Lewis in a series of publications (for references, see [9–14]). Revival of this field emerged simultaneously with the capsaicin era [12–14] after the first direct evidence for neurogenic inflammation was obtained by electrical stimulation of capsaicin-sensitive nerves [9].

Thermoregulation

Systemic application of histamine or capsaicin induced a fall in body temperature in mice and rats, which had been used by Jancsó to determine the level of respective desensitization [5]. Being a close coworker in his capsaicin studies, I conducted several series of experiments involving intracerebral microinjections and heating the preoptic area; these led to the conclusions that capsaicin stimulates and desensitizes not only peripheral sensory nerve terminals but also the central warmth sensors of the preoptic area [15, 16]. In further studies it turned out that capsaicin pretreatment prevents for months not only physiological [15] but also behavioral thermoregulation in a warm environment, but not against cold [4, 17, 18]. These impaired functions were accompanied by similar long-lasting ultrastructural changes in small neurons of the hypothalamic preoptic area, as well as in B-type of neurons in dorsal root ganglia [19–21]. These combined functional and morphological data, obtained more than three decades ago, provided the first examples both for an action of capsaicin in the central nervous system and for the regulatory function of these neurons affected by capsaicin exposure [16, 18, 21].

Prediction of the capsaicin receptor

Nicholas Jancsó published no full paper on capsaicin in well-recognized, refereed journals. Completion of our common works in this field resulted in four posthumous publications [9, 15, 16, 22] and a review for a conference book, which I coauthored (according to his views) with his wife Aurelia Jancsó-Gábor [23].

Several questions remained unanswered, however; particularly how could this compound be neuroselective when nociceptive and inflammatory effects of all noxious chemicals seemed to be abolished for weeks by capsaicin pretreatment (at least after instillation into the eye). By the late 1960s, data from four different experimental approaches, which I conducted together with Aurelia Jancsó-Gábor and Ferenc Joó, convinced me that capsaicin could be a lead molecule for sensory pharmacology, similar to the role played by ergotoxine or atropine in the field of autonomic nervous system pharmacology. The approaches were as follows:

1 In non-mammalian species like the frog, chicken or pigeon, a nocifensive reaction was not evoked by capsaicin [9]. Putting frogs into a 1% capsaicin solution did not elicit reactions or an impairment in behavior, and after a month the apparently healthy animals reacted to acid, like the controls.
2 The loss of sensation evoked by chemical agents was highly selective in humans after topical capsaicin pretreatment. Testings were conducted on ourselves by putting the tongue 10 times into a 1% capsaicin solution. Sensory thresholds to different stimuli were tested before and after the treatment (Fig. 1) [18, 24]. Note that the burning pain sensation to capsaicin, mustard oil and zingerone was abolished but the sensation of taste stimuli and chemical stimulation of cold receptors by menthol remained unimpaired. These personal experiences suggested that only the pain-sensory nerve endings lose their chemosensitivity. Likewise, temperature discrimination was impaired in the warm range but not in the cold range. These changes lasted for several hours but were fully reversible.
3 A clear structure-activity relationship for vanilloid compounds was obtained in behavioral studies. It turned out that pungency is not proportional to the desensitizing effect of capsaicin analogs or, in other words, the pharmacophores for the excitatory and blocking actions of the vanilloids were partially different. On the basis of these results the existence of a 'capsaicin receptor' was postulated, and some crucial requirements in vanilloid structures for their stimulatory and desensitizing effects were defined [25, 26].
4 Complementary to the long-lasting functional impairment induced by capsaicin, ultrastructural studies revealed long-lasting mitochondrial damage of B-type sensory neurons in trigeminal, nodosal and dorsal root ganglia, as well as in small-type neurons of the preoptic area. The characteristic changes were highly selective. Ultrastructural features of A-type cells, sympathetic pre- and postsynaptic elements, Schwann cells, glial cells and epithelial cells all remained unchanged [19–21].

Breakthroughs in the late 1970s

In spite of the fact that the sensory blockage induced by capsaicin and direct evidence for neurogenic inflammation were published for the first time in a widely distributed high-ranking journal in 1967 [9], this paper and additional publications relating to capsaicin remained abandoned for about 10 years. Figure 2 shows the number of papers published per year on capsaicin. A full list of papers until the end

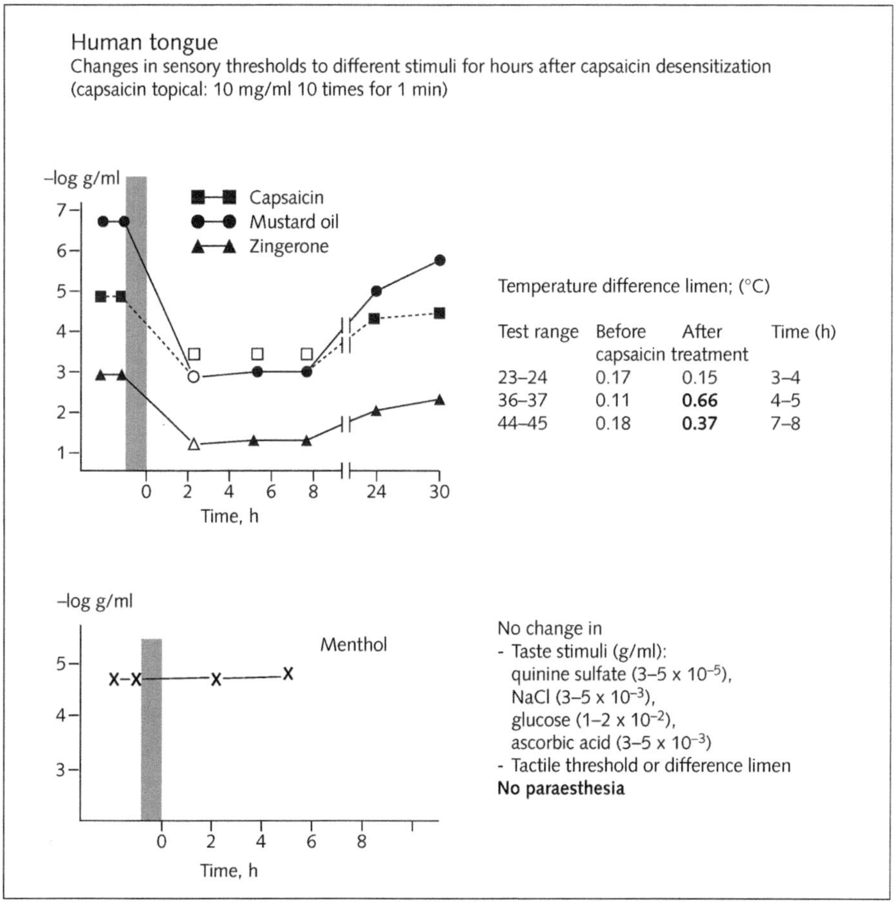

Figure 1
Effect of capsaicin desensitization of the human tongue on sensory thresholds to different stimuli. Open symbols indicate subthreshold concentrations. At the shaded column the tongue was put ten times for one minute in a 1% capsaicin solution. Modified figure from [18] with permission, see also [24].

of the 1970s, including abstracts of N. Jancsó, was collected in my review from 1982 [4] and recent figures were obtained from the PubMed database, where at present under the keyword 'capsaicin' more than 7000 articles are listed. According to my judgement, the following six different types of breakthrough discussed below and summarized in chronological order in Table 1 thrust capsaicin into the limelight of interest.

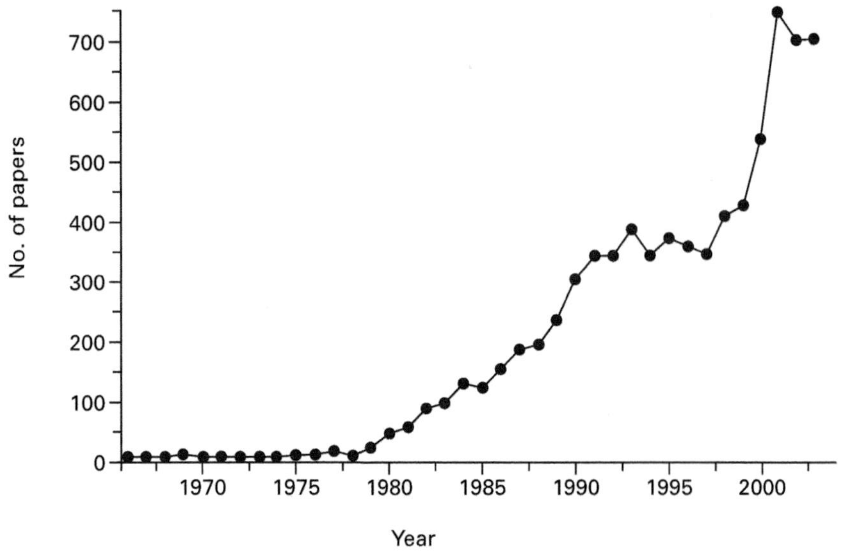

Figure 2
Number of papers published per year on capsaicin during the period 1966–2003.

Breakthrough 1

Single-unit recordings from cutaneous nerves revealed that there is a substantial proportion of C-afferents, which respond to high mechanical stimuli, noxious heat stimuli and irritants applied to intact skin. Activation of these C-polymodal nociceptors in the cat's saphenous nerve collided with the slowest-conducting (C_2) peak of the compound action potential, leaving unchanged its faster first component [27]. This evidence was verified also for close arterial injection of capsaicin, and the thermodependence of the capsaicin response was shown by complete blockade of burst activity of the rat's saphenous nerve by cooling the receptive field of injection to 20–22 °C [24]. Single-unit recordings from the great auricle nerve of the rabbit (which I performed in Ed Perl's laboratories during 1977–1978) [28], then from the rat, cat, monkey and humans in different laboratories, proved unequivocally that capsaicin excites with high selectivity the C-polymodal nociceptors and subsequently desensitizes them to their natural stimuli [29–32]. After close arterial injection, the responsiveness of the C-polymodal nociceptors to bradykinin, noxious heat or mechanical stimuli was diminished or abolished for more than 30 min but blockade

Table 1 - Breakthroughs in capsaicin research of the late 1970s

1	Selective action of capsaicin on C-polymodal nociceptors [24, 28]
2	Cell death of B-type neurons and irreversible loss of neurogenic inflammation [37]
3	Discovery of sensory-efferent responses of capsaicin-sensitive nerve endings in smooth-muscle preparation *in vitro* [47, 48]
3	Depletion of substance P from sensory neurons [53]
5	Substance P released from capsaicin-sensitive nerve endings induces neurogenic inflammation [54]
6	Selective thermal and chemical analgesia induced by intrathecal capsaicin application [58]

to one type of stimulus was often matched with unchanged responses of the same unit to the other stimuli. This observation showed clearly that, in these experiments, transduction processes were impaired at the sense organs without blockade of the axonal conduction [30, 33]. Similar enhanced thresholds of C-polymodal nociceptor units were observed in the rat after systemic capsaicin treatment, and in this survey Aδ-polymodal nociceptors were also excited by capsaicin [34]. After systemic capsaicin pretreatment of adult rats, electrophysiological and ultrastructural evidence showed that the nerve terminals of capsaicin-sensitive neurons are degenerated, but that the number of axons in the dorsal roots and nerve trunks remain unchanged [30, 35, 36]. On the other hand, five instillations of 1% capsaicin into the eye of rats induced ultrastructural changes only in free nerve endings, with swollen mitochondria and significant decreases in the number of vesicles [20]. Recent reports have shown that long-lasting exposure of mucosal areas with a high concentration of capsaicin induces degeneration of terminal arborizations of the sensory nerve fibers [30, 36].

Breakthrough 2

Pretreatment of newborn rats with capsaicin induced irreversible effects, attributed to an acute neurotoxic effect of the compound *in vivo*. Necrosis-like cell death was shown within 30 min [37, 38]. This type of pretreatment became popular for studying the effects of permanent loss of C-polymodal nociceptors induced by a 'highly selective sensory neurotoxin' [39–42].

However, in most of the above studies, two important caveats were not taken into account. First, there are qualitative differences between the long-term effects of neonatal and adult treatments in the rats and mice. For instance, there is an indiscriminate loss of C-afferents in adult rats after neonatal treatment, whereas after adult treatment only the proportion of C-polymodal nociceptors is decreased

[34, 36, 43]. The substantial loss of B-type neurons and C-afferent fibers is accompanied by variable loss (up to 33%) of A-type neurons and myelinated fibers [42, 44]. After neonatal treatment all types of sensory neuropeptide were depleted from tissues or sensory ganglia, whereas after adult treatment vasointestinal polypeptide (VIP), galanin and bombesin were not depleted [40, 42]. Reorganization of the pain pathway developed particularly in the dorsal horn [39–41]. The indiscriminate loss of C-afferents seems to be due to phenotypic changes of the primary afferent neurons after the pronounced loss of C-polymodal nociceptive neurons. Secondly, re-evaluation of the early morphological data is needed in the light of our recent quantitative morphometric study, which showed that no loss of trigeminal neurons occurred for 5 days after treatment of newborn rats with capsaicin (50 mg/kg, subcutaneous), even though capsaicin induced long-lasting mitochondrial swelling in B-type trigeminal neurons, similar the case in adults [45]. The significant loss of B-type neurons observed after 3 weeks was not due to acute necrosis-like or apoptotic cell death, but instead to the deprivation of nerve growth factor (NGF) uptake from the damaged nerve terminals. Similar results were obtained in rats treated with anandamide [46]. Thus, in contrast to the neonatal rat, NGF is not necessary for survival of sensory neurons in the adult animal and this is the reason why irreversible effects with loss of neurons are observed only after neonatal capsaicin pretreatment.

Breakthrough 3

Under *in vitro* conditions, the first evidence for a neuroselective site of action of capsaicin was obtained on the isolated ileum of guinea pig. In this classical preparation capsaicin elicited a new type of neural contraction without affecting smooth muscle responses to electrical stimulation of sympathetic postganglionic or vagal preganglionic fibers or myenteric cholinergic or intrinsic substance P-containing neurons [10–12, 40, 47, 48]. We suggested that the mediator of this neural smooth muscle response is released from sensory nerve endings expressing the putative capsaicin receptor. Therefore these nerve endings were defined as a 'capsaicin-sensitive chemoceptive neural system with dual sensory-efferent function' and the neuroselective action of capsaicin was attributed to the chemoceptive nature and high temperature coefficient (Q_{10}) of the excitatory process of these neurons [49] and not to its characteristic neuropeptide content [29–31, 42]. A similar new type of capsaicin-sensitive neural response was subsequently described by ourselves [50] and by Jan Lundberg and Alois Saria [51] in the trachea and main bronchi of the guinea pig. The impact of these studies was remarkable: sensory-efferent regulations mediated by tachykinins and calcitonin-gene-related peptide (CGRP) released from capsaicin-sensitive fibers was discovered in different organs of several species [10–13].

Breakthroughs 4, 5 and 6

The most fertile field in capsaicin research was related to its selective effect on sensory neuropeptides. Although some earlier data indicated that capsaicin depletes substance P from the spinal cord [1], the paper of Jessell, Iversen and Cuello [53] provided the first well-documented evidence for depletion of substance P by capsaicin, and in broader sense for depletion of a neuropeptide from sensory neurons by pharmacological means. Soon after this discovery Fred Lembeck, with Peter Holzer, showed that neurogenic inflammation is mediated by substance P released from capsaicin-sensitive nerve terminals [54]. Subsequently an impressive amount of information has been accumulated about the storage, release and tissue responses of tachykinins and other sensory neuropeptides in peripheral tissues. These studies were initiated mainly by research groups headed by Fred Lembeck in Graz, Carlo Maggi in Florance and Jan Lundberg in Stockholm [10–13, 52]. The protective role of sensory-efferent function of capsaicin-sensitive nerve endings in the rat gastric mucosa was shown by our group [55], and subsequently analysed in detail in the rat by Peter Holzer [56] and in humans and rats by ourselves in collaboration with the group of Gyula Mózsik [57]. From the point of view of pain and nociception, a paper by Tony Yaksh and colleagues [58] opened new avenues by showing that capsaicin could produce analgesic effects and depletion of substance P when it is applied intrathecally to the central terminals of capsaicin-sensitive sensory neurons.

Cellular mechanisms of capsaicin's actions

In order to reveal the intracellular mechanisms of the excitatory, sensory-neuron blocking and neurotoxic effects of capsaicin, the cell-culture technique has been exploited since around the late 1980s. However, the first electrophysiological evidence for the neuroselective action of capsaicin on a putative nociceptive subpopulation of cultured sensory neurons – which also responded to bradykinin – had been already reported in 1983 [59].

Capsaicin-gated cation channel

Quantitative analysis of ion fluxes [60] and studies on voltage- and capsaicin-activated currents, as well as on single-channel currents in isolated membrane patches from capsaicin-sensitive dorsal root ganglion neurons [61], revealed the existence of a novel capsaicin-gated cation channel. Gating of single patches of neuronal membrane by capsaicin supported the view that this compound acts through a membrane receptor protein [61]. Detailed patch-clamp analysis of the capsaicin-gated cation channel was reported few years later [62]. These early studies outlined also intracellular mechanisms for the explanation of the sensory neuron blocking and neurotoxic effects of capsaicin. The term sensory-neuron blocking effect was introduced [20, 63] to denote a functional impairment of responsiveness of capsaicin-sensitive sensory nerve terminals or the whole neuron to various stimuli, without degeneration of the affected neural elements. This functional blockade is not restricted to the capsaicin receptor protein but could be accompanied by long-term ultrastructural changes [20, 29, 31, 63]. On the basis of the described effects of capsaicin, influx of Na^+ and Ca^{2+} through the cation channel elicits depolarization, spike generation for nociceptive signals and neurotransmitter release from the nociceptive nerve endings, respectively [61]. Prolonged gating of these cation channels was proposed to produce the sensory-neuron blocking effect by enhancement of intracellular Ca^{2+}, which inhibited voltage-gated Ca^{2+} channels, and induced mitochondrial swelling. Enhanced intracellular NaCl was suggested to induce swelling of the neuron or nerve terminals, which together with Ca^{2+}-activated proteases might induce cell lysis (neurotoxic effect) under extreme gating conditions [61]. More details about recent achievements and further mechanisms that initiate the functional inactivation of these neurons are discussed in the chapters of Sándor and Szállási (Chapter 6) and Bley and Malmberg (Chapter 10) in this volume.

Resiniferatoxin (RTX): a new lead molecule

Identification of RTX as a capsaicin-like agent can be attributed to Peter Blumberg and his associates, particularly to Árpád Szállási [64, 65]. RTX is the irritant principal compound of a cactus-like plant, *Euphorbia resinifera*. It shares a common structural moiety, a 3-methoxy-4-hydroxy benzyl group, with capsaicin, the former being an ester of homovanillic acid and the latter an amide of vanillylamine (Fig. 3). The activities of this compound were soon appreciated by other research groups [66, 67]. RTX and capsaicin gated the same single channel in patch-clamp experiments [61]. RTX induced Ca^{2+} uptake by dorsal root ganglion cells, induced pungency, mimicked several *in vitro* effects of capsaicin and induced the same neurogenic inflammatory and hypothermic responses that were absent after capsaicin pretreatment. Furthermore, pretreatment with RTX prevented various effects of capsaicin both *in vitro* and *in vivo*. The potency of RTX for excitation and induction a functional blockade was in most cases several orders higher (up to 20 000-fold) relative to capsaicin (for references, see [41]). This high potency of RTX made this compound suitable for specific binding of [^{3}H]RTX by dorsal root ganglion membranes [41, 68], whereas a similar approach with [^{3}H]capsaicin was unsuccessful [39]. In rat dorsal root ganglion membranes, isolated neurons, spinal cord, sciatic nerve, urinary bladder, vagal nerve and dorsal vagal complex, the K_d value of RTX and the K_i value of capsaicin were within the ranges of 17–87 pM and 500–5400 nM, respectively [41]. Owing to the marked mismatches of the potencies for activation

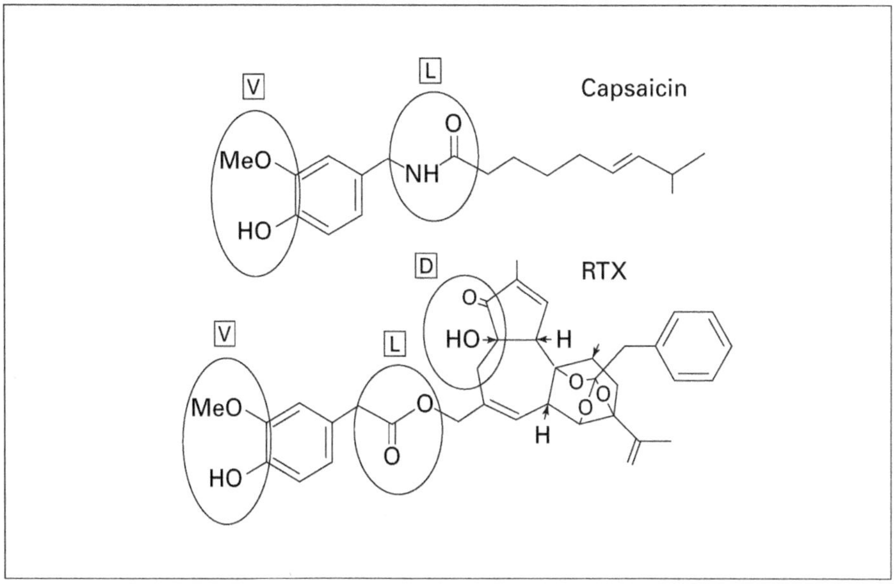

Figure 3
Chemical structure of capsaicin and resiniferatoxin (RTX). Encircled moieties are polar groups. V, vanilloid; L, linker; D, daphnane diterpene parts of the molecules.

and desensitization in different tissues and affinities to replace [^{3}H]RTX from binding sites, two types of receptor were proposed: the R-type receptor was characterized by its high affinity for RTX, which mediates desensitization without obvious agonism, whereas the C-type receptor was proposed to mediate the pronounced Ca^{2+} uptake and initiate the functional responses of the capsaicin-sensitive nerve terminals or neurons. The R- and C-subclasses of receptors showed distinct structure-activity relations for Ca^{2+} uptake and for high-affinity binding and were collectively denoted as vanilloid receptors [69].

Capsazepine and ruthenium red

Capsazepine and ruthenium red antagonize the effects of capsaicin and RTX on different tissues. Capsazepine, the first competitive antagonist of capsaicin, was discovered by Stuart Bevan, Humphrey Rang and coworkers [70] as a result of a systemic structure-activity study [71] at the former Sandoz (now Novartis) Institute for Medical Research in London. The blockade of capsaicin-induced Ca^{2+} uptake by ruthenium red was also discovered in these laboratories [60] and it turned out later that this polycationic dye antagonizes the effect of capsaicin and RTX by blocking the ion channel part of the capsaicin receptor [72]. These two compounds were useful under *in vitro* conditions, although their selective antagonist effect against capsaicin and RTX *in vitro* and efficacy *in vivo* are limited [32, 41, 42].

The TRPV1 (VR1) capsaicin receptor

An expression cloning strategy based on Ca^{2+} influx to isolate functional cDNA encoding the postulated capsaicin receptor from sensory neurons led to a remarkable breakthrough in capsaicin research. Michael Caterina, Mark Schumacher, Makoto Tominaga and David Julius with their coworkers cloned the capsaicin receptor in 1997 [72]. This capsaicin/vanilloid receptor was named by the authors VR1, for vanilloid receptor subtype 1. This nomenclature referred to the terminology of Peter Blumberg but became misleading because it refers in the name of a cloned receptor to a chemical moiety of some exogenous agonists, which is missing in the putative endogenous ligand [32, 73]. Structural features of VR1 that predict six transmembrane domains with a pore region between the fifth and sixth segments and both C- and N-termini located intracellularly also characterize the transient receptor potential (TRP) family, the first member of which was described in the retina of *Drosophila*. A further five members of this subfamily were identified and the nomenclature committee of the International Union of Pharmacology replaced the name VR1 with TRPV1 in 2003 [74]. The TRPV1 capsaicin receptor/cation channel is a membrane protein of 838 amino acids and became the first member of the so-called temperature-gated ion channels. It has a threshold temperature at near-noxious range of 43 °C (for further details see other chapters in this volume). With thermal transducer features, combined with responsiveness to noxious chemical agents and protons, TRPV1 is expressed in polymodal nociceptive neurons, indicating an important integrative nociceptor function for this protein [75] and forming a promising target for development of new types of analgesics [32, 76, 77].

It should be kept in mind that out of the six cloned vanilloid receptors (TRPV1–6) only the TRPV1 can be activated by vanilloids (capsaicin and RTX). Furthermore, the marked mismatch between R-type RTX binding and the C-type Ca^{2+} uptake response, formerly taken as an indication of two vanilloid receptors [41, 69], was also reproduced in cell lines transfected with rat TRPV1 or human TRPV1. On the one hand K_d values for RTX determined in several laboratories were within the range of 0.018–0.13 nM and its EC_{50} values for Ca^{2+} uptake were in the higher range of 1–6.5 nM, in both transfected cell lines and native neurons. On the other hand, in the case of capsaicin a reversed correlation was observed where its K_i values of 600–4000 nM were paradoxically much higher than its EC_{50} values for Ca^{2+} uptake in transfected cell lines and dorsal root ganglion neurons, where these figures were 6–38 and 200–340 nM, respectively (for references, see

[32]). Furthermore, a competitive antagonist (IBTU) of capsaicin and RTX on Ca^{2+} response did not antagonize the [^{3}H]RTX binding [78]. Point mutations in the TRPV1 molecule have not yet resolved this controversial issue.

In the TRPV1 protein, positions of key amino acids for vanilloid-gated responses and agonist binding favor multiple recognition sites for conformational changes to open the ion channel. They were localized in the N- and C-cytosolic tails on Arg-114 and Glu-761 [79], in the transmembrane domains 2 and 3, in the channel–lipid interface and its cytosolic region of Tyr-511, with additional residues of Ser-512 and Arg-491 [80], and on Met-547 and Thr-556 in transmembrane domains 3/4 [81]. Surprisingly, point mutation of Met-547 to Leu, or of Thr-550 to Ile, caused reduced binding of RTX without changing its potency for evoking Ca^{2+} responses. Differential influence of point mutation on responses to capsaicin and RTX was also observed [81]. These highly interesting data are far from complete but indicate multiple binding and recognition sites of vanilloids in the intracellular part of the protein.

Taking another approach to conventional structure-activity/affinity relationships of capsaicin and RTX congeners, a theory of differential recognition sites with partial overlap and possible involvement of protein–lipid interface was proposed [32]. In Figure 3 the encircled polar moieties prone for hydrogen-bonding interactions with the TRPV1 protein are at the vanilloid (V) and linker (L) groups and the α-hydroxy ketone moiety in positions C-3 and C-4 in the daphnane diterpene (D) skeleton. In the case of 'ultrapotent' RTX the vanilloid part seems to play a minor role in binding and the Ca^{2+} response, since replacement the 4-OH residue with a methoxy group at the vanilloid group induced a less-than 10-times fall in affinity and potency [32, 82]. In striking contrast the same change in the capsaicin molecule abolished its potency completely. The key structural moiety in the molecule of RTX for binding and action on TRPV1 seems to be the α-hydroxy ketone group at the daphnane diterpene part of the daphnane skeleton. Replacement of this 3-keto group by a 3-OH group diminished by more than 97% both the potency and affinity of RTX. Furthermore, omission at the daphnane diterpene group of the 5-membered substituted ring from the RTX molecule resulted in a completely inactive compound, even though the vanilloid and linker moieties, the two planar aromatic rings and the remaining part of the apolar diterpene skeleton are situated in identical positions with RTX in three-dimensional molecular overlay models [32, 82].

The TRPV1 ion channel of nociceptors is a promising target molecule for development of a new class of analgesic drugs but it is also a fascinating membrane protein as a molecular sensor, and thus constitutes a challenging field for basic research. Its conformational changes for activation and inactivation triggered by various chemical ligands, thermal stimuli and protons – as well as modulation of these responses by phosphorylation, thermal changes and by agonists of other receptors on the capsaicin-sensitive nociceptors – are discussed in other chapters of this volume.

Role of capsaicin-sensitive nociceptive neurons

Sensory nerve terminals, which express TRPV1 and are identical to capsaicin-sensitive afferents [4, 30, 32], subserve beyond their nociceptive sensory function as effector nerve terminals. The neuropeptide mediators of tachykinins and CGRP released from these sensors elicit neurogenic inflammation and enhancement of local microcirculation, respectively. These sensory-efferent dual functions are characteristic features of this substantial portion of primary afferent neurons, which comprise about 50% of the total neural population of sensory ganglia [10–12, 49].

Recent data have revealed a further 'unorthodox' [12] role of these TRPV1-expressing sensory neurons. Somatostatin released from these endings reaches into the circulation and elicits systemic analgesic and antiinflammatory effects [12, 77, 83, 84]. This novel endocrine-like systemic neurophumoral response mediated by somatostatin released from the activated capsaicin-sensitive nociceptors was denoted as 'sensocrine' role of these sensors [84]. Somatostatin agonism at sst4 receptors has turned out to be a promising target for development of new antiinflammatory analgesic drugs [77, 83, 84]. Furthermore, the unique multiple function of the capsaicin-sensitive, TRPV1-expressing neurons in respect of tissue responses of enhanced mircocirculation evoked by CGRP and systemic antiinflammatory effects due to somatostatin release is seen already at subnoxious level of stimulation of these sensors. For example, 0.1 Hz stimulation of capsaicin-sensitive afferents, which in humans does not evoke pain, induces maximum cutaneous vasodilatation and evokes systemic antiinflammatory effects in the rat [12, 83, 84]. Thus both the TRPV1 capsaicin receptor and the capsaicin-sensitive primary afferent nociceptive neuron population, together with the receptors of their sensory neuropeptides, are promising targets for drug research.

Acknowledgements
Recent research has been supported by OTKA TS-040753 and T-034911, and the Hungarian Academy of Sciences.

Notes

1 Langley JN, Dickinson WL (1889) On the local paralysis of peripheral ganglia, and on the connection of different classes of nerve fibers with them. *Proc Roy Soc* 46: 423–431
2 Dale HH (1906) On some physiological actions of ergot. *J Physiol London* 34: 163–206
3 Loewi O (1921) Über humorale Übertragbarkeit der Herznervenwirkungen. 1 Mitteilung. *Arch Ger Physiol* 189: 239–242
4 Bruce AN (1910) Über die Beziehungen der sensiblen Nervenendigungen zum Entzündungsvorgang. *Arch Exp Path Pharmak* 63: 424–433

5 Stricker S (1876) Untersuchungen über die Gefasserwurzeln des Ischiadicus. *S.-B., Kaiserl Akad Wiss Wien* 3: 173–185

6 Bayliss WM (1901) On the origin from the spinal cord of the vaso-dilatator fibres of the hind-limb, and on the nature of these fibres. *J Physiol London* 26: 173–209; Bayliss WM (1923) *The Vaso-motor system*. Longmans, Green, London

References

1 Szolcsányi J (1984) Capsaicin and neurogenic inflammation: history and early findings. In: LA Chahl, J Szolcsányi, F Lembeck (eds): *Antidromic vasodilatation and neurogenic inflammation*. Akadémiai Kiadó, Budapest, 7–26

2 Hőgyes A (1878) Beiträge zur physiologischen Wirkung der Bestandteile des *Capsicum* annuum. *Arch Exp Pathol Pharmakol* 9: 117–130

3 Molnár J (1965) Die pharmakologischen Wirkungen des Capsaicins, des scharf schmeckenden Wirkstoffes im Paprika. *Arzneim Forsch (Drug Res)* 15: 718–727

4 Szolcsányi J (1982) Capsaicin type pungent agents producing pyrexia. In: AS Milton (ed): *Handbook of Experimental Pharmacology, Vol. 60: Pyretics and antipyretics*. Springer-Verlag, Berlin, 437–478

5 Jancsó N (1955) *Speicherung. Stoffanreicherung im Retikuloendothel und in der Niere*. Akadémiai Kiadó, Budapest

6 Jancsó N (1960) Role of nerve terminals in the mechanism of inflammatory reactions. *Bull Millard Fillmore Hosp Buffalo, NY* 7: 53–77

7 Jancsó N, Jancsó-Gábor A, Takáts I (1961) Pain and inflammation induced by nicotine, acetylcholine and structurally-related compounds and their prevention by desensitizing agents. *Acta Physiol Hung* 19: 113–132

8 Makara GB, Stark E, Mihály K (1967) Sites at which formalin and capsaicin act to stimulate corticotropin secretion. *Can J Physiol Pharmacol* 45: 669–674

9 Jancsó N, Jancsó-Gábor A, Szolcsányi J (1967) Direct evidence for neurogenic inflammation and its prevention by denervation and by pretreatment with capsaicin. *Br J Pharmacol* 31: 138–151

10 Maggi CA (1995) Tachykinins and calcitonin gene-related peptide (CGRP) as co-transmitters released from peripheral endings of sensory nerves. *Prog Neurobiol* 45: 1–98

11 Holzer P (1992) Peptidergic sensory neurones in the control of vascular function: mechanisms and significance in the cutaneous and splanchnic vascular beds. *Rev Physiol Biochem Pharmacol* 121: 49–146

12 Szolcsányi J (1996) Capsaicin-sensitive sensory nerve terminals with local and systemic efferent functions: facts and scopes of an unorthodox neuroregulatory mechanism. *Prog Brain Res* 113: 343–359

13 Geppetti P, Holzer P (eds) (1996) *Neurogenic inflammation*. CRC Press, Boca Raton

14 Brain SD, Moore PK (eds) (1999) *Pain and neurogenic inflammation*. Birkhäuser Verlag, Basel

15 Jancsó-Gábor A, Szolcsányi J, Jancsó N (1970) Irreversible impairment of thermoregulation induced by capsaicin and similar pungent substances in rats and guinea-pigs. *J Physiol* 206: 495–507
16 Jancsó-Gábor A, Szolcsányi J, Jancsó N (1970) Stimulation and desensitization of the hypothalamic heat-sensitive structures by capsaicin in rats. *J Physiol* 208: 449–459
17 Szolcsányi J (1983) Disturbances of thermoregulation induced by capsaicin. *J Therm Biol* 8: 207–212
18 Szolcsányi J, Jancsó-Gábor A (1973) Capsaicin and other pungent agents as pharmacological tools in studies on thermoregulation. In: E Schönbaum, P Lomax (eds): *The pharmacology of thermoregulation.* Karger, Basel, 395-409
19 Joó F, Szolcsányi J, Jancsó-Gábor A (1969) Mitochondrial alterations in the spinal ganglion cells of the rat accompanying the long-lasting sensory disturbance induced by capsaicin. *Life Sci* 8: 621–626
20 Szolcsányi J, Jancsó-Gábor A, Joó F (1975) Functional and fine structural characteristics of the sensory neuron blocking effect of capsaicin. *Naunyn-Schmiedeberg's Arch Pharmacol* 287: 157–169
21 Szolcsányi J, Joó F, Jancsó-Gábor A (1971) Mitochondrial changes in preoptic neurones after capsaicin desensitization of the hypothalamic thermodetectors in rats. *Nature* 229: 116–117
22 Jancsó N, Jancsó-Gábor A, Szolcsányi J (1968) The role of sensory nerve endings in neurogenic inflammation induced in human skin and in the eye and paw of the rat. *Br J Pharmacol* 33: 32–41
23 Jancsó N (1968) Desensitization with capsaicin and related acylamides as a tool for studying the function of pain receptors. In: K Lim, D Armstrong, ED Pardo (eds.): Pharmacology of pain. Pergamon Press, Oxford, 33–55
24 Szolcsányi J (1977) A pharmacological approach to elucidate the role of different nerve fibres and receptor endings in mediation of pain. *J Physiol (Paris)* 73: 251–259
25 Szolcsányi J, Jancsó-Gábor A (1975) Sensory effects of capsaicin congeners I. Relationship between chemical structure and pain-producing potency. *Arzneim Forsch (Drug Res)* 25: 1877–1881
26 Szolcsányi J, Jancsó-Gábor A (1976) Sensory effects of capsaicin congeners II. Importance of chemical structure and pungency in desensitizing activity of capsaicin-type compounds. *Arzneim Forsch (Drug Res)* 26: 33–37
27 Bessou P, Perl ER (1969) Response of cutaneous sensory units with unmyelinated fibers to noxious stimuli. *J Neurophysiol* 32: 1025–1043
28 Szolcsányi J (1980) Effect of pain producing chemical agents on the activity of slowly conducting afferent fibres. *Acta Physiol Hung* 56: 86
29 Szolcsányi J (1990) Capsaicin, irritation and desensitization. Neurophysiological basis and future perspectives.In: BR Green, JR Mason, MR Kare (eds): *Chemical senses, Vol. 2. Irritation.* Marcel Dekker, New York, 141–168
30 Szolcsányi J (1993) Actions of capsaicin on sensory receptors. In: JN Wood (ed): *Capsaicin in the study of pain.* Academic Press, London, 1–26

31 Szolcsányi J (1991) Perspectives of capsaicin-type agents in pain therapy and research. In: WCV Parris (ed): *Contemporary issues in chronic pain management.* Kluwer Academic Press, Boston, 97–122
32 Szolcsányi J (2002) Capsaicin receptors as target molecules on nociceptors for development of novel analgesic agents. In: G Kéri, J Toth (eds): *Molecular pathomechanisms and new trends in drug research.* Taylor and Francis, London, 319–333
33 Szolcsányi J (1987) Selective responsiveness of polymodal nociceptors of the rabbit ear to capsaicin, bradykinin and ultra-violet irradiation. *J Physiol* 388: 9–23
34 Szolcsányi J, Anton F, Reeh PW, Handwerker HO (1988) Selective excitation by capsaicin of mechano-heat sensitive nociceptors in rat skin. *Brain Res* 446: 262–268
35 Chung K, Klein CM, Coggeshall RF (1990) The receptive part of the primary afferent axon is most valnerable to systemic capsaicin in adult rats. *Neurosci Lett* 53: 221–226
36 Winter J, Woolf C, Lynn B (1993) Degenerative and regenerative responses of sensory neurones to capsaicin-induced damage. In: JN Wood (ed): *Capsaicin in the study of pain.* Academic Press, London, 139–160
37 Jancsó G, Király E, Jancsó-Gábor A (1977) Pharmacologically induced selective degeneration of chemosensitive primary sensory neurons. *Nature* 270: 741–743
38 Jancsó G, Király E (1981) Sensory neurotoxins: chemically induced selective destruction of primary sensory neurons. *Brain Res* 210: 83–89
39 Buck SH, Burks TF (1986) The neuropharmacology of capsaicin: a review of some recent observations. *Pharmacol Rev* 38: 179–226
40 Holzer P (1991) Capsaicin: cellular targets, mechanism of action, and selectivity for thin sensory neurons. *Pharmacol Rev* 43: 143–201
41 Szállási Á, Blumberg PM (1999) Vanilloid (capsaicin) receptors and mechanisms. *Pharmacol Rev* 51: 159–211
42 Szolcsányi J, Pórszász R, Pethő G (1994) Capsaicin and pharmacology of nociceptors. In: JM Besson, G Guilbaud, H Ollat (eds): *Peripheral neurons in nociception. Physiopharmacological aspects.* John Libbey Eurotext, Paris 109–124
43 Welk E, Fleischer E, Petsche U, Handwerker HO (1984) Afferent C-fibers in rats after neonatal capsaicin treatment. *Pfügers Arch* 400: 66–71
44 Lawson SN (1987) The morphological consequences of neonatal treatment with capsaicin on primary afferent neurones in adult rats. *Acta Physiol Hung* 69: 315–321
45 Szőke É, Seress L, Szolcsányi J (2002) Neonatal capsaicin treatment results in prolonged mitochondrial damage and delayed cell death of B cells in the rat trigeminal ganglia. *Neurosci* 113: 925–937
46 Szőke É, Czéh G, Szolcsányi J, Seress L (2002) Neonatal anandamide treatment results in prolonged mitochondrial damage in the vanilloid receptor type 1-immunoreactive B-type neurons of the rat trigeminal ganglion. *Neuroscience* 115: 805–814
47 Barthó L, Szolcsányi J (1978) The site of action of capsaicin on the guinea-pig isolated ileum. *Naunyn-Schmiedeberg's Arch Pharmacol* 305: 75–81
48 Szolcsányi J, Barthó L (1978) New type of nerve-mediated cholinergic contractions of

the guinea-pig small intestine and its selective blockade by capsaicin. *Naunyn-Schmiedeberg's Arch Pharmacol* 305: 83–90
49 Szolcsányi J (1984) Capsaicin-sensitive chemoceptive neural system with dual sensory-efferent function. In: LA Chahl, J Szolcsányi, F Lembeck (eds): *Antidromic vasodilatation and neurogenic inflammation.* Akadémiai Kiadó, Budapest, 27–56
50 Szolcsányi J, Barthó L (1982) Capsaicin-sensitive non-cholinergic excitatory innervation of the guinea-pig tracheobronchial smooth muscle. *Neurosci Lett* 34: 247–251
51 Lundberg JM, Saria A (1982) Bronchial smooth muscle contraction induced by stimulation of capsaicin-sensitive sensory neurons. *Acta Physiol Scand* 116: 473–476
52 Lundberg JM (1996) Pharmacology of cotransmission in the autonomic nervous system: integrative aspects on amines, neuropeptides, adenosin triphosphate, amino acids and nitric oxid. *Pharmacol Rev* 48: 113–178
53 Jessell TM, Iversen LL, Cuello AC (1978) Capsaicin-induced depletion of substance P from primary sensory neurones. *Brain Res* 152: 183–188
54 Lembeck F, Holzer P (1979) Substance P as neurogenic mediator of antidromic vasodilatation and neurogenic plasma extravasation. *Naunyn-Schmiedeberg's Arch Pharmacol* 310: 175–183
55 Szolcsányi J, Barthó L (1981) Impaired defense mechanism to peptic ulcer in the capsaicin-desensitized rat. In: G Mózsik, O Hänninen, T Jávor (eds): *Gastrointestinal defense mechanisms.* Adv Physiol Sci Vol 29. Akadémiai Kiadó. Pergamon Press, Oxford, 39–51
56 Holzer P, Sametz W (1986) Gastric mucosal protection against ulcerogenic factors in the rat mediated by capsaicin-sensitive afferent neurons. *Gastroenterology* 91: 975–981
57 Szolcsányi J, Barthó L (2001) Capsaicin-sensitive afferents and their role in gastroprotection: an update. *J Physiol (Paris)* 95: 181–188
58 Yaksh TL, Farb DH, Leeman SE, Jessell TM (1979) Intrathecal capsaicin depletes substance P in the rat spinal cord and produces prolonged thermal analgesia. *Science* 206: 481–483
59 Baccaglini PJ, Hogan PG (1983) Some rat sensory neurons in culture express characteristic of differentiated pain sensory cells. *Proc Natl Acad Sci USA* 80: 594–598
60 Wood JN, Winter J, James IF, Rang HP, Yeats J, Bevan S (1988) Capsaicin-induced ion fluxes in dorsal root ganglion cells in culture. *J Neurosci* 8: 3208–3220
61 Bevan S, Szolcsányi J (1990) Sensory neuron-specific actions of capsaicin: mechanisms and applications. *Trends Pharmacol Sci* 11: 330–333
62 Oh U, Hwang SW, Kim D (1996) Capsaicin activates a nonselective cation channel in cultured neonatal rat dorsal root ganglion neurons. *J Neurosci* 16: 1659–1667
63 Szolcsányi J (1985) Sensory receptors and the antinociceptive effects of capsaicin. In: R Hakanson, F Sundler (ed.): *Tachykinin antagonists.* Elsevier, Amsterdam, 45–54
64 Szállási Á, Blumberg PM (1989) Resiniferatoxin, a phorbol-related diterpene, acts as an ultrapotent analog of capsaicin, the irritant constituent in red pepper. *Neuroscience* 30: 515–520
65 Szolcsányi J, Szállási Á, Szállási Z, Joó F, Blumberg PM (1990) Resiniferatoxin: An

ultrapotent selective modulator of capsaicin-sensitive primary afferent neurons. *J Pharmacol Exp Ther* 255: 923–928
66 Winter J, Dray A, Wood JN, Yeats JC, Bevan S (1990) Cellular mechanism of action of resiniferatoxin: A potent sensory neuron excitotoxin. *Brain Res* 520: 131–140
67 Maggi CA, Patacchini R, Tramontana M, Amann R, Giuliani S, Santicioli P (1990) Similarities and differences in the action of resiniferatoxin and capsaicin on central and peripheral endings of primary sensory neurons. *Neuroscience* 37: 531–539
68 Szállási Á, Blumberg PM (1990) Specific binding of resiniferatoxin, an ultrapotent capsaicin analog, by dorsal root ganglion membranes. *Brain Res* 524: 106–111
69 Ács G, Lee J, Marquez V, Blumberg PM (1996) Distinct structure-activity relations for stimulation of 45Ca uptake and for high affinity binding in cultured rat dorsal root ganglion neurons and dorsal root ganglion membranes. *Mol Brain Res* 35: 173–182
70 Bevan S, Hothi S, Hughes GA, James IF, Rang HP, Shah K, Walpole CSJ, Yeats JC (1992) Capsazepine: A competitive antagonist of the sensory neurone excitant capsaicin. *Br J Pharmacol* 107: 544–552
71 Walpole CSJ, Wrigglesworth R (1993) Structural requirements for capsaicin agonists and antagonists. In: JN Wood (ed): *Capsaicin in the study of pain.* Academic Press, San Diego, 63–82
72 Caterina MJ, Schumacher MA, Tominaga M, Rosen TA, Levine JD, Julius D (1997) The capsaicin receptor: a heat-activated ion channel in the pain pathway. *Nature* 389: 816–824
73 Szolcsányi J (2000) Are cannabinoids endogenous ligands for the VR1 capsaicin receptor? *Trends Pharmacol Sci* 21: 41–42
74 Clapham DE, Montell C, Schultz G, Julius D (2003) International Union of Pharmacology. XLIII. Compendium of voltage-gated ion channels: transient receptor potential channels. *Pharmacol Rev* 55: 591–596
75 Tominaga M, Caterina MJ, Malmberg AB, Rosen TA, Gilbert H, Skinner K, Raumann BE, Basbaum AJ, Julius D (1998) The cloned capsaicin receptor integrates multiple pain-producing stimuli. *Neuron* 21: 531–543
76 Szállási Á, Appendino G (2004) Vanilloid receptor TRPV1 antagonists as the next generation of painkillers. Are we putting the cart before the horse? *J Med Chem* 47: 2717–2723
77 Helyes Z, Pintér E, Németh J, Szolcsányi J (2003) Pharmacological targets for the inhibition of neurogenic inflammation. *Curr Med Chem Anti-Inflammatory and Anti-Allergy Agents* 2: 191–218
78 Tóth A, Blumberg PM, Chen Z, Kozikowski AP (2004) Design of a high-affinity competitive antagonist of the vanilloid receptor selective for the calcium entry-linked receptor population. *Mol Pharmacol* 65: 282–291
79 Jung J, Lee S-Y, Hwang SW, Cho H, Shin J, Kang Y-S, Kim S, Oh U (2002) Agonist recognization sites in the cytosolic tails of vanilloid receptor 1. *J Biol Chem* 277: 44448–44454

80 Jordt S-E, Julius D (2002) Molecular basis for species-species sensitivity to "hot" chili peppers. *Cell* 108: 421–430

81 Gavva NR, Klionsky L, Qu Y, Shi L, Tamir R Edenson S, Zhang TJ, Viswanadhan VN, Toth A, Pearce LV et al (2004) Molecular determinats of vanilloid sensitivity in TRPV1. *J Biol Chem* 279: 20283–20295

82 Walpole CSJ, Bevan S, Bloomfield G, Breckensidge R, James IF, Ritchie T, Szállási A, Winter J, Wrigglesworth R (1996) Similarities and differences in the structure-activity relationships of capsaicin and resiniferatoxin analogues. *J Med Chem* 39: 2939–2952

83 Helyes Z, Szabó Á, Németh J, Jakab B, Pintér E, Bánvölgi Á, Kereskai L, Kéri G, Szolcsányi J (2004) Antiinflammatory and analgesic effects of somatostatin released from capsaicin-sensitive nerve terminals in a Freund's adjuvant-induced chronic arthritis model in the rat. *Arthritis Rheum* 50: 1677–1685

84 Szolcsányi J, Pintér E, Helyes Z (2004) Sensocrine function of capsaicin-sensitive nociceptors mediated by somatostatin regulates against inflammation and hyperalgesia. In: HO Handwerker, K Brune (eds): *Hyperalgesias: molecular mechanisms and clinical implications.* IASP Press, Seattle, 113–128

Part II
Molecular and cellular properties of vanilloid receptors

Structural determinants of TRPV1 functionality

Makoto Tominaga

Section of Cell Signaling, Okazaki Institute for Integrative Bioscience, National Institutes of Natural Sciences, Okazaki 444-8787, Japan

Multimerization

The capsaicin receptor TRPV1 (Fig. 1) is predicted to have six transmembrane (TM) domains and a short, pore-forming hydrophobic stretch between the fifth and sixth TM domains [1]. A similar topological structure has been reported for many ion channels, including the cyclic-nucleotide-gated cation channels, the Shaker-related voltage-gated K^+ channels and the hyperpolarization-activated channels [2–4]. TRPV1 belongs to the TRPV subfamily of the large transient receptor potential (TRP) ion-channel superfamily, whose prototypical member, TRP, was found to be deficient in a *Drosophila* mutant exhibiting abnormal responsiveness to continuous light [5–7]. Like many other TRP channels, TRPV1 has a long N-terminus containing three ankyrin-repeat domains and a C-terminus containing a TRP domain close to the sixth TM. The ankyrin repeats consist of an approx. 33-residue motif named after the cytoskeletal protein ankyrin, which contains 24 copies of these repeats. In other membrane proteins, ankyrin repeats are known to bind to many cytosolic proteins [8]. One protein, calmodulin (CaM), has so far been reported to bind to the ankyrin-repeat domain of TRPV1 (the first ankyrin repeat) (Fig. 2) [9].

The *Drosophila* TRP channels have been shown to form a heteromultimeric channel complex. These complexes in turn form part of a larger signaling complex that also contains a G-protein-coupled receptor (rhodopsin), an effector [phospholipase C (PLC)], regulators [protein kinase C (PKC) and CaM], and the scaffolding protein INAD (inactivation no-after potential D) [10]. The capsaicin receptor is also presumed to consist of a multimeric channel, though experimental evidence along these lines has been relatively limited. Oligomer formation analysis using perfluorooctanoic acid polyacrylamide gel revealed that TRPV1 forms multimers, with a homotetramer as a predominant form (Fig. 1) [11]. Tetrameric stoichiometry for the native capsaicin receptor was further suggested by the behavior of a dominant-negative TRPV1 mutant bearing amino acid substitutions in the sixth TM domain [12].

Turning up the Heat on Pain: TRPV1 Receptors in Pain and Inflammation, edited by Annika B. Malmberg and Keith R. Bley

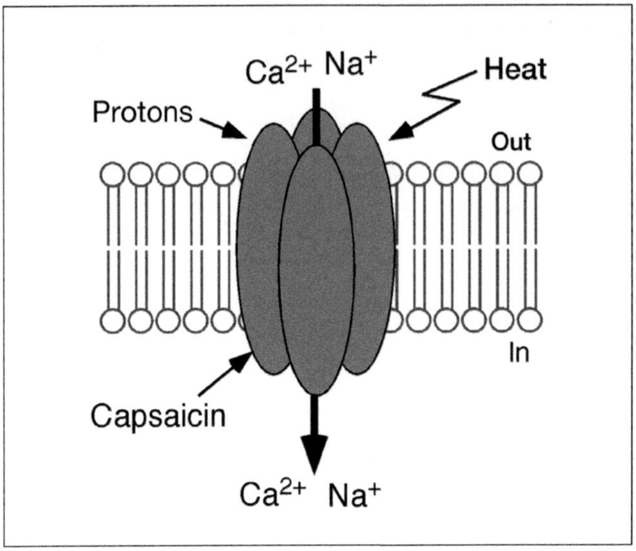

Figure 1
Proposed tetrameric structure of TRPV1 in the plasma membrane.

TRPV1 has also been suggested to form hetero-oligomers with TRPV3, another heat-sensitive TRP channel, based on the observations that TRPV3 is transcribed from a gene adjacent to TRPV1, is co-expressed in dorsal root ganglion neurons with TRPV1, co-precipitates with TRPV1 in heterologous expression systems, and may reduce TRPV1 responsiveness to capsaicin [13]. However, TRPV1 expressed alone in HEK-293 cells or *Xenopus* oocytes can account for the majority of the electrophysiological properties exhibited by native capsaicin receptors in sensory neurons, including ligand affinity, permeability sequence, current–voltage relationship, conductance, and open probability at both the single-channel and whole-cell levels. These results suggest either that TRPV1 can form homo-multimers without other subunits [1, 14–17] or that incorporation of subunits other than TRPV1 does not influence the functional properties.

Capsaicin binding

Because capsaicin and its analogues such as resiniferatoxin (RTX) are lipophilic, it is quite possible that they pass through the cell membrane and act on binding sites present in the intracellular surface of TRPV1. An apparent time lag between capsaicin uptake and pungent sensation might be partially explained by such a process.

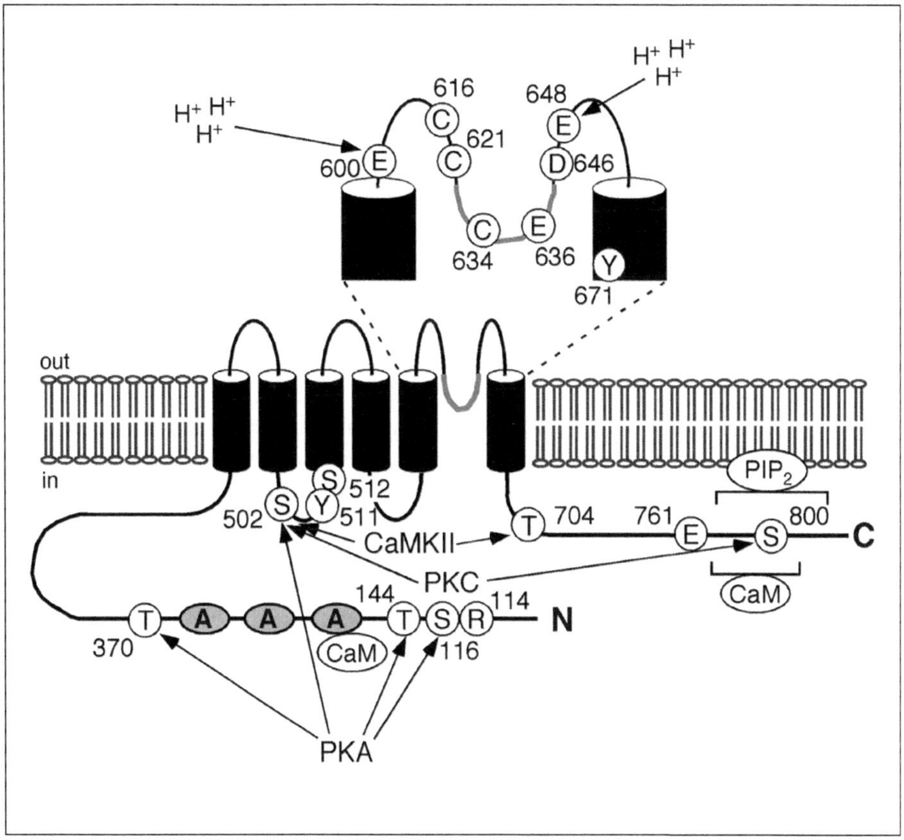

Figure 2
Regions and amino acids involved in TRPV1 function. Phosphatidylinositol (4,5)-bisphosphate (PIP_2) binds to the indicated region in the C-terminus. Calmodulin (CaM) binds to both C- and N-termini (the first ankyrin repeat; A). Protein kinase A (PKA), protein kinase C (PKC), or Ca^{2+}/CaM-dependent kinase II (CaMKII) phosphorylate overlapping serine or threonine residues, indicated by arrows. Protons act on the two glutamate residues in the extracellular loop, indicated by arrows.

Indeed, the existence of capsaicin-binding sites in the cytosolic domain of TRPV1 was proved using a synthetic water-soluble capsaicin analog (Fig. 1) [18]. Capsaicin is structurally related to putative endogenous TRPV1 agonists, such as anandamide [19], 12-hydroperoxyeicosatetraenoic acid (12-HPETE) [20], and *N*-arachidonoyl-dopamine (NADA) [21]. Therefore, there is significant pharmacological and physiological interest in identifying regions of the channel that transduce the effects of these molecules. In particular, the fact that the three-dimensional structures of capsaicin and 12-HPETE can be superimposed in the energy-minimized state suggests that active site of capsaicin are not a single amino acid [20]. Comparison of rat TRPV1 with its avian ortholog from chicken sensory neurons, together with mutational analysis, revealed that Tyr-511 and Ser-512, located at the transition between the second intracellular loop and the third TM domain, might interact with vanilloid ligands at the intracellular face of the membrane (Fig. 2) [22]. In addition, Arg-114 and Glu-761 in the N- and C-termini, respectively, were found to be involved in agonist recognition, based on studies involving stepwise deletions of TRPV1 and chimera construction between TRPV1 and its capsaicin-insensitive homolog, TRPV2 (VRL-1; Fig. 2) [23]. The apparently wide distribution of residues necessary for capsaicin binding is consistent with the fact that TRPV1 can be activated by compounds such as capsaicin and 12-HPETE (which are related to one another in their three-dimensional structures), and also suggests that these critical residues are relatively close to each other in native channel. It is also conceivable that there are regions other than the two loci in the cytosolic tails and third TM domain that control TRPV1 ligand binding. Indeed, mutations of the sixth TM domain were found to reduce [^{3}H]RTX binding affinity [12], and mutants of the pore domain (Glu-636, Asp-646, and Glu-648) were significantly more sensitive to capsaicin than wild-type TRPV1 (Fig. 2) [24].

As described above, TRPV1 shares structural similarity with the voltage-gated K^+ channels, including six-TM-domain topology. According to the current helix-packing models of the voltage-gated K^+ channels (derived from helical periodicity analysis and crystallographic approaches), the first, second, and third TM domains are located on the lipid-facing periphery of the tetrameric channel complex, whereas the fifth and sixth TM domains are located closer to the pore-forming channel core. Assuming similar helix packing for TRPV1, the lipophilic moiety of capsaicin may bind to the second and third TM domains on the channel–lipid interface, while the vanilloid moiety may interact with residues around Tyr-511 in the cytosolic region, thus linking the two TM domains together with a cytosolic tail.

Proton action

Acidification of the extracellular milieu has two primary effects on TRPV1 function. First, extracellular protons increase the potency of heat or capsaicin as TRPV1 agonists, in part by lowering the threshold for channel activation by either stimulus. Second, extracellular protons can themselves be viewed as agonists because further acidification (to pH <6.0) leads to channel opening at room temperature [17]. Extracellular protons are believed to act primarily by increasing the probability of channel opening [17, 25], rather than by altering unitary conductance or interacting directly with the vanilloid-binding site. Acidic solution evoked ionic currents with a EC_{50} value of about pH 5.4 when applied to outside-out, but not inside-out, membrane patches excised from HEK-293 cells expressing TRPV1 [17], suggesting that protons act on amino acids in the extracellular domain of TRPV1 having side-chain pK_a values in the physiologically relevant range. Mutational analyses revealed that Glu-600, located within a putative extracellular domain, serves as an important regulator site for proton potentiation of TRPV1 activity, whereas Glu-648 is involved in direct proton-evoked activation of TRPV1 (Fig. 2) [26]. These data indicate the existence of stimulus-specific steps in the TRPV1 activation process. A stimulus-specific gating mechanism is also supported by the existence of mutants exhibiting stimulus-specific reduction of their activity [12, 24].

Heat activation

Heat-evoked TRPV1 currents show properties similar to those of capsaicin-evoked currents. Heat-evoked single-channel openings were observed in inside-out membrane patches excised from HEK-293 cells expressing TRPV1, suggesting that TRPV1 is, itself, a heat sensor. It is not clear how and where heat acts to open the TRPV1 channel. However, it is now known that several TRP-family ion channels (TRPV1, TRPV2, TRPV3, TRPV4, TRPM8, TRPA1) are thermosensitive, suggesting that temperature-sensor domains are present in these channel proteins [27, 28]. The distal half of the TRPV1 C-terminus was reported to be partially involved in thermal sensitivity [29]. Furthermore, the observations that certain mutations and phosphorylation by protein kinase A (PKA) or PKC lead to a reduction of the threshold temperature for TRPV1 activation [26, 30, 31] suggests more global effects of heat on TRPV1. TRPV1 is known to have a voltage-dependent gating property [15]. It was very recently reported that temperature sensing in TRPV1 and TRPM8 (activated by cold stimulus and menthol) is tightly linked to voltage-dependent gating [32]. TRPV1 is activated upon depolarization, and changes in temperature result in graded shifts in its voltage-dependent activation curve. This result suggests that amino acids responsible for voltage-dependence are also involved in thermosensing, although the fourth TM domain of TRPV1 lacks the multiple positively charged residues typical of voltage-gated channels.

Desensitization

Capsaicin not only causes pain, but also seems to exhibit analgesic properties, particularly when used to treat pain associated with diabetic neuropathies or rheumatoid arthritis [33]. This paradoxical effect may relate to the ability of capsaicin to desensitize nociceptive terminals to capsaicin, as well as to other noxious stimuli, following prolonged exposure. At the molecular level, an extracellular Ca^{2+}-dependent reduction of TRPV1 responsiveness upon continuous vanilloid exposure (electrophysiological desensitization) may partially underlie this phenomenon [1, 33], although physical damage to the nerve terminal probably contributes to this effect as well. Ca^{2+}- and voltage-dependent desensitization of capsaicin-activated currents has also been observed in rat dorsal root ganglion neurons [34–37]. This inactivation of nociceptive neurons by capsaicin has generated extensive research on the possible therapeutic effectiveness of capsaicin as a clinical analgesic tool [38, 39].

Desensitization to capsaicin is a complex process with varying kinetic components: a fast component that appears to depend on Ca^{2+} influx through TRPV1 [34–37] and a slow component that does not. Calcineurin inhibitors reduce TRPV1 desensitization (the slow component), indicating the involvement of Ca^{2+}-dependent phophorylation/dephosphorylation process [37]. In addition, PKA-dependent phosphorylation of TRPV1 has been reported to mediate the slow component of TRPV1 desensitization [40]. TRPV1 becomes dephosphorylated upon exposure to capsaicin and this phosphorylation can be restored by 8-bromo-cAMP. Ser-116 in the N-terminus was found to be a substrate for PKA-dependent phosphorylation (Fig. 2). Thr-370 was also reported to be responsible for PKA-dependent reduction of TRPV1 desensitization (Fig. 2) [41].

CaM has also been reported to be involved in Ca^{2+}-dependent desensitization of TRPV1 [42]. CaM was found to bind to a 35-amino-acid segment in the C-terminus of TRPV1 (positions 767–801; Fig. 2). Disruption of the CaM-binding segment prevented extracellular Ca^{2+}-dependent TRPV1 desensitization to brief capsaicin application, although some desensitization was still observed upon more prolonged capsaicin application in cells expressing the mutant. It has also been reported that CaM binds to the first ankyrin-repeat domain in the N-terminus of TRPV1 (positions 189–222) and to be involved in desensitization (Fig. 2) [9]. Which of the N- or C-termini is predominantly involved in Ca^{2+}-dependent desensitization by CaM is not known. Interestingly, neither domain contains CaM-binding sites that are obvious from analysis of the primary sequence, such as a consensus isoleucine-glutamine motif. Ca^{2+}-dependent desensitization is a relatively common feature of many cation channels, including L-type Ca^{2+} channels, *N*-methyl-D-aspartate (NMDA) receptor channels and TRP channels. It may represent a physiological safety mechanism against a harmful Ca^{2+} overload in the cell, especially during large Ca^{2+} influx through the channels.

Permeability

The region important for the cation permeability of TRPV1 has not been well defined. Replacement of Asp-646, an amino acid that is involved in changing ligand affinity [24], as described above, with asparagine was found to decrease 10-fold ruthenium red blockade efficacy and reduce 4-fold the relative permeability of the divalent cation Mg^{2+} with respect to Na^{+} without changing the selectivity of monovalent cations (Fig. 2) [43]. It is easy to predict that a change in Ca^{2+} permeability modulates Ca^{2+}-dependent desensitization. Indeed, a TRPV1 mutant where Tyr-671 in the sixth TM domain was replaced with Lys caused a robust reduction of Ca^{2+} permeability from $P_{Ca/Na} = 9.0 \pm 1.3$ to 0.8 ± 0.1, leading to loss of desensitization in the presence of extracellular Ca^{2+} (Fig. 2) [44].

Phosphorylation

Like other ion channels, TRPV1 can be phosphorylated by several kinases including PKA [40, 45–47], PKC [31, 48–51], Ca^{2+}/CaM-dependent kinase II (CaMKII) [52], and Src kinase [53]. There has been extensive work demonstrating that activation of a PKA-dependent pathway by inflammatory mediators such as prostaglandins influences capsaicin- or heat-mediated actions in sensory neurons, probably by acting on TRPV1. These results suggest that PKA plays a pivotal role in the development of hyperalgesia and inflammation by inflammatory mediators. Ser-116 and Thr-370 in the N-terminus were reported to be phosphorylated by PKA and involved in desensitization (Fig. 2) [40, 41]. Phosphorylation of Ser-116 by PKA was found to inhibit dephosphorylation of TRPV1 caused by capsaicin exposure. Thr-144, Thr-370, and Ser-502 were also found to be involved in sensitization of heat-evoked TRPV1 responses upon PKA-dependent phosphorylation (Fig. 2) [47].

PKC-dependent phosphorylation of TRPV1 occurs downstream of activation of G_q-coupled receptors by several inflammatory mediators, including ATP, bradykinin, and trypsin or tryptase [31, 50, 51, 54]. PKC-dependent phosphorylation of TRPV1 caused not only potentiation of capsaicin- or proton-evoked responses, but also reduced the temperature threshold for TRPV1 activation so that normally non-painful temperatures in the range of normal body temperature were capable of activating TRPV1, thereby leading to the sensation of pain. These phenomena were also confirmed in native sensory neurons. Capsaicin potency was modulated by PKC-dependent phosphorylation, suggesting the interaction between phosphorylation and capsaicin binding described above. Direct phosphorylation of TRPV1 by PKC was proved using a biochemical approach [30], and two target serine residues (Ser-502 and Ser-800) were identified (Fig. 2) [30, 49]. In the mutant in which the two serine residues were replaced with alanine, sensitization or potentiation of TRPV1 activity induced by any of three different stimuli (capsaicin, proton, or heat) was abolished. Ser-502 and Ser-800 were also found to be involved in potentiation of NADA-induced TRPV1 activation [55], oleoylethanolamide-induced TRPV1 activation [56], and rephosphorylation of TRPV1 after Ca^{2+}-dependent desensitization [57].

As described above, calcineurin inhibits desensitization of TRPV1, indicating that a phosphorylation/dephosphorylation process is important for TRPV1 activity.

Indeed, CaMKII was reported to control TRPV1 activity upon phosphorylation of TRPV1 at Ser-502 and Thr-704 by regulating capsaicin binding (Fig. 2) [52]. Thus, phosphorylation of TRPV1 by three different kinases seems to control TRPV1 activity through the dynamic balance between the rate of phosphorylation and dephosphorylation. These mechanisms appear to converge, in part, on Ser-502, given that Ser-502 was found to be phosphorylated by all three kinases. If this is the case, Ser-502 represents a key site for regulation of the excitability of sensory neurons.

Modulation by lipids

Membrane-derived lipids are also known to regulate the function of some ion channels, including TRPV1. For example, TRPV1 was shown to be activated by anandamide, oleoylethanolamide, and some lipoxygenase products [19, 20, 56]. TRPM7 and G-protein-coupled inwardly rectifying K^+ channels can be inhibited or potentiated, respectively, by the binding of phosphatidylinositol (4,5)-bisphosphate (PIP_2) [58, 59]. PIP_2 was also found to inhibit TRPV1 activity and such PIP_2-mediated inhibition was released upon activation of PLC by metabotropic receptors [60] and the consequent hydrolysis of PIP_2 to diacylglycerol and inositol (1,4,5)-trisphosphate. A region of TRPV1 (amino acids 777–820) having eight positively charged residues was identified as a motif involved in regulation by PIP_2 (Fig. 2) [61], possibly because it interacts with the negatively charged head group of the phospholipid. Interestingly, this region includes Ser-800, one of the substrates for PKC-dependent phosphorylation and overlaps with the 35-amino-acid segment necessary for CaM binding. This region might thus be very important for regulation of TRPV1 function.

Other functions

Dithiothreitol, an agent that maintains -SH groups of cysteines in a reduced state, was reported to greatly facilitate membrane currents induced by noxious heat or capsaicin in a concentration-dependent manner [62]. This result, together with the existence of three cysteine residues located at positions 616, 621, and 634 in the loop between the fifth and sixth TM domains (Fig. 2), suggests that the sensitivity of TRPV1 is pre-set by the redox state of the thiol groups of these cysteine residues.

Conclusion

Many regions and amino acids involved in TRPV1 function have been identified. Given that TRPV1 is a key molecule in peripheral nociception, these regions and amino acids could prove useful for the development of novel anti-nociceptive or anti-inflammatory agents. Crystallographic analysis promises to provide even more information about the structural determinants of TRPV1 functionality.

Acknowledgement
I thank Dr M.J. Caterina (Johns Hopkins University School of Medicine) for his critical reading of the manuscript.

References

1 Caterina MJ, Schumacher MA, Tominaga M, Rosen TA, Levine JD, Julius D (1997) The capsaicin receptor: a heat-activated ion channel in the pain pathway. *Nature* 389: 816–824
2 Robinson RB, Siegelbaum SA (2003) Hyperpolarization-activated cation currents: from molecules to physiological function. *Annu Rev Physiol* 65: 453–480
3 Sansom MS, Shrivastava IH, Bright JN, Tate J, Capener CE, Biggin PC (2002) Potassium channels: structures, models, simulations. *Biochim Biophys Acta* 1565: 294–307
4 Kaupp UB, Seifert R (2002) Cyclic nucleotide-gated ion channels. *Physiol Rev* 82: 769–824
5 Montell C, Rubin GM (1989) Molecular characterization of the *Drosophila* trp locus: a putative integral membrane protein required for phototransduction. *Neuron* 2: 1313–1323
6 Montell C, Birnbaumer L, Flockerzi V (2002) The TRP channels, a remarkably functional family. *Cell* 108: 595–598
7 Clapham DE (2003) TRP channels as cellular sensors. *Nature* 426: 517–524
8 Sedgwick SG, Smerdon SJ (1999) The ankyrin repeat: a diversity of interactions on a common structural framework. *Trends Biochem Sci* 24: 311–316
9 Rosenbaum T, Gordon-Shaag A, Munari M, Gordon SE (2004) Ca^{2+}/calmodulin modulates TRPV1 activation by capsaicin. *J Gen Physiol* 123: 53–62
10 Harteneck C (2003) Proteins modulating TRP channel function. *Cell Calcium* 33: 303–310
11 Kedei N, Szabo T, Lile JD, Treanor JJ, Olah Z, Iadarola MJ, Blumberg PM (2001) Analysis of the native quaternary structure of vanilloid receptor 1. *J Biol Chem* 276: 28613–28619
12 Kuzhikandathil EV, Wang H, Szabo T, Morozova N, Blumberg PM, Oxford GS (2001) Functional analysis of capsaicin receptor (vanilloid receptor subtype 1) multimerization and agonist responsiveness using a dominant negative mutation. *J Neurosci* 21: 8697–8706
13 Smith GD, Gunthorpe MJ, Kelsell RE, Hayes PD, Reilly P, Facer P, Wright JE, Jerman

JC, Walhin JP, Ooi L et al (2002) TRPV3 is a temperature-sensitive vanilloid receptor-like protein. *Nature* 418: 186–190

14 Liu L, Wang Y, Simon SA (1996) Capsaicin activated currents in rat dorsal root ganglion cells. *Pain* 64: 191–195

15 Gunthorpe MJ, Harries MH, Prinjha RK, Davis JB, Randall A (2000) Voltage- and time-dependent properties of the recombinant rat vanilloid receptor (rVR1). *J Physiol* 525: 747–759

16 Premkumar LS, Agarwal S, Steffen D (2002) Single-channel properties of native and cloned rat vanilloid receptors. *J Physiol* 545: 107–117

17 Tominaga M, Caterina MJ, Malmberg AB, Rosen TA, Gilbert H, Skinner K, Raumann BE, Basbaum AI, Julius D (1998) The cloned capsaicin receptor integrates multiple pain-producing stimuli. *Neuron* 21: 531–543

18 Jung J, Hwang SW, Kwak J, Lee SY, Kang CJ, Kim WB, Kim D, Oh U (1999) Capsaicin binds to the intracellular domain of the capsaicin-activated ion channel. *J Neurosci* 19: 529–538

19 Zygmunt PM, Petersson J, Andersson DA, Chuang H, Sorgard M, Di Marzo V, Julius D, Hogestatt ED (1999) Vanilloid receptors on sensory nerves mediate the vasodilator action of anandamide. *Nature* 400: 452–457

20 Hwang SW, Cho H, Kwak J, Lee SY, Kang CJ, Jung J, Cho S, Min KH, Suh YG, Kim D et al (2000) Direct activation of capsaicin receptors by products of lipoxygenases: endogenous capsaicin-like substances. *Proc Natl Acad Sci USA* 97: 6155–6160

21 Huang SM, Bisogno T, Trevisani M, Al-Hayani A, De Petrocellis L, Fezza F, Tognetto M, Petros TJ, Krey JF, Chu CJ et al (2002) An endogenous capsaicin-like substance with high potency at recombinant and native vanilloid VR1 receptors. *Proc Natl Acad Sci USA* 99: 8400–8405

22 Jordt SE, Julius D (2002) Molecular basis for species-specific sensitivity to "hot" chili peppers. *Cell* 108: 421–430

23 Jung J, Lee SY, Hwang SW, Cho H, Shin J, Kang YS, Kim S, Oh U (2002) Agonist recognition sites in the cytosolic tails of vanilloid receptor 1. *J Biol Chem* 277: 44448–44454

24 Welch JM, Simon SA, Reinhart PH (2000) The activation mechanism of rat vanilloid receptor 1 by capsaicin involves the pore domain and differs from the activation by either acid or heat. *Proc Natl Acad Sci USA* 97: 13889–13894

25 Baumann TK, Martenson ME (2000) Extracellular protons both increase the activity and reduce the conductance of capsaicin- gated channels. *J Neurosci* 20: RC80

26 Jordt SE, Tominaga M, Julius D (2000) Acid potentiation of the capsaicin receptor determined by a key extracellular site. *Proc Natl Acad Sci USA* 97: 8134–8139

27 Jordt SE, McKemy DD, Julius D (2003) Lessons from peppers and peppermint: the molecular logic of thermosensation. *Curr Opin Neurobiol* 13: 487–492

28 Patapoutian A, Peier AM, Story GM, Viswanath V (2003) ThermoTRP channels and beyond: mechanisms of temperature sensation. *Nat Rev Neurosci* 4: 529–539

29 Vlachova V, Teisinger J, Susankova K, Lyfenko A, Ettrich R, Vyklicky L (2003) Func-

tional role of C-terminal cytoplasmic tail of rat vanilloid receptor 1. *J Neurosci* 23: 1340–1350

30 Numazaki M, Tominaga T, Toyooka H, Tominaga M (2002) Direct phosphorylation of capsaicin receptor VR1 by protein kinase Cepsilon and identification of two target serine residues. *J Biol Chem* 277: 13375–13378

31 Tominaga M, Wada M, Masu M (2001) Potentiation of capsaicin receptor activity by metabotropic ATP receptors as a possible mechanism for ATP-evoked pain and hyperalgesia. *Proc Natl Acad Sci USA* 98: 6951–6956

32 Voets T, Droogmans G, Wissenbach U, Janssens A, Flockerzi V, Nilius B (2004) The principle of temperature-dependent gating in cold- and heat-sensitive TRP channels. *Nature* 430: 748–754

33 Szallasi A, Blumberg PM (1999) Vanilloid (Capsaicin) receptors and mechanisms. *Pharmacol Rev* 51: 159–212

34 Piper AS, Yeats JC, Bevan S, Docherty RJ (1999) A study of the voltage dependence of capsaicin-activated membrane currents in rat sensory neurones before and after acute desensitization. *J Physiol* 518: 721–733

35 Liu L, Simon SA (1996) Capsaicin-induced currents with distinct desensitization and Ca^{2+} dependence in rat trigeminal ganglion cells. *J Neurophysiol* 75: 1503–1514

36 Koplas PA, Rosenberg RL, Oxford GS (1997) The role of calcium in the desensitization of capsaicin responses in rat dorsal root ganglion neurons. *J Neurosci* 17: 3525–3537

37 Docherty RJ, Yeats JC, Bevan S, Boddeke HW (1996) Inhibition of calcineurin inhibits the desensitization of capsaicin-evoked currents in cultured dorsal root ganglion neurones from adult rats. *Pflugers Arch* 431: 828–837

38 Bernstein JE (1987) Capsaicin in the treatment of dermatologic disease. *Cutis* 39: 352–353

39 Maggi CA (1991) Capsaicin and primary afferent neurons: from basic science to human therapy? *J Auton Nerv Syst* 33: 1–14

40 Bhave G., Zhu W, Wang H, Brasier DJ, Oxford GS, Gereau RWT (2002) cAMP-dependent protein kinase regulates desensitization of the capsaicin receptor (VR1) by direct phosphorylation. *Neuron* 35: 721–731

41 Mohapatra DP, Nau C (2003) Desensitization of capsaicin-activated currents in the vanilloid receptor TRPV1 is decreased by the cyclic AMP-dependent protein kinase pathway. *J Biol Chem* 278: 50080–50090

42 Numazaki M, Tominaga T, Takeuchi K, Murayama N, Toyooka H, Tominaga M (2003) Structural determinant of TRPV1 desensitization interacts with calmodulin. *Proc Natl Acad Sci USA* 100: 8002–8006

43 Garcia-Martinez C, Morenilla-Palao C, Planells-Cases R, Merino JM, Ferrer-Montiel A (2000) Identification of an aspartic residue in the P-loop of the vanilloid receptor that modulates pore properties. *J Biol Chem* 275: 32552–32558

44 Mohapatra DP, Wang SY, Wang GK, Nau C (2003) A tyrosine residue in TM6 of the

vanilloid receptor TRPV1 involved in desensitization and calcium permeability of capsaicin-activated currents. *Mol Cell Neurosci* 23: 314–324
45 De Petrocellis L, Harrison S, Bisogno T, Tognetto M, Brandi I, Smith GD, Creminon C, Davis JB, Geppetti P, Di Marzo V (2001) The vanilloid receptor (VR1)-mediated effects of anandamide are potently enhanced by the cAMP-dependent protein kinase. *J Neurochem* 77: 1660–1663
46 Hu HJ, Bhave G, Gereau RWT (2002) Prostaglandin and protein kinase A-dependent modulation of vanilloid receptor function by metabotropic glutamate receptor 5: potential mechanism for thermal hyperalgesia. *J Neurosci* 22: 7444–7452
47 Rathee PK, Distler C, Obreja O, Neuhuber W, Wang GK, Wang SY, Nau C, Kress M (2002) PKA/AKAP/VR-1 module: a common link of Gs-mediated signaling to thermal hyperalgesia. *J Neurosci* 22: 4740–4745
48 Premkumar LS, Ahern GP (2000) Induction of vanilloid receptor channel activity by protein kinase C. *Nature* 408: 985–990
49 Bhave G, Hu HJ, Glauner KS, Zhu W, Wang H, Brasier DJ, Oxford GS, Gereau RWT (2003) Protein kinase C phosphorylation sensitizes but does not activate the capsaicin receptor transient receptor potential vanilloid 1 (TRPV1). *Proc Natl Acad Sci USA* 100: 12480–12485
50 Sugiura T, Tominaga M, Katsuya H, Mizumura K (2002) Bradykinin lowers the threshold temperature for heat activation of vanilloid receptor 1. *J Neurophysiol* 88: 544–548
51 Dai Y, Moriyama T, Higashi T, Togashi K, Kobayashi K, Yamanaka H, Tominaga, M, Noguchi K (2004) Proteinase-activated receptor 2-mediated potentiation of transient receptor potential vanilloid subfamily 1 activity reveals a mechanism for proteinase-induced inflammatory pain. *J Neurosci* 24: 4293–4299
52 Jung J, Shin JS, Lee SY, Hwang SW, Koo J, Cho H, Oh U (2004) Phosphorylation of vanilloid receptor 1 by Ca^{2+}/calmodulin-dependent kinase II regulates its vanilloid binding. *J Biol Chem* 279: 7048–7054
53 Jin X, Morsy N, Winston J, Pasricha PJ, Garrett K, Akbarali HI (2004) Modulation of TRPV1 by nonreceptor tyrosine kinase, c-Src kinase. *Am J Physiol Cell Physiol* 287: C558–C563
54 Moriyama T, Iida T, Kobayashi K, Higashi T, Fukuoka T, Tsumura H, Leon C, Suzuki N, Inoue K, Gachet C et al (2003) Possible involvement of P2Y2 metabotropic receptors in ATP-induced transient receptor potential vanilloid receptor 1-mediated thermal hypersensitivity. *J Neurosci* 23: 6058–6062
55 Premkumar LS, Qi ZH, Van Buren J, Raisinghani M (2004) Enhancement of potency and efficacy of NADA by PKC-mediated phosphorylation of vanilloid receptor. *J Neurophysiol* 91: 1442–1449
56 Ahern GP (2003) Activation of TRPV1 by the satiety factor oleoylethanolamide. *J Biol Chem* 278: 30429–30434
57 Mandadi S, Numazaki M, Tominaga M, Bhat MB, Armati PJ, Roufogalis BD (2004) Activation of protein kinase C reverses capsaicin-induced calcium-dependent desensitization of TRPV1 ion channels. *Cell Calcium* 35: 471–478

58 Runnels LW, Yue L, Clapham DE (2002) The TRPM7 channel is inactivated by PIP(2) hydrolysis. *Nat Cell Biol* 4: 329–336
59 Huang CL, Feng S, Hilgemann DW (1998) Direct activation of inward rectifier potassium channels by PIP_2 and its stabilization by Gbetagamma. *Nature* 391: 803–806
60 Chuang HH, Prescott ED, Kong H, Shields S, Jordt SE, Basbaum AI, Chao MV, Julius D (2001) Bradykinin and nerve growth factor release the capsaicin receptor from PtdIns(4,5)P2-mediated inhibition. *Nature* 411: 957–962
61 Prescott ED, Julius D (2003) A modular PIP2 binding site as a determinant of capsaicin receptor sensitivity. *Science* 300: 1284–1288
62 Vyklicky L, Lyfenko A, Susankova K, Teisinger J, Vlachova V (2002) Reducing agent dithiothreitol facilitates activity of the capsaicin receptor VR-1. *Neuroscience* 111: 435–441

TRPV1 distribution and regulation

Janet Winter

Novartis Institute for Medical Sciences, 5 Gower Place, London WC1E 6BN, UK

Introduction

Long before the cloning of the capsaicin receptor, TRPV1, capsaicin-sensitive sensory neurons were characterized and studied by measuring capsaicin-induced depolarization of afferent fibres and of individual sensory neuron cell bodies. The ability of capsaicin to permanently ablate a subset of sensory neurons enabled characterization of those surviving: peptidergic afferents were underrepresented in the remaining population [1], suggesting that many capsaicin-sensitive sensory neurons normally express neuropeptides. Not surprisingly, capsaicin evoked the release of the neuropeptide substance P from primary sensory neurons [2]. A silver-staining technique for degenerating fibres identified capsaicin-sensitive afferent terminals in the dorsal horn of the spinal cord, following systemic capsaicin treatment. Silver staining also revealed widespread capsaicin-sensitive structures in the central nervous system [3]. Individual cells with functional capsaicin-sensitive channels were identified by a cobalt-uptake method in dissociated cultures of primary sensory neurons [4] and a high-affinity radioligand for the capsaicin receptor, tritiated resiniferotoxin ([^{3}H]RTX), localized capsaicin-binding sites autoradiographically in tissue sections [5]. Recent studies used a fluorescent dye that permeates activated TRPV1 channels to label capsaicin-sensitive cells *in vitro* or *in vivo* [6].

However, there was an explosion in receptor-localization studies following cloning and sequencing of the TRPV1 receptor, which allowed production of specific *in situ* hybridization probes and sensitive antibodies and led to the confirmation of expression of TRPV1 beyond the sensory afferent [7, 8]. This review attempts to summarize findings on the widespread distribution of TRPV1 receptors both within, and intriguingly outside, the nervous system and, where known, suggests multiple functions for these receptors at multiple sites.

What little is known about the regulation of TRPV1 protein levels is also discussed and this is almost entirely limited to the plasticity of the sensory nervous sys-

Turning up the Heat on Pain: TRPV1 Receptors in Pain and Inflammation, edited by
Annika B. Malmberg and Keith R. Bley

tem in pain models, plus a few examples in humans. This chronic regulation is separate from the acute regulation (i.e. sensitization by post-translational modifications), which will be dealt with in depth in other chapters in this volume.

TRPV1 in the peripheral nervous system

Sensory ganglia

By far the greatest interest in capsaicin over the years has focused on the role of capsaicin receptors in nociceptive sensory neurons; originally sensory neurons were thought to be more or less the exclusive site of action of capsaicin and its analogues. Neonatal capsaicin treatment permanently destroyed most C-fibres, but also some A-fibres [9]. Labelling of cultured adult rat sensory dorsal root ganglion (DRG) neurons with cobalt during capsaicin stimulation confirmed that most of the 'large, light' neurofilament-positive A-fibre neurons were capsaicin-insensitive, while many of the 'small, dark' C-fibre neurons, some of which are peptidergic, were sensitive, as they accumulated cobalt [4].

Sensory neurons in sections of rat DRG express TRPV1 immunoreactivity [7, 10]. In naïve animals, most TRPV1 immunolabelling overlaps with either the calcitonin-gene-related peptide (CGRP) or the isolectin B4 (IB4) population, with a small percentage in the neurofilament-positive A-fibres [11]. Trigeminal ganglia – the cranial sensory ganglia – also express TRPV1 in rat [12, 13] and human [14]. The petrosal, but not the geniculate, ganglia neurons express mRNA for TRPV1 [15]. Nodose TRPV1 mRNA [12, 16] and protein [18] are expressed in most nodose ganglion sensory neurons (Fig. 1). Nodose neurons tend to sense feelings of fullness and nausea from the stomach rather than painful sensations. Superior cervical ganglion (SCG) sympathetic neurons do not express TRPV1 (Fig. 1) and do not respond to capsaicin with calcium uptake [19].

Innervation of sensory organs

Many free nerve endings in skin express TRPV1 [20] but recently Meissner corpuscles in glabrous skin of monkeys have been shown to be innervated by TRPV1-expressing C-fibres in addition to the innervation by A-fibres [21]. Meissner corpuscles are widely regarded as low-threshold mechanoreceptors but the presence of TRPV1-positive innervation suggests a further role in nociception, possibly in allodynia in pathological conditions. TRPV1 is also found in high concentrations in the innervation of taste papillae, with some TRPV1-positive fibres penetrating the taste buds. TRPV1 is not expressed on taste buds themselves; hence the sensation of hot peppers is not included in the taste modalities. Capsaicin rather pro-

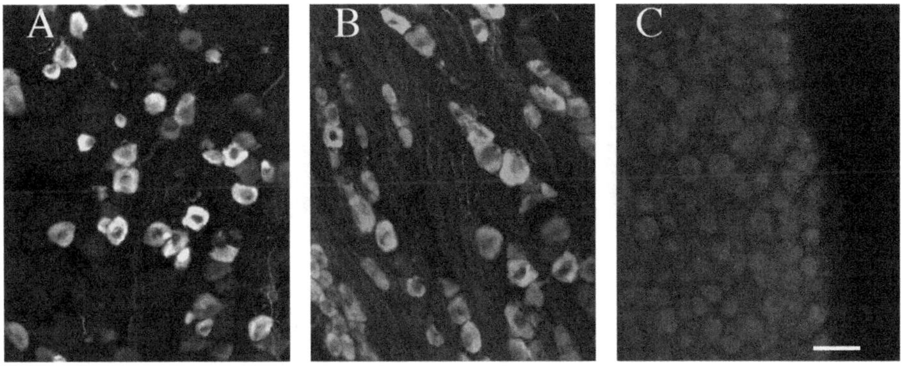

Figure 1
TRPV1 in sensory but not sympathetic ganglia. (A–C) Rabbit anti-TRPV1 C-terminal antibody. Sections of rat (A) dorsal root ganglion (DRG), (B) nodose ganglion and (C) superior sympathetic ganglion. Scale bar, 100 μm.

duces a burning sensation in the oral cavity by direct activation of sensory afferents [22]. The cochlear vasculature is innervated by TRPV1-positive afferents from the trigeminal ganglion [23] and the authors suggest that perturbation of these afferents may play a role in vertigo, tinnitus and hearing deficits associated with migraine.

Spinal distribution

TRPV1 is expressed on sensory afferent terminals in the spinal cord predominantly in laminae I and II, consistent with C- and Aδ-fibre expression [24, 25]. As expected, this immunostaining is largely depleted following dorsal rhizotomy. However, there has been one report showing that spinal cord astrocytes also express TRPV1 [26] and the same workers show that some dorsal horn NK-1-expressing cells also colocalize with TRPV1 [27].

Innervation of other organs

Capsaicin evokes vasodilatation in the dura mater of the rat [28], presumably by stimulation of capsaicin-sensitive afferents and the resultant release of vasoactive neuropeptides. TRPV1, which can be stimulated by alcohol [29], may be a major player in ethanol-evoked headaches and migraines; these are strongly associated with dural vasodilatation which can be evoked by CGRP release from TRPV1-

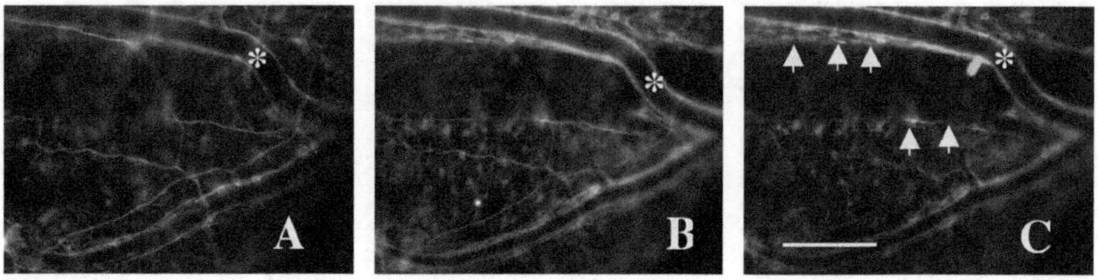

Figure 2
*TRPV1 in the dura. (A–C) Whole mounts of rat dura stained with (A) mouse anti-β-III tubulin to reveal all nerve fibres, (B) sheep anti-calcitonin-gene-related peptide (CGRP) and (C) TRPV1. *Denotes blood vessels; arrows mark nerve fibres. Scale bar, 50 μm.*

expressing trigeminal afferents. TRPV1 and CGRP immunoreactivity in the dura mater is shown in Figure 2.

Teeth are also innervated by trigeminal sensory fibres and tooth-pulp afferents express TRPV1 in samples of both human and rat tissue [30, 31].

The viscera have a higher density of capsaicin-sensitive afferents than either skin or skeletal muscle [32] with 60% of bladder afferents expressing sensitivity compared to 40% of muscle and only 20–30% of skin afferents. TRPV1 is expressed on extrinsic sensory nerve fibres distributed in all the regions of the gastrointestinal tract in myenteric ganglia, the muscle layer and mucosa. High expression of TRPV1-expressing fibres has been detected in biopsies from patients with several inflammatory diseases of the colon and ileum [33], in vulvodynia [34] and in familial rectal pain [35], and TRPV1 activation seems to play a role in intestinal motility disorders [36] as well as pain in the gastrointestinal tract. However, TRPV1-positive sensory afferents in the gastric mucosa also play a protective role against ulceration [37, 38].

Numerous TRPV1-immunoreactive fibres are found in the mucosa and muscular layer of the entire urinary tract except the kidney [39]. As well as detecting pain the TRPV1-expressing endings may also encode the tone of the bladder smooth muscle. The heart is innervated by TRPV1-expressing afferent nerves and these afferent nerves are involved in the cardiogenic sympathoexcitatory reflex during myocardial ischaemia [40]. Thyroglobulin-producing follicles in thyroid tissue are innervated by β-III tubulin-expressing nerve fibres, some of which are also TRPV1-positive (Fig. 3). TRPV1 function in the thyroid is unknown but may have a neuroendocrine role. Lungs are innervated by capsaicin-sensitive afferents and activation of TRPV1 is an important mechanism in cough [41]. As ethanol can activate TRPV1 channels, ethanol-induced cough may be mediated via TRPV1 expressed on sensory afferents.

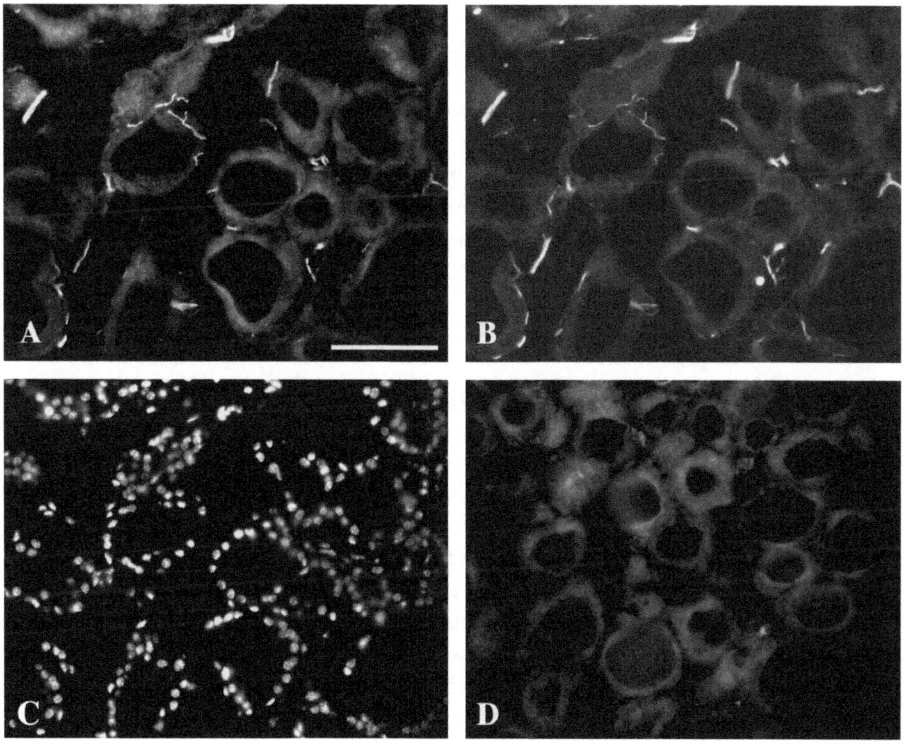

Figure 3
TRPV1 on wild-type and knockout mouse thyroid. (A–C) The same section of wild-type mouse thyroid labelled with (A) rabbit anti-TRPV1 C-terminal antibody, (B) mouse anti-β-III tubulin to reveal all nerve fibres, (C) 4,6-diamidino-2-phenylindole (DAPI) to reveal cell nuclei. The rings of nuclei are thyroid follicular cells. (D) Rabbit anti-TRPV1, on knockout mouse thyroid; no specific staining is seen on nerve fibres. Scale bar, 100 μm.

Central nervous system distribution

Multiple lines of evidence point to TRPV1 expression in the brain. The direct action of capsaicin on the preoptic/anterior hypothalamus (POAH) is responsible for the hypothermia evoked by systemic capsaicin. This functional effect is abolished in TRPV1-knockout animals, supporting a TRPV1-evoked mechanism [10]. Capsaicin causes ultrastructural damage to some hypothalamic neurons and [^{3}H]RTX-binding sites [42], TRPV1 mRNA and TRPV1 immunoreactivity [43] have all been detected in the hypothalamus. These TRPV1-expressing neurons may function as warm receptors sensing core body temperature [44]. A wide range of other brain structures has been shown to express TRPV1; interestingly, some neocortical and other brain regions have higher density of [^{3}H]RTX binding than even the DRG or the spinal cord [42, 43]. The function of these channels is unknown, but they are more likely to respond to anandamide and eicosenoids than temperatures above the threshold of activation (42 °C) [10]. A recent study has suggested a neuroprotective role for TRPV1 in the brain, with an endovanilloid protecting against ouabain-induced excitotoxicity [45]. Moreover, in a model of Huntington's disease, TRPV1 agonists produced antihyperkinetic activity [46].

TRPV1 expression outside the nervous system

As well as being expressed on the sensory innervation of the airways, TRPV1 is also expressed on airways epithelial cells, and activation of TRPV1 by particulate matter induces apoptosis, which may be an important mechanism for particulate matter damage to the airways [47]. Air pollutants are thought to cause a reduced immune function and lowered resistance to infection as well as hypersensitivity to allergens and may exacerbate asthma and chronic obstructive pulmonary disease (COPD). Environmental particulate matter can act like synthetic particles made from negatively charged polymers, which are thought to act by concentrating protons near proton-gated receptors. Particulate matter causes a slowly desensitizing activation of TRPV1, and the calcium influx leads to apoptosis in sensory neurons and lung epithelial cells [47]. The authors suggest that blocking TRPV1 receptors would be useful in alleviating particulate matter-induced inflammation and toxicity.

In the skin, activation of TRPV1-expressing afferents can release the neuropeptides substance P and neurokinin A, which induce cutaneous nerve growth factor (NGF) production. Although high doses of capsaicin can destroy peptideric endings and impair wound healing by depleting neuropeptides, low-level TRPV1 activation may release neuropeptides and have trophic effects. This trophic action of sensory neuropeptides may have important consequences during inflammation and wound healing [48]. However, rat and human keratinocytes themselves express TRPV1, and activation by capsaicin caused an upregulation of cyclooxygenase 2 (COX-2) and an increased release of interleukin-8 and prostaglandin E2 [49]. These results suggest that TRPV1 expression in keratinocytes may have a role in the inflammation that occurs secondary to epidermal damage or insult, and thus may function as a sensor for noxious cutaneous stimulation [49–51]. In puritic skin of patients with prurigo nodularis there is an increased expression of TRPV1 in keratinocytes and nerve fibres. The same study describes TRPV1 expression in skin mast cells [52]. This distribution, in nerve cells, keratinocytes and skin mast cells, along with the efficacy of capsaicin in pruritus, suggests TRPV1 modulators may be useful antipruritics. An anti-proliferative role for capsaicin on melanoma cells has also been shown [53].

Bladder epithelial cells also express TRPV1. The urothelium itself is a responsive structure capable of detecting stretch and it can release signaling molecules such as ATP, which then triggers reflex bladder contraction. TRPV1-null mice exhibit reduced responses to bladder stretch. Moreover, excised urothelial cells from TRPV1-null mice release less ATP when stimulated by hypotonic-evoked swelling. This finding is currently not fully understood, as recombinant TRPV1 cannot be activated by hypotonic solutions. One explanation is that TRPV1 is coexpressed with a 'stretch-sensitive' subunit in the bladder. Whatever underlies the molecular basis for TRPV1 and stretch responses, there is a clear involvement of TRPV1 in bladder function that involves TRPV1 expressed on both sensory fibres and urothelial cells [7] and infusion of capsaicin into bladders of patients with hyperactive bladders can reduce micturition [54].

As mentioned above, TRPV1-positive sensory fibres can have a gastroprotective role. There is evidence of TRPV1 expression also in epithelial cells of the gastrointestinal tract [36] and these authors suggest that TRPV1 blockers may be useful drugs for treatment of gut inflammation. Finally, a possible neuroendocrine role is suggested by the presence of TRPV1 receptors in pancreatic islet β-cells [55].

Regulation of TRPV1

Most of the information in the literature on regulation of TRPV1 relates to studies of rat DRG sensory neurons, either *in vitro* or *in vivo*. In two neuropathic models, partial sciatic nerve ligation and spinal nerve ligation, the DRG neurons with injured axons lose the majority of the TRPV1 protein in their cell bodies, although accumulation is seen proximal to the ligation site following nerve ligation [56]. TRPV1 mRNA also declines in DRG cell bodies following axotomy [57]. However, an increase in TRPV1 protein and mRNA expression is evident in DRG cell bodies with uninjured axons; that is, these neurons are still connected to the peripheral targets [56, 58]. Moreover, there is a pronounced increase of TRPV1 expression in some A-fibres, which normally express very little TRPV1 [56]. It is possible that some of these newly expressing A-fibres may be Aβ and Aδ nociceptors [59]. Of particular interest is that this increase in TRPV1 in A-fibres following nerve injury is even more obvious in mice that have been neonatally treated with capsaicin and have very few C-fibres or TRPV1 protein left in the DRG. Neonatally capsaicin-pretreated animals have decreased responsiveness to capsaicin and heat. When these animals undergo a subsequent partial nerve ligation, there is a dramatic increase in TRPV1 in the A-fibres and a concomitant increase in capsaicin responsiveness of the animals, but thermal responsiveness is not changed, suggesting that these newly expressing fibres are not thermally sensitive [60]. A similar increase in TRPV1 expression in A-fibres is observed in diabetic mice (induced by streptozotocin) [61].

Functional TRPV1 protein expression is also increased in DRG neurons in an inflammatory pain model [20, 62]. No upregulation of TRPV1 mRNA is seen in the DRG when measured by RNase protection, and the authors suggest transcription-independent upregulation involving the kinase p38, which is activated by NGF produced in the inflamed tissue [20]. However, a different study showed increase in TRPV1 protein by immunohistochemistry and TRPV1 mRNA, by *in situ* hybridization, especially in neurofilament-positive A-fibres in DRG from complete Freund's adjuvant (CFA)-treated rats [63]. One possible explanation for this discrepancy is that methods for measuring mRNA in whole ganglia may not be sensitive enough to see changes in small populations of cells innervating the inflamed tissue, whereas such differences can be detected by *in situ* hybridization. Carageenan-induced inflammation can also cause increased transport of TRPV1 mRNA along primary afferents [64].

Regulation of functional TRPV1 protein has also been studied in cultured sensory neurons from adult rats. Functional TRPV1 protein expression is upregulated in DRG neurons by NGF and glial cell line-derived neurotrophic factor (GDNF) [65–67], whereas brain-derived neurotrophic factor (BDNF) controls capsaicin sensitivity in nodose ganglion neuron cultures [68]. NGF also regulates TRPV1 mRNA levels in such cultures [66, 69]. The signalling pathways involved in TRPV1 regulation are now partially understood. NGF is known to signal through its high-affinity receptor, TRKA, via a mitogen-activated protein (MAP) kinase cascade involving Ras, extracellular-signal-regulated kinases (ERKs) and MAP kinase/ERK kinase (MEK) [70]. In DRG neurons, growth factor-mediated upregulation of capsaicin sensitivity and functional TRPV1 protein can be partially blocked by MEK inhibitors [67, 71] and microinjection of constitutively active Ras protein into cultured DRG neurons is sufficient for increased TRPV1 expression in the absence of growth factors. Constitutively active ERK, which is downsteam of Ras, is not sufficient for TRPV1 expression, suggesting that activation of other pathways downstream of Ras may also be involved. One of these pathways could be via activation of p38 kinase.

In summary, plasticity of TRPV1 expression in chronic pain conditions may constitute one of the mechanisms responsible for hyperalgesia.

Acknowledgements

I would like to thank Glen Wotherspoon for providing Figures 2 and 3, and Stuart Bevan, Peter McIntyre and Marzia Malcangio for reading the manuscript.

References

1 Buck SH, Burks TF (1986) The neuropharmacology of capsaicin: review of some recent observations. *Pharmacol Rev* 38: 179–226
2 Theriault E, Otsuka M, Jessell T (1979) Capsaicin-evoked release of substance P from primary sensory neurons. *Brain Res* 170: 209–213
3 Ritter S, Dinh TT (1988) Capsaicin-induced neuronal degeneration: silver impregnation of cell bodies, axons, and terminals in the central nervous system of the adult rat. *J Comp Neurol* 271: 79–90
4 Winter J (1987) Characterization of capsaicin-sensitive neurones in adult rat dorsal root ganglion cultures. *Neurosci Lett* 80: 134–140
5 Winter J, Walpole CS, Bevan S, James IF (1993) Characterization of resiniferatoxin binding sites on sensory neurons: co-regulation of resiniferatoxin binding and capsaicin sensitivity in adult rat dorsal root ganglia. *Neuroscience* 57: 747–757
6 Meyers JR, MacDonald RB, Duggan A, Lenzi D, Standaert DG, Corwin JT, Corey DP (2003) Lighting up the senses: FM1-43 loading of sensory cells through nonselective ion channels. *J Neurosci* 23: 4054–4065
7 Caterina MJ (2003) Vanilloid receptors take a TRP beyond the sensory afferent. *Pain* 105: 5–9
8 Szallasi A (2002) Vanilloid (capsaicin) receptors in health and disease. *Am J Clin Pathol* 118: 110–121
9 Lawson SN (1987) The morphological consequences of neonatal treatment with capsaicin on primary afferent neurones in adult rats. *Acta Physiol Hung* 69: 315–321
10 Caterina MJ, Schumacher MA, Tominaga M, Rosen TA, Levine JD, Julius D (1997) The capsaicin receptor: a heat-activated ion channel in the pain pathway. *Nature* 389: 816–824
11 Ma QP (2001) Vanilloid receptor homologue, VRL1, is expressed by both A- and C-fiber sensory neurons. *Neuroreport* 12: 3693–3695
12 Ichikawa H, Sugimoto T (2001) VR1-immunoreactive primary sensory neurons in the rat trigeminal ganglion. *Brain Res* 890: 184–188
13 Dinh QT, Groneberg DA, Mingomataj E, Peiser C, Heppt W, Dinh S, Arck PC, Klapp BF, Fischer A (2003) Expression of substance P and vanilloid receptor (VR1) in trigeminal sensory neurons projecting to the mouse nasal mucosa. *Neuropeptides* 37: 245–250
14 Hou M, Uddman R, Tajti J, Kanje M, Edvinsson L (2002) Capsaicin receptor immunoreactivity in the human trigeminal ganglion. *Neurosci Lett* 330: 223–226
15 Matsumoto I, Emori Y, Ninomiya Y, Abe K (2001) A comparative study of three cranial sensory ganglia projecting into the oral cavity: in situ hybridization analyses of neurotrophin receptors and thermosensitive cation channels. *Brain Res* Mol *Brain Res* 93: 105–112
16 Ichikawa H, Sugimoto T (2003) The co-expression of VR1 and VRL-1 in the rat vagal sensory ganglia. *Brain Res* 980: 293–296
17 Helliwell RJ, McLatchie LM, Clarke M, Winter J, Bevan S, McIntyre P (1998) Capsaicin

sensitivity is associated with the expression of the vanilloid (capsaicin) receptor (VR1) mRNA in adult rat sensory ganglia. *Neurosci Lett* 250: 177–180

18 Zhang L, Jones S, Brody K, Costa M, Brookes SJ (2004) Thermosensitive transient receptor potential channels in vagal afferent neurons of the mouse. *Am J Physiol Gastrointest Liver Physiol* 286: G983–G991

19 Wood JN, Winter J, James IF, Rang HP, Yeats J, Bevan S (1988) Capsaicin-induced ion fluxes in dorsal root ganglion cells in culture. *J Neurosci* 8: 3208–3220

20 Ji RR, Samad TA, Jin SX, Schmoll R, Woolf CJ (2002) p38 MAPK activation by NGF in primary sensory neurons after inflammation increases TRPV1 levels and maintains heat hyperalgesia. *Neuron* 36: 57–68

21 Pare M, Elde R, Mazurkiewicz JE, Smith AM, Rice FL (2001) The Meissner corpuscle revised: a multiafferented mechanoreceptor with nociceptor immunochemical properties. *J Neurosci* 21: 7236–7246

22 Ishida Y, Ugawa S, Ueda T, Murakami S, Shimada S (2002) Vanilloid receptor subtype-1 (VR1) is specifically localized to taste papillae. *Brain Res* Mol *Brain Res* 107: 17–22

23 Vass Z, Dai CF, Steyger PS, Jancso G, Trune DR, Nuttall AL (2004) Co-localization of the vanilloid capsaicin receptor and substance P in sensory nerve fibers innervating cochlear and vertebro-basilar arteries. *Neuroscience* 124: 919–927

24 Guo A, Vulchanova L, Wang J, Li X, Elde R (1999) Immunocytochemical localization of the vanilloid receptor 1 (VR1): relationship to neuropeptides, the P2X3 purinoceptor and IB4 binding sites. *Eur J Neurosci* 11: 946–958

25 Guo A, Simone DA, Stone LS, Fairbanks CA, Wang J, Elde R (2001) Developmental shift of vanilloid receptor 1 (VR1) terminals into deeper regions of the superficial dorsal horn: correlation with a shift from TrkA to Ret expression by dorsal root ganglion neurons. *Eur J Neurosci* 14: 293–304

26 Doly S, Fischer J, Salio C, Conrath M (2004) The vanilloid receptor-1 is expressed in rat spinal dorsal horn astrocytes. *Neurosci Lett* 357: 123–126

27 Doly S, Fischer J, Conrath M (2004) The vanilloid receptor-1 (TRPV1) is expressed in some rat dorsal horn NK1 cells. *Brain Res* 1004: 203–207

28 Dux M, Santha P, Jancso G (2003) Capsaicin-sensitive neurogenic sensory vasodilatation in the dura mater of the rat. *J Physiol* 552: 859–867

29 Trevisani M, Smart D, Gunthorpe MJ, Tognetto M, Barbieri M, Campi B, Amadesi S, Gray J, Jerman JC, Brough SJ et al (2002) Ethanol elicits and potentiates nociceptor responses via the vanilloid receptor-1. *Nat Neurosci* 5: 546–551

30 Renton T, Yiangou Y, Baecker PA, Ford AP, Anand P (2003) Capsaicin receptor VR1 and ATP purinoceptor P2X3 in painful and nonpainful human tooth pulp. *J Orofac Pain* 17: 245–250

31 Stenholm E, Bongenhielm U, Ahlquist M, Fried K (2002) VRl- and VRL-l-like immunoreactivity in normal and injured trigeminal dental primary sensory neurons of the rat. *Acta Odontol Scand* 60: 72–79

32 Hu-Tsai M, Winter J, Woolf CJ (1992) Regional differences in the distribution of cap-

saicin-sensitive target-identified adult rat dorsal root ganglion neurons. *Neurosci Lett* 143: 251–254
33 Yiangou Y, Facer P, Dyer NH, Chan CL, Knowles C, Williams NS, Anand P (2001) Vanilloid receptor 1 immunoreactivity in inflamed human bowel. *Lancet* 357: 1338–1339
34 Tympanidis P, Casula MA, Yiangou Y, Terenghi G, Dowd P, Anand P (2004) Increased vanilloid receptor VR1 innervation in vulvodynia. *Eur J Pain* 8: 129–133
35 Chan CL, Facer P, Davis JB, Smith GD, Egerton J, Bountra C, Williams NS, Anand P (2003) Sensory fibres expressing capsaicin receptor TRPV1 in patients with rectal hypersensitivity and faecal urgency. *Lancet* 361: 385–391
36 Geppetti P, Trevisani M (2004) Activation and sensitisation of the vanilloid receptor: role in gastrointestinal inflammation and function. *Br J Pharmacol* 141: 1313–1320
37 Kato S, Aihara E, Nakamura A, Xin H, Matsui H, Kohama K, Takeuchi K (2003) Expression of vanilloid receptors in rat gastric epithelial cells: role in cellular protection. *Biochem Pharmacol* 66: 1115–1121
38 Horie S, Yamamoto H, Michael GJ, Uchida M, Belai A, Watanabe K, Priestley JV, Murayama T (2004) Protective role of vanilloid receptor type 1 in HCl-induced gastric mucosal lesions in rats. *Scand J Gastroenterol* 39: 303–312
39 Avelino A, Cruz C, Nagy I, Cruz F (2002) Vanilloid receptor 1 expression in the rat urinary tract. *Neuroscience* 109: 787–798
40 Zahner MR, Li DP, Chen SR, Pan HL (2003) Cardiac vanilloid receptor 1-expressing afferent nerves and their role in the cardiogenic sympathetic reflex in rats. *J Physiol* 551: 515–523
41 Morice AH, Geppetti P (2004) Cough. 5: The type 1 vanilloid receptor: a sensory receptor for cough. *Thorax* 59: 257–258
42 Roberts JC, Davis JB, Benham CD (2004) [3H]Resiniferatoxin autoradiography in the CNS of wild-type and TRPV1 null mice defines TRPV1 (VR-1) protein distribution. *Brain Res* 995: 176–183
43 Mezey E, Toth ZE, Cortright DN, Arzubi MK, Krause JE, Elde R, Guo A, Blumberg PM, Szallasi A (2000) Distribution of mRNA for vanilloid receptor subtype 1 (VR1), and VR1-like immunoreactivity, in the central nervous system of the rat and human. *Proc Natl Acad Sci USA* 97: 3655–3660
44 Jancso-Gabor A, Szolcsanyi J, Jancso N (1970) Irreversible impairment of thermoregulation induced by capsaicin and similar pungent substances in rats and guinea-pigs. *J Physiol* 206: 495–507
45 Veldhuis WB, van der Stelt SM, Wadman MW, van Zadelhoff G, Maccarrone M, Fezza F, Veldink GA, Vliegenthart JF, Bar PR, Nicolay K et al (2003) Neuroprotection by the endogenous cannabinoid anandamide and arvanil against *in vivo* excitotoxicity in the rat: role of vanilloid receptors and lipoxygenases. *J Neurosci* 23: 4127–4133
46 Lastres-Becker I, de Miguel R, De Petrocellis L, Makriyannis A, Di Marzo V, Fernandez-Ruiz J (2003) Compounds acting at the endocannabinoid and/or endovanilloid systems

reduce hyperkinesia in a rat model of Huntington's disease. *J Neurochem* 84: 1097–1109
47 Agopyan N, Head J, Yu S, Simon SA (2004) TRPV1 receptors mediate particulate matter-induced apoptosis. *Am J Physiol Lung Cell Mol Physiol* 286: L563–L572
48 Burbach GJ, Kim KH, Zivony AS, Kim A, Aranda J, Wright S, Naik SM, Caughman SW, Ansel JC, Armstrong CA (2001) The neurosensory tachykinins substance P and neurokinin A directly induce keratinocyte nerve growth factor. *J Invest Dermatol* 117: 1075–1082
49 Southall MD, Li T, Gharibova LS, Pei Y, Nicol GD, Travers JB (2003) Activation of epidermal vanilloid receptor-1 induces release of proinflammatory mediators in human keratinocytes. *J Pharmacol Exp Ther* 304: 217–222
50 Inoue K, Koizumi S, Fuziwara S, Denda S, Inoue K, Denda M (2002) Functional vanilloid receptors in cultured normal human epidermal keratinocytes. *Biochem Biophys Res Commun* 291: 124–129
51 Denda M, Fuziwara S, Inoue K, Denda S, Akamatsu H, Tomitaka A, Matsunaga K (2001) Immunoreactivity of VR1 on epidermal keratinocyte of human skin. *Biochem Biophys Res Commun* 285: 1250–1252
52 Stander S, Moormann C, Schumacher M, Buddenkotte J, Artuc M, Shpacovitch V, Brzoska T, Lippert U, Henz BM, Luger TA et al (2004) Expression of vanilloid receptor subtype 1 in cutaneous sensory nerve fibers, mast cells, and epithelial cells of appendage structures. *Exp Dermatol* 13: 129–139
53 Patel PS, Yang S, Li A, Varney ML, Singh RK (2002) Capsaicin regulates vascular endothelial cell growth factor expression by modulation of hypoxia inducing factor-1alpha in human malignant melanoma cells. *J Cancer Res Clin Oncol* 128: 461–468
54 Cruz F (2004) Mechanisms involved in new therapies for overactive bladder. *Urology* 63: 65–73
55 Akiba Y, Kato S, Katsube K, Nakamura M, Takeuchi K, Ishii H, Hibi T (2004) Transient receptor potential vanilloid subfamily 1 expressed in pancreatic islet beta cells modulates insulin secretion in rats. *Biochem Biophys Res Commun* 321: 219–225
56 Hudson LJ, Bevan S, Wotherspoon G, Gentry C, Fox A, Winter J (2001) VR1 protein expression increases in undamaged DRG neurons after partial nerve injury. *Eur J Neurosci* 13: 2105–2114
57 Michael GJ, Priestley JV (1999) Differential expression of the mRNA for the vanilloid receptor subtype I in cells of the adult rat dorsal root and nodose ganglia and its down-regulation by axotomy. *J Neurosci* 19: 1844–1854
58 Fukuoka T, Tokunaga A, Tachibana T, Dai Y, Yamanaka H, Noguchi K (2002) VR1, but not P2X(3), increases in the spared L4 DRG in rats with L5 spinal nerve ligation. *Pain* 99: 111–120
59 Ritter AM, Mendell LM (1992) Somal membrane properties of physiologically identified sensory neurons in the rat: effects of nerve growth factor. *J Neurophysiol* 68: 2033–2041
60 Rashid MH, Inoue M, Kondo S, Kawashima T, Bakoshi S, Ueda H (2003) Novel expres-

sion of vanilloid receptor 1 on capsaicin-insensitive fibers accounts for the analgesic effect of capsaicin cream in neuropathic pain. *J Pharmacol Exp Ther* 304: 940–948
61 Rashid MH, Inoue M, Bakoshi S, Ueda H (2003) Increased expression of vanilloid receptor 1 on myelinated primary afferent neurons contributes to the antihyperalgesic effect of capsaicin cream in diabetic neuropathic pain in mice. *J Pharmacol Exp Ther* 306: 709–717
62 Nicholas RS, Winter J, Wren P, Bergmann R, Woolf CJ (1999) Peripheral inflammation increases the capsaicin sensitivity of dorsal root ganglion neurons in a nerve growth factor-dependent manner. *Neuroscience* 91: 1425–1433
63 Amaya F, Oh-hashi K, Naruse Y, Iijima N, Ueda M, Shimosato G, Tominaga M, Tanaka Y, Tanaka M (2003) Local inflammation increases vanilloid receptor 1 expression within distinct subgroups of DRG neurons. *Brain Res* 963: 190–196
64 Tohda C, Sasaki M, Konemura T, Sasamura T, Itoh M, Kuraishi Y (2001) Axonal transport of VR1 capsaicin receptor mRNA in primary afferents and its participation in inflammation-induced increase in capsaicin sensitivity. *J Neurochem* 76: 1628–1635
65 Bevan S, Winter J (1995) Nerve growth factor (NGF) differentially regulates the chemosensitivity of adult rat cultured sensory neurons. *J Neurosci* 15: 4918–4926
66 Ogun-Muyiwa P, Helliwell R, McIntyre P, Winter J (1999) Glial cell line derived neurotrophic factor (GDNF) regulates VR1 and substance P in cultured sensory neurons. *Neuroreport* 10: 2107–2111
67 Bron R, Klesse LJ, Shah K, Parada LF, Winter J (2003) Activation of Ras is necessary and sufficient for upregulation of vanilloid receptor type 1 in sensory neurons by neurotrophic factors. *Mol Cell Neurosci* 22: 118–132
68 Winter J (1998) Brain derived neurotrophic factor, but not nerve growth factor, regulates capsaicin sensitivity of rat vagal ganglion neurones. *Neurosci Lett* 241: 21–24
69 Winston J, Toma H, Shenoy M, Pasricha PJ (2001) Nerve growth factor regulates VR-1 mRNA levels in cultures of adult dorsal root ganglion neurons. *Pain* 89: 181–186
70 Huang EJ, Reichardt LF (2003) Trk receptors: roles in neuronal signal transduction. *Annu Rev Biochem* 72: 609–642
71 Ganju P, O'Bryan JP, Der C, Winter J, James IF (1998) Differential regulation of SHC proteins by nerve growth factor in sensory neurons and PC12 cells. *Eur J Neurosci* 10: 1995–2008

Part III
Pharmacology and physiology of vanilloid receptors

Insights into TRPV1 pharmacology provided by non-capsaicin ligands

Peter M. Blumberg, Derek C. Braun, Noemi Kedei, Jozsef Lazar, Vladimir Pavlyukovets and Larry V. Pearce

Laboratory of Cellular Carcinogenesis and Tumor Promotion, National Cancer Institute, Bethesda, MD 20892, USA

Resiniferatoxin as an ultrapotent capsaicin analog with a unique pattern of activity

The identification of resiniferatoxin as a capsaicin analog had important conceptual and practical impact on the field. Hecker and coworkers were analyzing the irritant constituents in the latex of *Euphorbia resinifera*, long used as a traditional medicine [1], as part of a much more extensive effort to identify congeners of the tumor-promoting phorbol esters, which are found in many members of the family Euphorbiaceae. One of the constituents they identified in this latex, resiniferatoxin (RTX), was unique [2]. Structurally, RTX was a typical diterpene ester of the daphnane class, thus being related to the phorbol esters (Fig. 1). Importantly, however, RTX differed in possessing a homovanillyl ester at the C-20 position. Already at that time, it was known that a free C-20 hydroxyl group was essential for typical phorbol ester activity, and, indeed, RTX was inactive as a tumor promoter and inactive in cellular assays responsive to typical phorbol esters. We now appreciate that the free C-20 hydroxyl of the typical phorbol ester inserts into a cleft in the binding domain of its target, the C1 domain of protein kinase C, and that substitution at the C-20 position sterically interferes with this insertion.

The bioassay used by the Hecker group for identification of irritant compounds was that of mouse ear reddening. In that bioassay, RTX was also unique. First, it was extraordinarily potent, being approximately 1000-fold more potent that the most potent of the phorbol esters, phorbol 12-myristate 13-acetate. Secondly, its irritancy was transient – unlike the much more persistent inflammation induced by the phorbol esters – and was followed by complete tachyphylaxis.

My group undertook to elucidate the mechanism of action of RTX as part of our broader program to understand the mechanisms accounting for the diversity in the patterns of response to phorbol ester congeners. We hypothesized that RTX might function as a capsaicin homolog based on the structural similarity between the homovanillyl side chain of RTX, which differentiated it from typical phorbol esters,

Turning up the Heat on Pain: TRPV1 Receptors in Pain and Inflammation, edited by
Annika B. Malmberg and Keith R. Bley

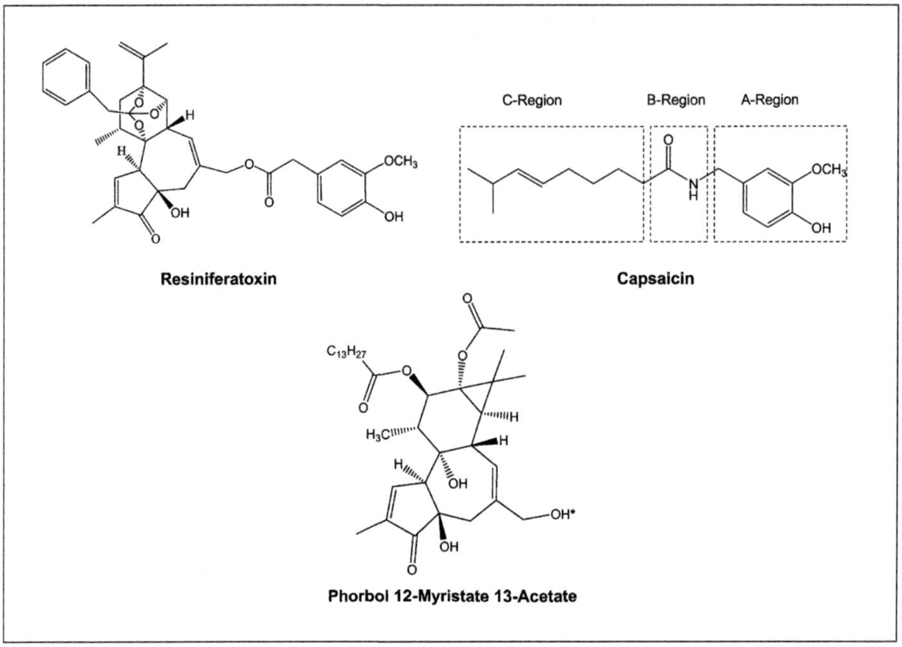

Figure 1
Comparison of the structures of resiniferatoxin (RTX), phorbol 12-myristate 13-acetate, and capsaicin. For capsaicin, the elements of the structure defining the A-, B-, and C-regions are indicated. For phorbol 12-myristate 13-acetate, an asterisk marks the primary C-20 hydroxyl group, which is a critical element of the pharmacophore required for binding to protein kinase C.

and the homovanillylamide moiety of capsaicin. We found that RTX indeed acted as a capsaicin analog [3]. RTX mimicked capsaicin in its effects on pungency, neurogenic inflammation, and thermoregulation. Like capsaicin, RTX induced desensitization/defunctionalization, and RTX showed cross-desensitization/defunctionalization with capsaicin. Secondly, RTX was ultrapotent. For thermoregulation and neurogenic inflammation, RTX was three or four orders of magnitude more potent than capsaicin; thus, it was five or six orders of magnitude more potent than the habañero pepper! Finally, RTX was not simply an ultrapotent capsaicin analog but showed selectivity between responses; RTX was only modestly more pungent than capsaicin. Since pungency represents the limiting toxicity for the therapeutic application of capsaicin, RTX thus had promise as a new therapeutic agent. Subsequent studies have confirmed and extended all of these observations [4].

This analysis of RTX made two important conceptual contributions. First, it showed that there were great, unexploited opportunities for enhancing ligand potency, as well as other properties. This had not been particularly evident from the extensive medicinal chemistry that had been done up to that time [5–8]. Second, it demonstrated that pungency was not intrinsically linked to potency for desensitization, complementing the findings with olvanil, which was appreciably less potent but likewise showed dissociation of desensitization and pungency [9].

[^{3}H]RTX binding demonstrated the existence of a capsaicin receptor

A receptor mechanism for the action of capsaicin had been postulated by Szolcsányi and coworkers based on structure-activity studies [8], but capsaicin was unsuitable for a radioligand binding assay to demonstrate its existence because of its high lipophilicity relative to its potency. Although RTX is also quite lipophilic, its dramatically enhanced potency made it correspondingly more suitable as a ligand for binding analysis. Using [^{3}H]RTX, we were able to demonstrate the existence of the vanilloid receptor, with appropriate structure-activity relations and species specificity consistent with the characterization of capsaicin action in biological systems [10]. The binding assay was further refined through the addition of α_1-acid glycoprotein, which reduced non-specific binding [11].

Although less widely used, 5-iodoRTX has also been used for binding analysis of the capsaicin receptor [12]. It has the advantage of higher potential specific activity but the disadvantages of weaker binding affinity (4 nM) and more complicated kinetics.

The demonstration of the vanilloid receptor provided strong motivation for the cloning of the receptor, which in turn has permitted its further molecular characterization [13]. The [^{3}H]RTX binding assay has further provided a convenient screening assay for the quantitative evaluation of vanilloid structure-activity relations. An intrinsic advantage of this assay is that it can be performed with long incubation times and on broken cell preparations. It thereby avoids the kinetic distortion of the measures of calcium uptake, which saturate rapidly, are subject to desensitization, and are affected by problems of penetration of ligand into the inner face of the membrane, where the site of vanilloid recognition of the receptor appears to be located [14]. On the other hand, careful analysis of the structure-activity relations emerging from the [^{3}H]RTX binding assay and from assays of calcium uptake suggested that the quantitative values were not identical [15]. Rather, some ligands were somewhat selective for the [^{3}H]RTX binding assay, whereas others were somewhat selective for the calcium-uptake assay. Although a plausible explanation was that there were multiple receptors for vanilloids – which we termed R-type and C-type, based on their selectivity – it is now clear from expression of TRPV1 in heterologous systems that the cloned TRPV1 can account for both activities [16]. Likewise, knockout mice for TRPV1 lose capsaicin responsiveness [17, 18]. The pharmacological reality remains, however, that different assays yield different views of vanilloid structure-activity relations. The plausible explanation, as will be discussed in more detail below, is that TRPV1 exists in different states with different ligand-recognition properties, whether these different states arise through association with other proteins or through the influence of different effectors.

Diversity of structures active on TRPV1

A dominant theme emerging from the analysis of vanilloid action is that every aspect shows great complexity. An implication, of course, is that this diversity provides extraordinary opportunities for pharmacological exploitation.

One aspect of this diversity is the great variation in structures that bind to TRPV1 through the capsaicin/RTX-recognition site. The standard concept of the structural features involved in binding is that presented in an elegant series of papers by Walpole et al. [5–7]. The capsaicin pharmacophore is postulated to be contributed by the A-region *p*-hydroxy, *m*-methoxyphenyl group, the B-region amide or ester, and the hydrophobic C-region (Fig. 1). RTX fits within this model, with the further insight that elaboration of the structure of the C-region can provide marked enhancement in potency and differential selectivity. Phorbol 12,13-esters can substitute for the resiniferonol 9,13,14-orthophenylacetate group in RTX, although with a substantial decrease in potency, as exemplified by phorbol 12-phenylacetate 13-acetate 20-homovanillate (PPAHV) [19]. Numerous derivatives built upon a structure, designed to embody the essential elements of the RTX C-region, have also been explored [20].

Like RTX, the natural products anandamide and arachidonyl-dopamine, which have been postulated to be endogenous regulators of TRPV1, fit within the standard pharmacophore model, although with a simplified A-region [21]. In contrast, some other natural products possess markedly divergent structures (Fig. 2). Evodiamine, a component of the pungent herbal medicine Wu-Zhu-Yu, binds TRPV1 competitively, functions as a TRPV1 agonist, and behaves as a typical capsaicin analog in multiple-tissue and whole-animal assays [22, 23]. Its potency is some 3–20-fold less than that of capsaicin. Thapsigargin, a potent inhibitor of the sarcoplasmic reticulum ATPase, inhibited [^{3}H]RTX binding with comparable affinity to capsaicin but through a mixed mechanism, and functioned as an antagonist of TRPV1 function [24]. Finally, the pungent dialdehyde sesquiterpenes such as isovelleral interact with TRPV1. However, since they are reactive molecules and affect many cellular processes, it is clear that they function through multiple mechanisms. These natural products demonstrate that a broad range of structural solutions exist which can confer affinity for TRPV1.

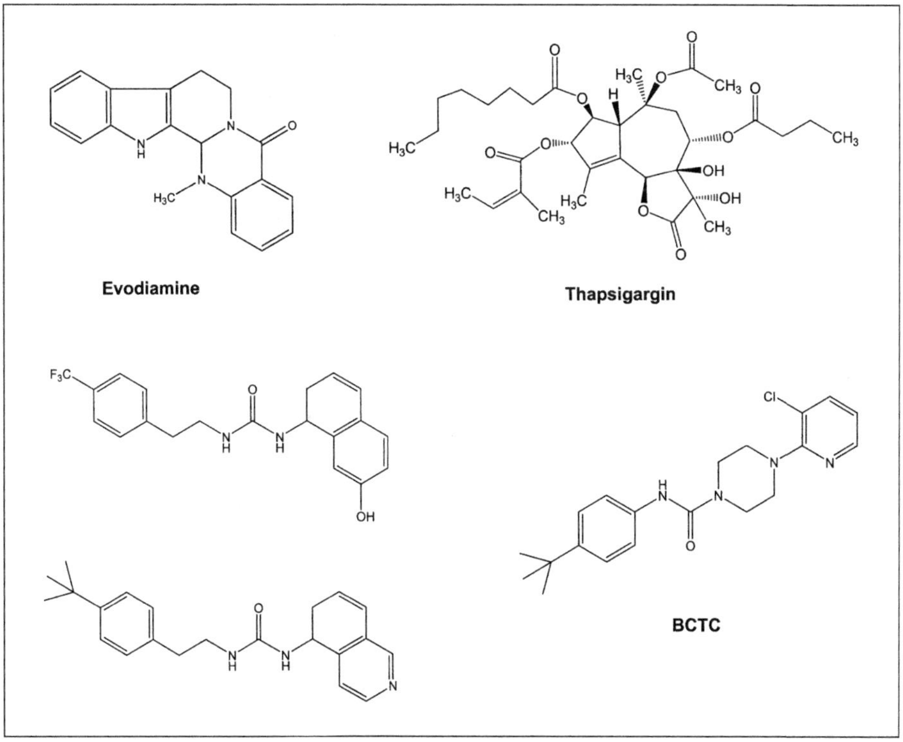

Figure 2
Diversity of structures active on TRPV1. Evodiamine and thapsigargin are natural products. The others are among the synthetic antagonists that have been described.

This same conclusion is emerging from the vigorous current efforts in medicinal chemistry directed at TRPV1 [25]. For example, a 3-pyridyl group can substitute for the A-region to generate an agonist, and an isoquinoline, 7-hydroxynapthalene, or a 4-(2-pyridyl)piperazine group as an A-region can generate an antagonist.

Diversity of responses of TRPV1 to ligand binding

Capsaicin functions as a full agonist at TRPV1. It is now clear, however, that antagonism and partial agonism (partial antagonism) are other possible outcomes to ligand binding. Each of these outcomes could represent a desirable therapeutic strategy, depending on circumstances.

Agonism

The rationale for agonists as therapeutic agents is that the agonism leads to desensitization/defunctionalization upon extended exposure. At least one mechanism for the desensitization/defunctionalization is calcium toxicity to the nerve terminals, leading to suppression of C-fiber function for a prolonged period. One advantage is thus that limited treatment could have a long-term therapeutic effect. This expectation is realized with RTX treatment of the overactive bladder, where the effect of treatment persists for months [26]. A second advantage of desensitization/defunctionalization is that it should extend to suppression of all responses mediated by the C-fiber neurons, whereas antagonism would be limited to those responses for which TRPV1 itself is involved.

Antagonism

The potential long-term effects of desensitization/defunctionalization, on the other hand, make it less appropriate where short-term suppression of C-fiber function is desired. Development of TRPV1 antagonists has consequently received great attention recently, with the identification of numerous antagonistic structures with high potency [25]. In contrast to capsazepine [27], the first TRPV1 antagonist reported that provided proof of principle but which was both weak in potency and blocked a range of other targets, the newer agents afford greater selectivity as well as greater potency. An important concept which has emerged from the characterization of such antagonists is that the breadth of antagonism differs between antagonists. For example, we characterized two compounds, *N*-(4-tert-butylbenzyl)-*N*'-[4-(methylsulfonylamino)benzyl]thiourea (KJM429) and *N*-(4-tert-butylbenzyl)-*N*'-[3-fluoro-4-(methylsulfonylamino)benzyl]thiourea (JYL1421), which both fully antagonized the stimulation of calcium uptake in response to capsaicin [28]. JYL1421 also fully antagonized the stimulation of calcium uptake induced by pH 6.0 or 44 °C. In contrast, KJM429 only partially antagonized these responses. *N*-(4-tert-butylphenyl)-4-(3-chloropyridin-2-yl)tetrahydropyrazine1(2*H*)-carbox-amide (BCTC), likewise, inhibited the activation of TRPV1 by acid as well as by capsaicin [29]. A plausible explanation is that antagonists blocking the response of TRPV1 to other stimuli are able to stabilize the closed conformation of the channel, whereas those that only block capsaicin action occupy the capsaicin-binding site but are less effective at stabilizing this conformation.

A critical issue is whether antagonists can block physiological responses. In fact, BCTC reduced both thermal and mechanical hyperalgesia in the Freund's complete adjuvant inflammation model and reduced mechanical hyperalgesia and tactile allodynia in a partial sciatic nerve injury model [30]. Likewise, *N*-[2-(2,3-dimethylbenzyl-3-(pivaloyloxy)propyl]-*N*'-[3-fluoro-4-(methylsulfonylamino)benzyl]thiourea was a potent inhibitor in the acetic acid writhing assay [31]. These results provide strong support for the therapeutic strategy of using TRPV1 antagonists.

Partial agonism/partial antagonism

A third pattern of behavior, which has received the least attention, is that of partial agonism. We characterized in detail two compounds, *N*-[2-(3,4-dimethylbenzyl)-3-(pivaloyloxy)propyl]-*N'*-[n-(methylsulfonylamino)benzyl]thiourea (JYL827) and *N*-(4-tert-butylbenzyl)-*N'*-[3-methoxy-4-(methylsulfonylamino)benzyl]thiourea (JYL1511), which showed partial agonism compared to capsaicin [32]. JYL827 induced only 6.8% of the maximal response to capsaicin. JYL1511 evoked 17.4% of the maximal response, and both compounds induced a reciprocal level of antagonism of the action of capsaicin. As expected, the extent of partial agonism depended on the level of TRPV1 expression. Less expectedly, the level of partial agonism was dependent on the extent of costimulators of TRPV1. Indeed, when assayed in the presence of pH 5.5, 44 °C, and phorbol 12-myristate 13-acetate, JYL827 functioned as almost a complete agonist (89% of the maximal capsaicin response).

The analysis of structure-activity relations for agonism, partial agonism, and antagonism emphasizes that the functional behavior of ligands depends on the structure as a whole rather than being attributable to only one region of the ligand, such as the A-region. This was clearly evident in the analog series of Lee and coworkers, where there was a continuum between agonism and antagonism as the A-region was varied, but the tendency toward antagonism also depended on the specific C-region [31]. For capsazepine, antagonism appeared to depend on the nature of the B-region, where a shift in orientation of the B-region relative to the A-region was necessary for the antagonism. On the other hand, it is also clear that certain A-regions favor antagonism. Examples include the 5-iodo substitution on the 3-methoxy-4-hydroxyphenyl A-region of RTX, generating the potent antagonist 5-iodoRTX [12], or replacement of the 3-methoxy-4-hydroxyphenyl A-region of more typical vanilloids with a 4-methylsulfonylaminophenyl group [31].

A critical implication of the analysis of partial agonists is that the nature of the response to ligands is dependent on the other signals impinging on TRPV1 [33, 34]. In addition to pH and temperature, TRPV1 has been shown to be modulated by calcium, protein kinase C, protein kinase A, Src kinase, phosphatidylinositol 4,5-bisphosphate, and calmodulin. It is subject to oligomerization, possibly to heterooligomerization, and to association with other proteins. These signals will be different between different cell types in which TRPV1 is found and will depend on the local tissue environment. It is thus entirely plausible that it may be possible to develop compounds that function as antagonists for TRPV1 in the normal setting, but which change their activity to agonism at sites of inflammation, thereby causing local desensitization/defunctionalization. Such compounds could thereby yield local effect with systemic application.

Diversity of cellular localization of TRPV1

The development of antibodies against TRPV1 and of fusion proteins between TRPV1 and green fluorescent protein has permitted the localization of TRPV1 within cells to be determined. The clear answer is that TRPV1 is not simply a ligand-gated channel located at the plasma membrane, permitting the flux of calcium from the outside [33]. Rather, most of the TRPV1 is located inside the cell [35]. This is true for endogenously expressed TRPV1 in dorsal root ganglion neurons as well as for heterologously expressed TRPV1 [36]. Not unexpectedly, the localization of TRPV1 depends somewhat on the specific cell type in which it is expressed. In the heterologously expressed systems, the internal TRPV1 can be shown to be largely associated with endoplasmic reticulum markers.

The strong prediction from the localization of TRPV1 is that stimulation will lead to release of calcium stores from the endoplasmic reticulum into the cytoplasm as well as the influx of external calcium. Indeed, this result has been confirmed by multiple laboratories. Interestingly, RTX appears to be more effective at causing this release of internal calcium than is capsaicin [35, 37]. Conversely, we have described a ligand which was potent as an antagonist of capsaicin-stimulated calcium uptake but was not able to antagonize the release of calcium from internal stores by RTX [38].

A speculative model which emerges is that the internal TRPV1 shows the structure-activity relations of the R-type receptor, whereas the plasma membrane-localized TRPV1 defines, of course, the C-type response. Since the TRPV1 is predominantly internal, the domination of the [^{3}H]RTX binding results by the internal TRPV1 would be expected. The development of compounds that preferentially interact with the internal TRPV1 and the determination of their biological activity should reveal the relevance of this pool of TRPV1 to therapeutic strategies.

Diversity in kinetics of response to ligands

Using the 'impermeable' TRPV1 ligand DA-5018, Oh and coworkers [14] have argued that the ligand-recognition site on TRPV1 is located on the internal face of the membrane. The same conclusion is supported from mutational analysis of TRPV1 (see below), although there are still some uncertainties. The expectation is thus that vanilloids need to penetrate into the cell in order to activate TRPV1 and that the kinetics of penetration will thereby influence the kinetics of response. This is of significance, because it is clear with the phorbol esters that they may penetrate cells with a half-time of many minutes. It would be expected that RTX would behave likewise. In fact, the time courses for elevation of internal calcium in response to increasing concentrations of capsaicin and RTX were very different [16]. With capsaicin, increasing doses of capsaicin caused increasing extents of elevation of internal calcium, with similar very rapid attainment of the plateau level. With RTX, in marked contrast, low concentrations of RTX caused a slow increase in internal calcium concentration extending over many minutes but with a similar plateau level to that caused by the concentration of RTX causing an immediate elevation [16]. Complementing these studies are those with 4-O-(aminoethyl)vanillylamide analogs structurally related to DA-5018 [39]. They were shown to retain their analgesic potency but to be virtually devoid of pungency. These compounds would be expected to penetrate slowly because only the small proportion that was uncharged at physiological pH would be expected to penetrate the membrane. The authors postulated that the loss of potency might lie in 'their *rate* of excitation of the sensory neuron'. A similar explanation might account for the low pungency relative to desensitization that is characteristic of RTX. This model, if correct, suggests a strategy for the design of non-pungent agonists for attainment of desensitization/defunctionalization.

Species differences in selectivity of TRPV1 to ligands

The early analysis of capsaicin sensitivity revealed marked differences in sensitivity. Whereas capsaicin is pungent for people or rats, chickens and rabbits do not respond. In addition, human and rat TRPV1 show differences in ligand recognition. Several groups compared the homologs of TRPV1 from these various species, identified the specific differences in TRPV1 sequence between them, and used site-directed mutagenesis to evaluate the role of specific residues in TRPV1 on ligand responsiveness [40–43].

The consensus of these studies is that the ligand binds to the third and fourth transmembrane segments of the receptor. Critical residues in the rat include Tyr-511, Met-547, and Thr-550. In the human, Met-547 is replaced by Leu- 547. In the rabbit, Thr-550 is replaced by Ile-550. Alteration of these residues had two dramatic effects. First, it changed the structure-activity relations. Second, the structure-activity relations show different dependence on these residues for the [^{3}H]RTX binding assay and for the calcium response, again emphasizing the distinction between R- and C-type receptor recognition. Position 547 has received the most attention. Reciprocal experiments mutating rat Met-547→Leu or human Leu-547→Met have been evaluated for RTX, 5-iodoRTX, PPAHV, and capsazepine. The former three compounds have higher potency on rat TRPV1; capsazepine has higher potency on human TRPV1. The mutations caused the expected qualitative shifts in the binding affinities, although the shifts were not quantitative. Although there is incomplete

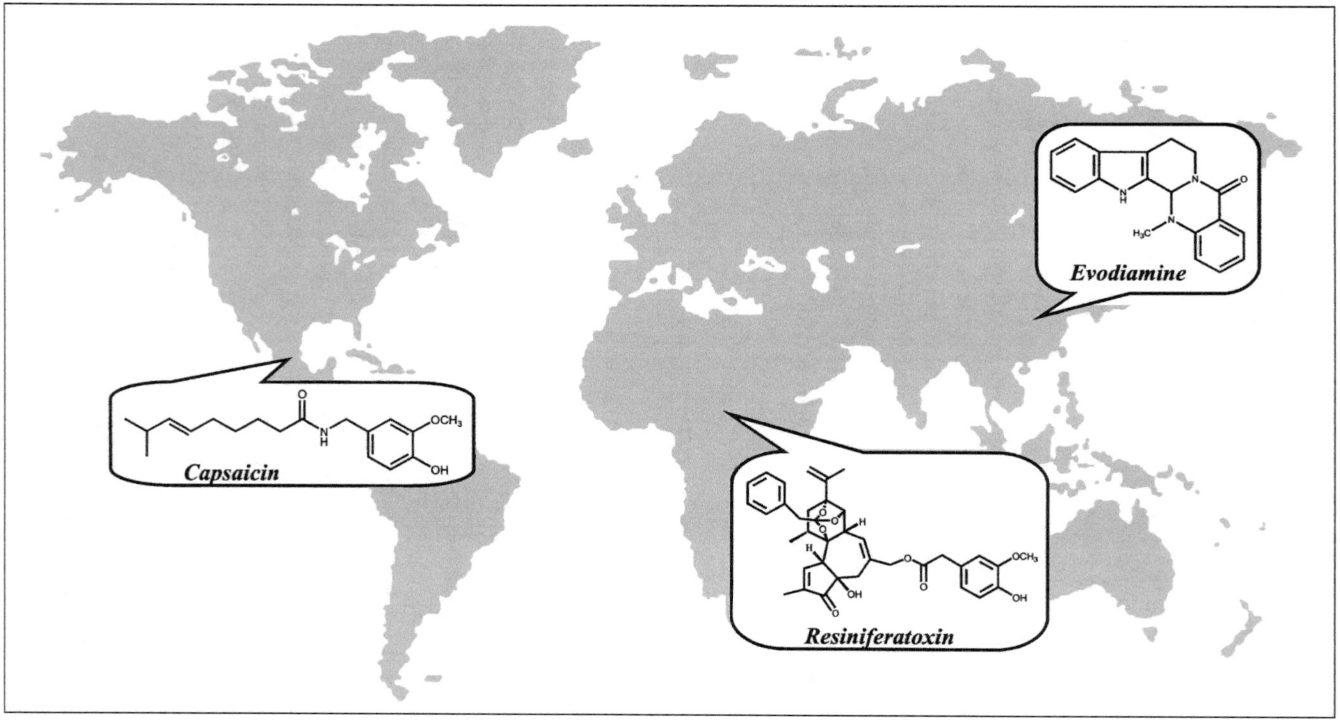

Figure 3
Traditional medicines from around the world have highlighted the capsaicin receptor as a therapeutic target. Capsaicin is of course the pungent constituent in red pepper. Resiniferatoxin (RTX) is one of multiple irritant compounds in euphorbium, the dried sap of Euphorbia resinifera. *Evodiamine is a major active principle compound in Wu-Zhu-Yu, a Chinese herbal medicine prepared from the dried fruits of* Evodia rutaecarpa. *Map from www.geographic.org, used with permission.*

agreement in the literature, two of three reports suggest that the Met-547→Leu mutation did not shift the agonist response to RTX, and the mutation likewise did not affect either the binding or agonist response to capsaicin. Interestingly, it did shift the agonist response to PPAHV. In the rabbit, mutation of the native Ile-550 to Thr, as found in the rat and human, rendered it fully responsive to agonism by capsaicin or RTX, highlighting the importance of this residue.

Diversity of tissues containing TRPV1

The early concept was that specific capsaicin responsiveness was limited to C-fiber sensory neurons. The use of [^{3}H]RTX binding, antibodies to TRPV1, and measurements of TRPV1 mRNA now make it clear that TRPV1 is much more widely distributed [44]. It is widespread in the central nervous system, being located within multiple brain regions [45] and found within both neurons and glial cells. Other tissues or tissue cell types that express TRPV1 include bladder, keratinocytes, and mast cells. The physiological role of TRPV1 in most of these tissues remains to be determined, as do its pharmacological characteristics in these different locations.

Summary

Natural products have long highlighted promising sites for therapeutic intervention, afforded pharmacological tools for delineating their mechanism of action, and provided lead molecules for further medicinal chemistry. The abundance of diverse structures active on a single target emphasizes the breadth of structural solutions for generating affinity and specificity. TRPV1 was introduced to Africa and Europe through RTX, to the New World through capsaicin, and to Asia through evodiamine (Fig. 3). The pungency associated with these lead molecules both initially directed attention to them and has been, to some degree, an impediment to their therapeutic exploitation. Our rapidly increasing understanding of the great diversity within their target system holds great promise that we will be able to exploit this diversity for development of new generations of selective therapeutic agents.

References

1 Appendino G, Szallasi A (1997) Euphorbium: modern research on its active principle, resiniferatoxin, revives an ancient medicine. *Life Sci* 60: 681–696
2 Adolph W, Sorg B, Hergenhahn M, Hecker E (1982) Structure-activity relations of polyfunctional diterpenes of the daphnane type I. Revised structure for resiniferatox-

in and structure-activity relations for resiniferonol and some of its esters. *J Nat Prod* 45: 347–354
3 Szallasi A, Blumberg PM (1989) Resiniferatoxin, a phorbol-related diterpene, acts as an ultrapotent analog of capsaicin, the irritant constituent in red pepper. *Neuroscience* 30: 515–520
4 Szallasi A, Blumberg PM (1999) Vanilloid (capsaicin) receptors and mechanisms. *Pharmacol Rev* 51: 159–212
5 Walpole CSJ, Wrigglesworth R, Bevan S, Campbell EA, Dray A, James IF, Perkins MN, Reid DJ, Winter J (1993) Analogues of capsaicin with agonist activity as novel analgesic agents: structure-activity studies. 1. The aromatic "A-region". *J Med Chem* 36: 2362–2372
6 Walpole CSJ, Wrigglesworth R, Bevan S, Campbell EA, Dray A, James IF, Masdin KJ, Perkins MN, Winter J (1993) Analogues of capsaicin with agonist activity as novel analgesic agents; structure-activity studies. 2. The amide bond "B-region". *J Med Chem* 36: 2373–2380
7 Walpole CSJ, Wrigglesworth R, Bevan S, Campbell EA, Dray A, James IF, Masdin KJ, Perkins MN, Winter J (1993) Analogues of capsaicin with agonist activity as novel analgesic agents; structure-activity studies. 3. The hydrophobic side-chain "C-region". *J Med Chem* 36: 2381–2389
8 Szolcsányi J, Jancsó-Gábor A (1975) Sensory effects of capsaicin congeners I. Relationship between chemical structure and pain-producing potency of pungent agents. *Arzneim Forsch* 25: 1877–1881
9 Janusz JM, Buckwalter BL, Young PA, LaHann TR, Farmer RW, Kasting GB, Loomans ME, Kerckaert GA, Maddin CS, Berman EF et al (1993) Vanilloids. 1. Analogs of capsaicin with antinociceptive and antiinflammatory activity. *J Med Chem* 36: 2595–2604
10 Szallasi A, Blumberg PM (1990) Specific binding of resiniferatoxin, an ultrapotent capsaicin analog, by dorsal root ganglion membranes. *Brain Res* 524: 106–111
11 Szallasi A, Lewin NE, Blumberg PM (1992) Identification of alpha1-acid glycoprotein (orosomucoid) as a major vanilloid-binding protein in serum. *J Pharmacol Exp Ther* 262: 883–888
12 Wahl P, Foged C, Tullin S, Thomsen C (2001) Iodo-resiniferatoxin, a new potent vanilloid receptor antagonist. *Mol Pharmacol* 59: 9–15
13 Caterina MJ, Julius D (2001) The vanilloid receptor: A molecular gateway to the pain pathway. *Annu Rev Neurosci* 24: 487–517
14 Jung J, Hwang SW, Kwak J, Lee SY, Kang CJ, Kim WB, Kim D, Oh U (1999) Capsaicin binds to the intracellular domain of the capsaicin-activated ion channel. *J Neurosci* 19: 529–538
15 Acs G, Lee J, Marquez VE, Blumberg PM (1996) Distinct structure activity relations for stimulation of ^{45}Ca uptake and for high affinity binding by cultured rat dorsal root ganglion neurons and dorsal root ganglion membranes. *Mol Brain Res* 35: 173–182

16 Szallasi A, Blumberg PM, Annicelli LL, Krause JE, Cortright DN (1999) The cloned rat vanilloid receptor VR1 mediates both R-type binding and C-type calcium response in dorsal root ganglion neurons. *Mol Pharmacol* 56: 581–587
17 Caterina MJ, Leffler A, Malmberg AB, Martin WJ, Trafton J, Petersen-Zeitz KR, Koltzenburg M, Basbaum AI, Julius D (2000) Impaired nociception and pain sensation in mice lacking the capsaicin receptor. *Science* 288: 306–313
18 Davis JB, Gray J, Gunthorpe MJ, Hatcher JP, Davey PT, Overend P, Harries MH, Latcham J, Clapham C, Atkinson K et al (2000) Vanilloid receptor-1 is essential for inflammatory thermal hyperalgesia. *Nature* 405: 183–187
19 Appendino G, Cravotto G, Palmisano G, Annunziata R, Szallasi A (1996) Synthesis and evaluation of phorboid 20-homovanillates: discovery of a class of ligands binding to the vanilloid (capsaicin) receptor with different degrees of cooperativity. *J Med Chem* 39: 3123–3131
20 Lee J, Lee J, Kim J, Kim SY, Chun MW, Cho H, Hwang SW, Oh U, Park YH, Marquez VE et al (2001) *N*-(3-acyloxy-2-benzylpropyl)-*N'*-(4-hydroxy-3-methoxybenzyl) thiourea derivative as potent vanilloid receptor agonists and analgesics. *Bioorg Med Chem* 9: 19–32
21 van der Stelt M, Di Marzo V (2004) Endovanilloids. Putative endogenous ligands of transient receptor potential vanilloid 1 channels. *Eur J Biochem* 271: 1827–1834
22 Pearce LV, Petukhov PA, Szabo T, Kedei N, Bizik F, Kozikowski AP, Blumberg PM (2004) Evodiamine functions as an agonist for the vanilloid receptor TRPV1. *Org Biomol Chem* 2: 2281–2286
23 Kobayashi Y (2002) The nociceptive and anti-nociceptive effects of evodiamine from fruits of Evodia rutaecarpa in mice. *Planta Medica* 69: 425–428
24 Toth A, Kedei N, Szabo T, Wang Y, Blumberg PM (2002) Thapsigargin binds to and inhibits the cloned vanilloid receptor 1. *Biochem Biophys Res Commun* 293: 777–782
25 Appendino G, Munoz E, Fiebich BL (2003) TRPV1 (vanilloid receptor, capsaicin receptor) agonists and antagonists. *Expert Opin Ther Patents* 13: 1825–1837
26 Cruz F (2004) Mechanisms involved in new therapies for overactive bladder. *Urology* 63 (suppl 3A): 65–73
27 Bevan S, Hothi S, Hughes G, James IF, Rang HP, Shah K, Walpole CSJ, Yeats JC (1992) Capsazepine: a competitive antagonist of the sensory neurone excitant capsaicin. *Br J Pharmacol* 107: 544–552
28 Wang Y, Szabo T, Welter JD, Toth A, Tran R, Lee J, Kang SU, Suh YG, Blumberg PM, Lee J (2002) High affinity antagonists of vanilloid receptor. *Mol Pharmacol* 62: 947–956
29 Valenzano KJ, Grant ER, Wu G, Hachicha M, Schmid L, Tafesse L, Sun Q, Rotshteyn Y, Francis J, Limberis J et al (2003) BCTC (*N*-(4-tertiarybutylphenyl)-4-(3-chloropyridin-2-yl)tetrahydropryazine-1(2H)-carbox-amide), a novel, orally-effective vanilloid receptor 1 antagonist with analgesic properties: I. *In vitro* characterization and pharmacokinetic properties. *J Pharmacol Exp Ther* 306: 377–386

30 Pomonis JD, Harrison JE, Mark L, Bristol DR. Valenzano KJ, Walker K (2003) BCTC (*N*-(4-tertiarybutylphenyl)-4-(3-chloropyridin-2-yl)tetrahydropryazine-1(2H)-carboxamide), a novel, orally-effective vanilloid receptor 1 antagonist with analgesic properties: II. *In vivo* characterization in rat models of inflammatory and neuropathic pain. *J Pharmacol Exp Ther* 306: 387–393

31 Lee J, Lee J, Kang M, Shin M, Kim JM, Kang SU, Lim JO, Choi HK, Suh YG, Park HG et al (2003) *N*-(3-acyloxy-2-benzylpropyl)-*N*-[4-(methylsulfonylamino)benzyl] thiourea analogues: novel potent and high affinity antagonists and partial antagonists of the vanilloid receptor. *J Med Chem* 46: 3116–3126

32 Wang Y, Toth A, Tran R, Szabo T, Welter JD, Blumberg PM, Lee J, Kang SU, Lim JO, Lee J (2003) High affinity partial agonists of the vanilloid receptor. *Mol Pharmacol* 64: 325–333

33 Cortright DN, Szallasi A (2004) Biochemical pharmacology of the vanilloid receptor TRPV1. An update. *Eur J Biochem* 271: 1814–1819

34 Ferrer-Montiel A, Garcia-Martinez C, Morenilla-Pallao C, Garcia-Sanz N, Fernandez-Carvajal A, Fernandez-Ballester G, Planells R (2004) Molecular architecture of the vanilloid receptor. Insights for drug design. *Eur J Biochem* 271: 1820–1826

35 Olah Z, Szabo T, Karai L, Hough C, Fields RD, Caudle RM, Blumberg PM, Iadarola MJ (2001) Ligand induced dynamic membrane changes and cell deletion conferred by vanilloid receptor 1. *J Biol Chem* 276: 11021–11030

36 Guo A, Vulchanova L, Wang J, Li X, Elde R (1999) Immunocytochemical localization of the vanilloid receptor 1 (VR1): relationship to neuropeptides, the $P2X_3$ purinoceptor and IB4 binding sites. *J Neurosci* 11:946–958

37 Marshall IC, Owen DE, Cripps TV, Davis JB, NcNulty S, Smart D (2003) Activation of vanilloid receptor 1 by resiniferatoxin mobilizes calcium from inositol 1,4,5-trisphosphate-sensitive stores. *Br J Pharmacol* 138: 172–176

38 Toth A, Blumberg PM, Chen Z, Kozikowski AP (2004) Design of a high affinity, competitive antagonist of the vanilloid receptor selective for the calcium entry linked receptor population. *Mol Pharmacol* 65: 282–291

39 Wrigglesworth R, Walpole CSJ, Bevan S, Campbell EA, Dray A, Hughes GA, James I, Masdin KJ, Winter J (1996) Analogues of capsaicin with agonist activity as novel analgesic agents: structure-activity studies. 4. Potent, orally active analgesics. *J Med Chem* 39: 4942–4951

40 Jordt SE, Julius D (2002) Molecular basis for species-specific sensitivity to "hot" chili peppers. *Cell* 108: 421–430

41 Gavva NR, Klionsky L, Qu Y, Shi L, Tamir R, Edenson S, Zhang TJ, Vellarkad NV, Toth A, Pearce LV et al (2004) Molecular determinants of vanilloid sensitivity in TRPV1. *J Biol Chem* 279: 20283–20295

42 Phillips E, Reeve A, Bevan S, McIntyre P (2004) Identification of species-specific determinants of the action of the antagonist capsazepine and the agonist PPAHV on TRPV1. *J Biol Chem* 279: 17165–17172

43 Chou MZ, Mtui T, Gao YD, Kohler M, Middleton RE (2004) Resiniferatoxin binds to the capsaicin receptor (TRPV1) near the extracellular side of the S4 transmembrane domain. *Biochemistry* 43: 2501–2511
44 Caterina MJ (2003) Vanilloid receptors take a TRP beyond the sensory afferent. *Pain* 105: 5–9
45 Acs G, Palkovits M, Blumberg PM (1996) Specific binding of [^{3}H]resiniferatoxin by human and rat preoptic area, locus ceruleus, medial hypothalamus, reticular formation, and ventral thalamus membrane preparations. *Life Sci* 59: 1899–1908

Endocannabinoids and vanilloid TRPV1 receptors

Ruth A. Ross

School of Medical Sciences, Institute of Medical Sciences, University of Aberdeen, Foresterhill, Aberdeen AB25 2ZD, Scotland, UK

Anandamide – endocannabinoid and endovanilloid?

The endocannabinoid system comprises two known receptors (CB_1 and CB_2), a family of endogenous ligands, and specific molecular machinery for the synthesis, transport, and inactivation of cannabinoid ligands [1]. *N*-Arachidonoyl-ethanolamide (anandamide) (Fig. 1) is known as an endocannabinoid, as defined by its ability to be produced endogenously and to bind to and activate cannabinoid CB_1 and CB_2 receptors (see [1]). Since its discovery, anandamide has been shown to have numerous physiological actions that encompass cardiovascular, immune, gastrointestinal, and nervous systems [2–5]. The pharmacology of anandamide is complex, its actions being mediated by cannabinoid CB_1 and CB_2 receptors and by putative non-CB_1, non-CB_2 receptors [1, 6, 7]. The search for endogenous TRPV1 receptor activators or endovanilloids is on-going and recent advances suggest that anandamide may be one such compound [8–11]. There is evidence that anandamide can activate TRPV1 receptors both directly and indirectly via lipoxygenase metabolites [12–15]. There is ample evidence that the interaction of anandamide with TRPV1 receptors is specific: TRPV1 actions are blocked by receptor- specific antagonists and not by antagonists of CB_1 and CB_2 receptors; desensitization of TRPV1 via capsaicin pretreatment abrogates the effects of anandamide; neonatal capsaicin treatment prevents anandamide activation of TRPV1; anandamide-mediated TRPV1 effects are absent from untransfected cells that do not express TRPV1 receptors; and anandamide displaces radiolabelled resiniferatoxin (RTX) from specific binding sites [12–14].

Whereas anandamide has similar affinity for TRPV1 to that of capsaicin, it has significantly lower potency [14]. In high-expression recombinant cell lines using various methods, the potency (EC_{50}) of anandamide has been measured in the range of 0.7–5 μM. In native systems, the potency of anandamide ranges from 0.3 to 0.8 μM in blood vessels (relaxation), compared with 6–10 μM in bronchus (contraction), dorsal root ganglion (DRG) neurons and trigeminal neurons (intracellular Ca^{2+} con-

Turning up the Heat on Pain: TRPV1 Receptors in Pain and Inflammation, edited by
Annika B. Malmberg and Keith R. Bley

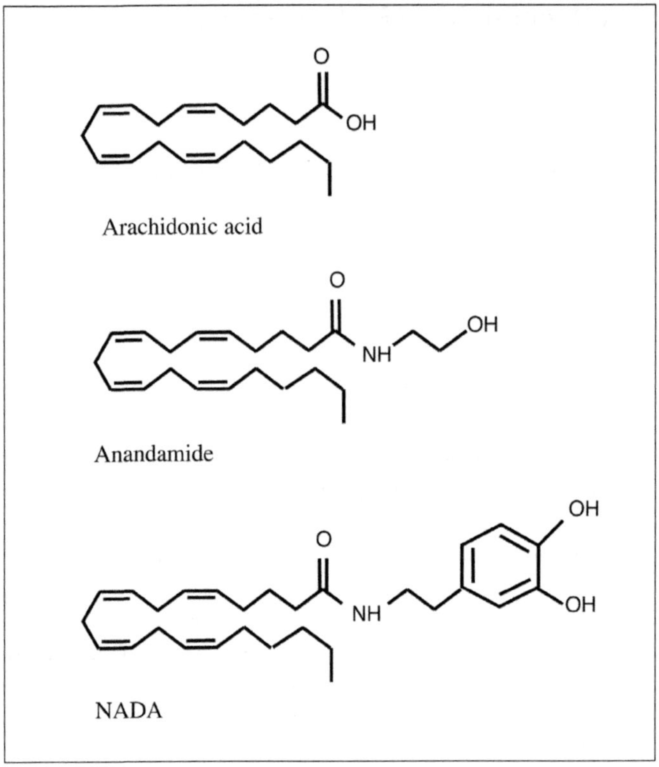

Figure 1
Structures of arachidonic acid, anandamide and N-arachidonoyl-dopamine (NADA).

centration, $[Ca^{2+}]_i$, and inward current), and dorsal horn neurons (miniature excitatory postsynaptic currents (mEPSCs)) [16, 17]. Significant differences emerge when comparisons are made of the efficacy of anandamide in native and recombinant receptor systems. There are also significant differences between tissues. Using measurement of $[Ca^{2+}]_i$ in high-expression recombinant cell lines expressing either rat or human TRPV1, anandamide appears to be a full agonist, as defined by an E_{max} value that is not significantly different from that of capsaicin [13, 14, 18–20]. Some authors find the E_{max} value of anandamide to be less than 100% in recombinant expression systems. Similar differences are found in tissues expressing native TRPV1 receptors and the efficacy of anandamide appears to be tissue-dependent. Thus, it is a full agonist in blood vessels and mesenteric arterial bed, and a partial agonist in the bronchus, DRG, trigeminal neurons, and substantia gelatinosa neurons of the dorsal horn [13, 17, 21, 22]. Ross et al. [14] found that, although anandamide has an E_{max} value of 100%, it displays a low K_i/EC_{50} ratio, as determined from radioligand binding and $^{45}Ca^{2+}$-uptake data; this is indicative of low intrinsic efficacy. As a consequence of the low intrinsic efficacy of anandamide at TRPV1, the E_{max} value will vary with the receptor reserve and the pharmacological endpoint that is measured. Thus, whereas elevation of $[Ca^{2+}]_i$ has a high degree of signal amplification, measurement of TRPV1 current has no downstream amplification.

The low intrinsic efficacy of anandamide at the TRPV1 receptor has important physiological implications. A low intrinsic efficacy agonist will attenuate the effects of a full agonist. Indeed, in trigeminal neurons, co-application of anandamide with capsaicin significantly reduces the currents produced by capsaicin [22]. If endogenous activators of TRPV1 exist that are full agonists, then compounds such as anandamide may have an inhibitory role and serve as anti-inflammatory and analgesic compounds.

Intracellular *versus* extracellular

It is important to give consideration to the location of the agonist-binding site on TRPV1, which is located intracellularly [23, 24]: the apparent low intrinsic efficacy of anandamide may be due to limited access to this binding site. As a consequence, the potency of exogenously applied anandamide will be affected by its ability to enter the cell (see [16]). Anandamide gains access to the intracellular environment either via the anandamide membrane transporter (AMT) or via accumulation sustained by a concentration gradient that is maintained through the rapid intracellular hydrolysis of anandamide by fatty acid amide hydrolase (FAAH; see [25, 26]). The finding of the enhanced potency of anandamide when administered intracellularly is in line with the demonstration that compounds that alter the function of the AMT modulate the potency of anandamide in HEK-293 cells expressing TRPV1 [27–29]. A recent study by Andersson et al. [21] provides further evidence that primary afferent fibers express the AMT and that variability in the expression levels of the transporter affect the potency of anandamide. Thus the AMT inhibitor, VDM13, causes a 2.3-fold rightward shift in the log concentration-response curve for vasodilatation of mesenteric arteries by anandamide but does not affect the contractile response to anandamide in the bronchus. Similarly, the potency of the putative endovanilloid *N*-arachidonoyl-dopamine (NADA; Fig. 1) varies between tissue preparations, an observation that has been attributed to differential expression of the AMT [30, 31]. As has been demonstrated that for CB_1 receptors [32] the affinity and potency of anandamide at TRPV1 is enhanced by inhibition of FAAH in certain tissues, including Chinese hamster ovary cells [14], HEK-293 cells, rat ileum, and neuroblastoma cells (see [16]). On the other hand, in the mesenteric bed, FAAH inhibitors attenuate the TRPV1 receptor-mediated relaxation by anandamide [33]. In the bronchus and trigeminal neurons, however, the potency of anandamide as a TRPV1 agonist is unaffected by FAAH inhibitors [21, 34, 35].

Recent data represent the first direct demonstration of a significant difference between the potency of anandamide when the compound is applied directly to the intracellular environment (Fig. 2). As previously demonstrated by others [12, 13, 22, 25], Evans et al. [11] confirm that, when applied to the extracellular environment, anandamide had low potency at the TRPV1 receptor in rat cultured DRG neurons. Consequently they found that, when applied extracellularly, 100 nM anandamide fails to produce inward currents in 93% of neurons. In contrast, when applied intracellularly, this relatively low concentration evoked inward currents in 60% of neurons, the current being significantly attenuated by the TRPV1 receptor antagonist, capsazepine (1 μM). In another demonstration of intracellular activation of TRPV1, Hwang et al. [35] recorded TRPV1 receptor-mediated inward currents in inside-out patches of cultured DRG neurons: anandamide activated TRPV1 receptors, but the efficacy was found to be much lower than that of capsaicin. Differences in compound preparation and vehicles may account for the variation in potency observed.

These data raise the possibility that anandamide may be a relatively high-potency TRPV1 agonist when it is released intracellularly, close to the TRPV1 receptor ligand-binding site. Mass spectrometric analysis shows that both capsaicin and depolarization (KCl) induce significant release of anandamide in DRG cultures [2, 36]. It is also notable that the capsaicin-evoked release of anandamide is significantly attenuated when the FAAH inhibitor, methylarachidonyl-fluorophosphonate (MAFP), is excluded from the buffer, demonstrating that anandamide is rapidly metabolized in DRG neurons. It would therefore appear that anandamide can activate both CB_1 and TRPV1 receptors in a similar concentration range, raising the possibility that it is indeed both an endocannabinoid and an endovanilloid.

CB_1/TRPV1 crosstalk

An intriguing possibility is that when applied to the extracellular environment, anandamide first activates CB_1 receptors, whose ligand-binding domain is extracellular, which in turn modulates activation of TRPV1 (see [16, 37]).

CB_1/TRPV1 signaling

Phosphorylation of TRPV1 by protein kinase C (PKC) and protein kinase A (PKA) sensitizes TRPV1 receptors to vanilloids, anandamide, heat, and protons (see [2]). In contrast, PKC-mediated phosphorylation of the CB_1 receptor prevents cannabinoid-mediated activation of K_{ir} currents and inhibition of P/Q-type Ca^{2+} channels

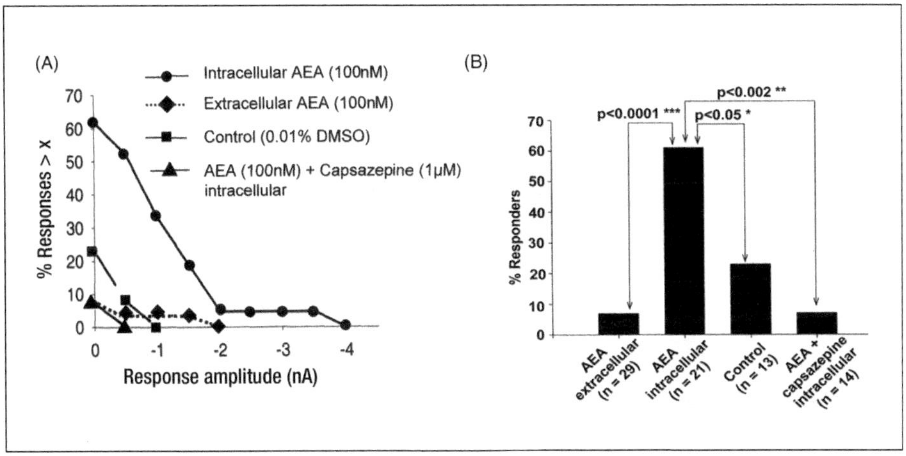

Figure 2
Intracellular versus extracellular application of anandamide in dorsal root ganglion (DRG) neurons. Using whole-cell patch-clamp electrophysiology to measure TRPV1 receptor-mediated inward currents; anandamide (AEA; 100 nM) is more potent as a TRPV1 agonist when applied directly to the intracellular environment (from [11]). (A) Population response distribution for neurons to which 100 nM anandamide was applied directly to the intracellular domain via the patch pipette solution; to which 100 nM anandamide was applied extracellularly; vehicle controls, where 0.01% DMSO was included in the patch pipette solution; and to which 100 nM anandamide and 1 µM capsazepine were co-applied intracellularly in the patch pipette solution. Both the number of responders and the amplitude of responses were greater when anandamide was applied intracellularly. (B) Anandamide (100 nM) applied intracellularly evoked inward currents in significantly more neurons (Fisher's exact test) than extracellular application or controls, and these currents were capsazepine-sensitive.

[38]. The CB_1 receptor inhibits adenylate cyclase via the α-subunit of G_i, resulting in inhibition of PKA activation. Thus, one may anticipate that PKC and PKA are pivotal in the regulating the balance of CB_1 and TRPV1 activation by anandamide.

In HEK-293 cells overexpressing both CB_1 and TRPV1 receptors, pre-treatment with a cannabinoid receptor agonist significantly enhances capsaicin-evoked increases in $[Ca^{2+}]_i$ [37]. This effect is inhibited by SR141716A and is absent from cells that express TRPV1 only, indicating that CB_1 receptor activation leads to an enhanced activation of TRPV1. Furthermore, anandamide produces significantly greater increase in intracellular calcium in CB_1/TRPV1-expressing HEK-293, as compared to TRPV1-expressing HEK-293 cells. The CB_1 receptor-mediated enhancement of TRPV1 activation is attenuated by inhibitors of phosphoinositide

3-kinase (PI-3K) and phospholipase C (PLC). PI-3K is responsible for the formation of phosphatidylinositol bisphosphate (PIP_2), whereas PI-PLC catalyzes PIP_2 hydrolysis; the result of inhibition of these enzymes will be attenuation of turnover and hydrolysis of PIP_2. The TRPV1 receptor is under the inhibitory control of PIP_2; both antibody sequestration of plasma membrane PIP_2 and PLC hydrolysis of PIP_2 potentate the TPRV1 channel activation [39]. CB_1 receptor activation is coupled to stimulation of PLC and PI-3K [40–42] and hence may release TRPV1 from tonic inhibition by PIP_2. PKC is also activated as a consequence of hydrolysis of PIP_2 to diacylglycerol. Hermann et al. [37] suggest that these mechanisms may underlie the CB_1 receptor-mediated enhancement of TRPV1 receptor activation. Furthermore, Hermann et al. [37] found that if cAMP-mediated signaling is activated, CB_1 agonists mediate inhibition and not enhancement of TRPV1 receptor activation: when PKA is activated via forskolin stimulation of adenylate cyclase, CB_1 receptor agonists inhibit TRPV1 receptor activation by capsaicin.

Thus, in cells in which there is co-expression of CB_1 and TRPV1 receptors, there are two possibilities for CB_1 receptor influence on the activation of TRPV1. Firstly, CB_1 receptor activation may lead to an enhanced TRPV1 receptor activation. One may predict such an outcome in cells in which the CB_1 receptor couples to pathways that facilitate gating of TRPV1. Alternatively, CB_1 receptor activation may lead to an inhibition of TRPV1 receptor activation. Such an outcome may be observed in cells in which CB_1 receptors are not coupled to PI-PLC or PI-3K or in which the TRPV1 receptor is phosphorylated by cAMP-dependent PKA activation.

Co-localization of CB_1 and TRPV1

Within the DRG and trigeminal ganglia there is a heterogeneous population of cells containing small-, intermediate-, and large-diameter neurons that give rise to C-, Aδ-, or Aβ-fiber axons. Both in culture and *in situ*, DRG neurons express both CB_1 and TRPV1 receptors. TRPV1 receptor expression is largely associated with small-diameter nociceptive primary afferent fibers, but the localization of CB_1 receptors is controversial (see [4]): some studies suggest a high degree of co-localization [43, 44] and others suggest little co-localization [45, 46]. A recent study combined *in situ* hybridization and immunohistochemistry (using a C-terminus antibody) to investigate the profile of expression of CB_1 receptors in DRG sections [47]. This group found that the CB_1 receptor is expressed by approx. 25% of DRG neurons, and that the majority (approx. 75%) of these receptors were expressed on medium- and large-diameter myelinated DRG cells that express neurofilament 200 kDa protein. Of CB_1-expressing cells, approx. 10% co-stained for calcitonin-gene-related peptide (CGRP) and approx. 15% co-stained for TRPV1. A further study in trigeminal neurons corroborates findings that CB_1 receptors do not predominate in nociceptive fibers: using *in situ* hybridization, Price et al. [48] find that 5% of CB_1 mRNA is co-localized with TRPV1 immunoreactivity or with peptidergic immunoreactivity. In contrast, 75% of CB_1 mRNA co-localized with N52 immunoreactivity, the marker for myelination. It is notable that – based on size distribution and co-localization analysis – the authors conclude that CB_1 receptors are found predominantly in cell bodies which give rise to large-diameter Aβ-fibers.

Modulation of TRPV1 by CB_1 receptor activation

A number of studies in native tissues have addressed the question of CB_1 receptor-mediated modulation of TRPV1 responses in nociceptive pathways. These studies have yielded evidence both supporting and negating CB_1 receptor-mediated modulation of TRPV1 receptor responses.

Certain studies provide functional data to suggest that CB_1 receptors may not be located on small-diameter nociceptive primary afferent fibers: the cannabinoid receptor agonist, CP55940 (100 nM), significantly inhibits KCl-evoked increases in $[Ca^{2+}]_i$ in intermediate (800–1500 μM^2) DRG neurons, but does not affect these responses in small-diameter neurons (<800 μM^2) [46]. In line with this finding, CP55940 fails to inhibit increases in $[Ca^{2+}]_i$ evoked by capsaicin, which activates only small-diameter neurons. Khasabova et al. [49] found differences in the populations of neurons in which cannabinoid agonists and μ-opioid receptor agonists inhibit depolarization-evoked increases in $[Ca^{2+}]_i$: while in intermediate neurons CP55940 inhibits evoked increases in $[Ca^{2+}]_i$ by 30%, it has no effect in small-diameter neurons; morphine displays the opposite pattern of activity. Price et al. [33] studied the effects of anandamide and NADA in trigeminal neurons *in vitro*. The compounds evoke CGRP release, an effect that is sensitive to TRPV1 receptor antagonists iodoRTX and capsazepine. This group found that anandamide does not inhibit either capsaicin or depolarization-evoked CGRP release. The authors did not investigate whether SR141716A, which selectively blocks the CB_1 receptor, influences the ability of anandamide to stimulate CGRP release. The inability of anandamide to inhibit evoked CGRP release is in line with mounting evidence that the CB_1 receptor is predominantly located in medium/large myelinated neurons in both trigeminal neurons and DRG [46–49].

In contrast, others find evidence that cannabinoids modulate depolarization and capsaicin-evoked neuropeptide release. Tognetto et al. [50] find that anandamide has a dual effect on neuropeptide release from cultured DRG neurons: at low concentrations anandamide induced a CB_1 -receptor mediated inhibition of electrically stimulated neuropeptide release (presumably by inhibition of voltage-gated Ca^{2+} channels), while at higher concentrations anandamide evoked a TRPV1 receptor-mediated release [51]. Anandamide also attenuates capsaicin-evoked neuropeptide release from DRG neurons and isolated paw skin [52, 53]. Furthermore, in the presence of a CB_1 receptor antagonist, anandamide becomes equipotent with capsaicin as a TRPV1 receptor agonist [52]. Similarly, in the guinea pig ileum, anandamide increases acetylcholine release in a capsazepine-sensitive manner and this effect is enhanced in the presence of the CB_1 receptor antagonist, SR141716A [54]. Finally, in human neuroblastoma cells, cannabinoid receptor antagonists potentiate the TRPV1 receptor-mediated cell death induced by anandamide [55]. The implication of such data is that activation of the CB_1 receptor by anandamide may attenuate its TRVP1 receptor-mediated action. In the spinal cord, superfusion of the CB_1 receptor antagonist significantly enhances the release of neuropeptide evoked by capsaicin [56]. Activation of the TRPV1 receptor may evoke release of endocannabinoids that subsequently activate CB_1 receptors to attenuate the TRPV1 receptor-mediated release of neuropeptide. This hypothesis is supported by the finding that capsaicin induces significant release of anandamide in DRG cultures [36]. Cannabinoid receptors are constitutively active and there are numerous examples of the CB_1 receptor antagonist SR141716A producing inverse cannabimimetic effects (see [57]). In DRG neurons, a cannabinoid agonist inhibits and antagonist enhances voltage-gated Ca^{2+} currents (VACCs) [51]. It is possible that in DRG neurons and the spinal cord, constitutively active CB_1 receptors may exert a tonic inhibitory effect on TRPV1 receptor by attenuating cAMP-dependent PKA activation. In this scenario, inverse agonists may enhance TRPV1 receptor activation. Ellington et al. [58] find that the CB_1 receptor agonist, CP55940, inhibits capsaicin-evoked CGRP release from paw skin of both control and diabetic animals, an effect that is blocked by a CB_1 receptor antagonist. Anandamide also has bi-directional effects on cough in conscious guinea pigs: when given by aerosol, anandamide induces cough, an effect that is prevented by pre-treatment with capsazepine and not CB_1 or CB_2 receptor antagonists; and, when airways are pretreated with the ligand for a longer time in experimental conditions, anandamide inhibits cough through CB_1 receptors [59].

Novel receptors

Non-CB_1 receptors

When considering crosstalk between cannabinoid receptors and TRPV1 receptors it is important to discuss the role of novel receptors. There is a growing body of evidence for non- CB_1, non-CB_2 receptors which are activated by the endogenous cannabinoid [1, 6, 7]. Anandamide and WIN55212 stimulate [^{35}S]guanosine 5'-[γ-thio]triphosphate ([^{35}S]GTP[S]) binding in both CB_1-knockout and wild-type mice [7, 60]. Indeed, in CB_1-knockout mice, anandamide retains the ability to induce the tetrad of cannabimimetic effects (see [6]). Evans et al. [11] found that anandamide inhibits VACCs in the presence of the CB_1 receptor antagonist, SR1417161A. Thus, in the presence of SR141716A (100 nM), anandamide (100 nM) retains the ability to inhibit VACCs in 70% of neurons. In the remaining neurons, anandamide does not inhibit VACCs in the presence of SR141716A. In the absence of antagonist, anandamide inhibits VACCs in a similar percentage of neurons. Consequently, the authors conclude from these studies that, in a certain population of DRG neurons, anandamide can inhibit VACCs via a non-CB_1 receptor. The nature of the receptors mediating inhibition of VACCs by anandamide is the subject of on-going investigations, and it may be that the effect is antagonized by higher concentrations of SR141716A. The failure of 100 nM SR141716A to block the action of anandamide is not unexpected. Anandamide (100 nM) inhibits capsaicin-evoked CGRP release from the rat isolated paw skin; an effect blocked by neither CB_1 nor CB_2 receptor antagonists (100 nM) [58]. In addition, a number of effects of anandamide are only antagonized by concentrations of SR141716A that are considerably higher than those expected for antagonism of the CB_1 receptor [1, 63]. Furthermore, in a recent study, Khasabova et al. [46] found that the inhibition of depolarization-induced Ca^{2+} influx in DRG neurons induced by cannabinoids is not affected by 100 nM SR141716A, but is significantly attenuated by the higher concentration of 3 μM. However, at concentrations above 1 μM, SR141716A may no longer be CB_1 receptor-selective [1]. Evans et al. [11] also find that anandamide has opposing effects on depolarization-evoked calcium influx in distinct populations of DRG neurons. In line with the inhibition of VACCs measured electrophysiologically, they found that in a population of small-/intermediate-diameter neurons, 1 μM anandamide inhibits increases in $[Ca^{2+}]_i$ evoked by 30 mM KCl. In contrast, in a population of intermediate/large neurons, this concentration of anandamide significantly enhances the level of $[Ca^{2+}]_i$ evoked by high-K^+-evoked depolarization. Furthermore, in the presence of pertussis toxin, the inhibitory effect of anandamide is apparently abolished: in cultures pre-treated with pertussis toxin, anandamide exhibited an enhancing effect in all the cells analyzed. The implication of these findings is that the inhibition of depolarization-evoked increases in $[Ca^{2+}]_i$ by anandamide in small-/intermediate-sized DRG neurons is a $G_{i/o}$-mediated event. The receptor in question may be a CB_1 receptor or a non-CB_1 G-protein-coupled receptor. It is notable that there is evidence for $G_{i/o}$-coupled (pertussis toxin-sensitive) effects of anandamide that are not blocked by SR141716A in the concentration range expected for the CB_1 receptor [61, 62]. Zygmunt et al. [12] demonstrated that, in isolated mesenteric vessels, anandamide-mediated vasorelaxation is dependent on TRPV1 receptor activation and is capsazepine-sensitive. However, activation of TRPV1 receptors in sensory nerves does not totally account for the vasorelaxation observed in response to anandamide; indeed, anandamide displays additional novel pharmacology in the cardiovascular system. Anandamide-induced vasorelaxation involves a novel non- CB_1 receptor which is sensitive to SR141716A, albeit at higher concentrations than those required to antagonize the CB_1 receptor (i.e. 1 μM and above). This receptor is antagonized by a novel cannabidiol analogue, O1918 [62].

ANKTM1 channel

It is known that the plant-derived cannabinoid, Δ^9-tetrahydrocannabinol (THC), has analgesic and anti-inflammatory properties, which may be mediated by CB_1 receptors present on primary afferent sensory neurons. However, THC appears to retain an analgesic effect in CB_1 $^{-/-}$ mice. In addition, cannabidiol is anti-inflammatory and analgesic and does not share the ability of THC to bind to and activate CB_1 receptors. These data raise the possibility of a non-CB_1 receptor-mediated action of plant-derived cannabinoids [63]. Furthermore, THC and cannabidiol induce a concentration-related relaxation of arterial segment – an effect that is insensitive to antagonism by CB_1 receptor antagonists [64]. The relaxant effect is absent in capsaicin-treated vessels, indicating a role for small-diameter sensory fibers, but is not antagonized by the TRPV1 receptor antagonist capsazepine. The relaxation is abolished by ruthenium red, which is a non-competitive antagonist of vanilloid receptors. This novel effect of THC and cannabidiol has been demonstrated to be mediated by activation of a TRP-related channel ANKTM1, which is a cold-activated receptor [65]. This channel is located in a small sub-population of nociceptive sensory neurons, which also express TRPV1. The receptor is differentially expressed in trigeminal and DRG neurons, with significantly greater expression in trigeminal neurons. It remains to be investigated fully whether endocannabinoids also activate this channel.

Anandamide metabolites

Anandamide and 2-arachidonoylglycerol (2-AG) are rapidly hydrolyzed by the microsomal enzyme FAAH. In addition, anandamide and 2-AG can also be metabolized by a range of oxygenase enzymes, which are already known to convert arachidonic acid (Fig. 1) to potent biologically active compounds (see [66]). These include cyclooxygenase (COX), lipoxygenase and cytochrome P450 enzymes. Such pathways have important implications for the pharmacology of endocannabinoids, which may be directed by tissue-specific differences in the balance of these metabolic enzymes. Oxygenation of anandamide and 2-AG leads to the production of a range of novel lipid products whose physiological role has yet to be elucidated. The metabolism of anandamide may influence the activation of TRPV1 receptors by this endocannabinoid.

Arachidonic acid and TRPV4

There is a large body of evidence demonstrating that the pharmacological effects of anandamide both *in vitro* and *in vivo* can be markedly enhanced by inhibition of the enzymatic hydrolysis of this fatty acid amide by FAAH [1, 32]. The FAAH inhibitor phenylmethanesulfonyl fluoride (PMSF) has been shown to have anti-nociceptive actions when administered alone [67]. Thus it would appear that these effects of anandamide are not generally due to its conversion to arachidonic acid and the subsequent production of COX and lipoxygenase arachidonic acid metabolites. However, the analgesic and motor-inhibitory actions of anandamide in the mouse tetrad *in vivo* are not blocked by SR141716A, and are present in CB_1 receptor-knockout mice [68]. In the presence of FAAH inhibitors and in FAAH-knockout mice, anandamide acts as a potent and selective CB_1 agonist and the behavioral and analgesic effects are reversed by SR141716A [32]. It may be that the non-CB_1 pharmacology of anandamide in this test is due to its rapid hydrolysis to arachidonic acid or other active metabolites, which are cannabamimetic, but not CB_1 receptor agonists. Certainly, some of the pharmacological effects of anandamide have been shown to remain after its levels in the brain have diminished [68], implicating the formation of active anandamide metabolites.

TRPV4 is located in a broad range of tissues including the sensory terminals of DRG neurons [69]. The non-CB_1 receptor-mediated analgesic actions of anandamide may be related to the recent finding that TRPV4 is activated by cytochrome P450 metabolites of arachidonic acid. In the low micromolar concentration range, anandamide and arachidonic acid increase intracellular Ca^{2+} and activate whole-cell currents in TRPV4-expressing cells [70]. Anandamide metabolism to arachidonic acid appears to be a prerequisite for TRPV4 activation: the metabolically stable anandamide analogue methanandamide is not a TRPV4 agonist and anandamide is inactive in the presence of the FAAH inhibitor PMSF.

COX metabolites

Yu et al. [71] demonstrate that anandamide is oxygenated by COX-2, while COX-1 does not display any detectable activity with anandamide as substrate. The major product of COX-2 metabolism of anandamide is prostaglandin E_2 (PGE_2) ethanolamide. Furthermore, anandamide is oxygenated by COX-2 in the physiological environment: in human foreskin fibroblast cells which express both COX isoforms, PG ethanolamides are produced as products of anandamide metabolism. However, in the non-P-glycoprotein-mediated resistant clone of parent HL60 cells, which express COX-1 only, there is no detectable metabolism. In RAW 264.7 macrophages, PGE_2 ethanolamide is synthesized from anandamide and pre-treatment of the cells with lipopolysaccharide – which induces COX-2 expression – leads to a significant enhancement of the production of this metabolite [72]. A recent study further investigates the nature of the PG ethanolamides that can be formed as a consequence of COX-2 metabolism of anandamide [73]. In HCA-7 cells that constitutively express COX-2, COX-2 oxygenation of anandamide leads to the production of the endoperoxide intermediate PGH_2 ethanolamide, the major metabolites of which are PGE_2 and PGD_2 ethanolamide.

PGE_2 ethanolamide produces only a modest displacement (approx. 24%) of $[^3H]$RTX from rat-TRPV1-Chinese hamster ovary cell membranes [14]. Matias et al. [74] find that $PGF_{2\alpha}$ did increase intracellular Ca^{2+} in HEK-293 cells overexpressing the TRPV1 receptor; albeit with a low potency, the IC_{50} and E_{max} values were 15 μM and 31%, respectively. In the same study, PGE_2 ethanolamide and PGD_2 ethanolamide were found to be inactive. Prostanoids are known both to directly activate sensory neurons and to sensitize sensory neurons to other potent nociceptive agents such as bradykinin. PGE_2 has been shown to activate a subpopulation of small-diameter capsaicin-sensitive DRG neurons and to potentiate the bradykinin-evoked increases in $[Ca^{2+}]_i$. Both these actions of PGE_2 appear to involve PKA-dependent mechanisms [75]. Smith et al. [75] found that 1 μM PGE_2 evoked an increase in $[Ca^{2+}]_i$ in 16% of capsaicin-sensitive DRG neurons. PGI_2 and $PGF_{2\alpha}$ (1 μM) also evoke calcium transients in 26 and 29% of DRG neurons, respectively. Similarly, Ross et al. [76] find that PGE2 ethanolamide (3 μM) evoked an increase in $[Ca^{2+}]_i$ in 21% of small-diameter capsaicin-sensitive DRG neurons (Fig. 3). Therefore, it is possible that PGE2 ethanolamide shares the ability of PGE2 to sensitize DRG neurons to capsaicin and bradykinin.

Lipoxygenase metabolites

In vitro, 12- and 15-lipoxygenase convert anandamide to 12- and 15-hydroperoxyeicosatetraenoyl acid (HPETE) ethanolamide respectively, the reaction rates being similar to those for arachidonic acid (see [66]). 12- and 15-HPETE ethanolamide are produced when anandamide is incubated with porcine leukocytes 12-lipoxygenase and rabbit reticulocyte 15-lipoxygenase, respectively [77]. In line with this, 12-lipoxygenase isolated from the rat pineal gland effectively metabolizes anandamide [78]. In these studies, the reaction rates for anandamide metabolism were found to be not significantly different from those obtained with arachidonic acid as substrate.

The lipoxygenase metabolites of arachidonic acid, particularly 12-(*S*)-HPETE, 5-(*S*)-hydroxyeicoatetraenoic acid (HETE), and leukotriene B_4 (LTB_4), are agonists of the TRPV1 receptor [79, 80]. Recent studies of the action of the potent inflammatory mediator bradykinin provide more compelling evidence for the role of lipoxygenase metabolites in activation of TRPV1. Thus, bradykinin activation of TRPV1 receptors in both cultured DRG neurons and skin is significantly attenuated by lipoxygenase inhibitors [81]. Furthermore, extracellular recording from C-fiber receptive fields in guinea pig isolated airways reveals that lipoxygenase inhibitors dramatically inhibit bradykinin-induced action potentials, which are TRPV1 receptor-mediated [82]. As an alternative substrate for lipoxygenase enzymes, anan-

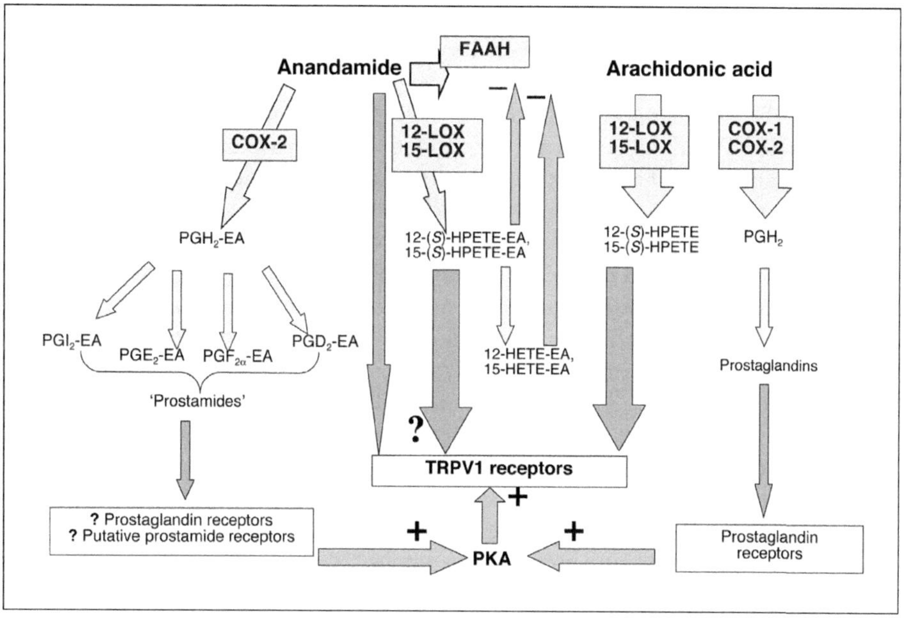

Figure 3
Metabolic pathways of anandamide relating to the TRPV1 receptor modulation. Anandamide is rapidly hydrolyzed by fatty acid amide hydrolase (FAAH) to yield arachidonic acid and ethanolamide. Arachidonic acid is oxygenated by lipoxygenase (LOX) enzymes: the products include 12-(S)- and 15-(S)-hydroperoxyeicosatetraenoyl acid (HPETE), 5-(S)-hydroxyeicoatetraenoic acid (HETE), and leukotriene B_4, which are TRPV1 agonists [26]. Anandamide is a substrate for cyclooxygenase 2 (COX-2), yielding prostaglandin ethanolamides (PG-EA) [66]. There is evidence that these compounds may activate TRPV1 possibly via protein kinase A (PKA) activation [76]. Anandamide is also a substrate for lipoxygenase [66], yielding equivalent HPETE ethanolamides (HPETE-EAs) and HETE ethanolamides (HETE-EAs), which may also be TRPV1 agonists [15]. The lipoxygenase products of anandamide are potent inhibitors of FAAH [86].

damide may attenuate the production of metabolites of arachidonic acid, which are TRPV1 receptor agonists. Alternatively, anandamide may activate TRPV1 via lipoxygenase metabolites of arachidonic acid formed subsequent to its metabolism by FAAH (Fig. 3). A further possibility is that anandamide lipoxygenase metabolites may themselves activate TRPV1 receptors. In the bronchus, the TRPV1 receptor-mediated contractile action of anandamide is little affected by FAAH inhibitors but is significantly attenuated by lipoxygenase inhibitors [15]. These data suggest that, in this tissue, anandamide may be metabolied to HPETE ethanolamides and LTB_4 ethanolamides which – like the hydroperoxy derivatives of arachidonic acid (HPETEs) and LTB_4 – may be vanilloid receptor agonists. In human-TRPV1-HEK-293 cells, the anandamide lipoxygenase metabolites 11-(*S*)-HPETE ethanolamide and 5-(*S*)-HPETE ethanolamide, do not induce an increase in $[Ca^{2+}]_i$, and 15-(*S*)-HPETE ethanolamide has only a modest effect at concentrations above 10 μM [29]. However, it is feasible that these compounds and related molecules are more potent as TRPV1 agonists if produced intracellularly via metabolism of arachidonic acid and/or anandamide close to the intracellular TRPV1 agonist-binding site. There is evidence that anandamide is neuroprotective against oubain-induced excitotoxicity, an action that is shared by capsaicin [83]. The neuroprotective effects of anandamide are not attenuated by concentrations of capsazepine that antagonize TRPV1. Intriguingly, the 12-lipoxygenase metabolite of anandamide had a neuroprotective effect while the 15-lipoxygnease metabolite enhanced the neuroprotective effect of anandamide. However, pretreatment with the non-selective lipoxygenase inhibitor, nordihydroguaiaretic acid (NDGA), does not reverse anandamide-mediated inhibition of cytotoxic edema, suggesting that lipoxygenase products are not implicated in this action of the endocannabinoid.

In the bronchus, lipoxygenase inhibitors modestly attenuate the contractile action of capsaicin [15], raising the possibility that the increase in intracellular calcium elicited by TRPV1 receptor activation leads to the release of arachidonic acid and/or anandamide, whose hydroxylation by lipoxygenase may lead to the formation of compounds that are themselves vanilloid agonists. Indeed, mass spectrometric analysis shows that capsaicin and depolarization (KCl) induce significant release of anandamide in DRG cultures [36]. It is perhaps notable that the capsaicin-evoked release of anandamide is significantly attenuated when the FAAH inhibitor, MAFP, is excluded from the buffer, demonstrating that anandamide is rapidly metabolized in DRG neurons. Metabolic products of anandamide have also been implicated in anandamide-induced depolarization of the guinea pig isolated vagus nerve, which is TRPV1 receptor-mediated [84]. In this preparation, depolarization by anandamide, but not capsaicin, is inhibited by lipoxygenase inhibitors but in only in the presence of calcium. It is not clear, however, whether these active metabolites are produced via direct lipoxygenase metabolism of anandamide or via metabolism of arachidonic acid, because these experiments did not include FAAH inhibitors. Potent FAAH inhibitors have recently been synthesized which enhance the levels of anandamide significantly and these compounds may be of considerable therapeutic benefit [85]. In the event of inhibition of FAAH metabolism of anandamide, increased levels of endogenous anandamide may lead to the production of significant levels of the HPETE ethanolamides. In addition, lipoxygenase metabolism to metabolites that are FAAH inhibitors [86] of anandamide may enhance TRPV1 receptor activation by increasing the levels of available anandamide.

NADA and TRVP1

NADA (Fig. 1) has been identified in nervous tissue [10]. This novel endovanilloid is produced in highest concentrations in the striatum, hippocampus, and cerebellum. In HEK-293 cells overexpressing TRPV1 receptors, NADA increases intracellular Ca^{2+} levels with an EC_{50} of approx. 50 nM that is comparable to that obtained with capsaicin. In cultured DRG neurons, NADA increases intracellular Ca^{2+} levels with an EC_{50} of approx. 800 nM, an effect that is inhibited by capsazepine and iodoR-TX. The compound stimulates release of neuropeptides from isolated DRG neurons and dorsal spinal cord slices. Furthermore, when administered intradermally, NADA produces a strong thermal hyperalgesia.

Price et al. [33] studied the effects of anandamide, arachidonyl-2-chloroethylamide, and NADA in trigeminal neurons *in vitro*. The compounds evoke CGRP release, an effect that is sensitive to TRPV1 receptor antagonists iodoRTX and capsazepine. In line with previous findings, anandamide has a low potency with an EC_{50} value of 11 μM, as compared to that of approx. 40 nM for capsaicin. NADA appears to be considerably more potent than anandamide, the EC_{50} being 857 nM. Both compounds behaved as full agonists. This study provides evidence that anandamide and arachidonyl-2-chloroethylamide are non-pungent, being inactive in the rat eye-wipe test. An influence of the CB_1 receptor was ruled out, as inclusion of SR141716A did not unmask a nociceptive effect of anandamide. This is in contrast to NADA, which exhibits significant nocifensive responses.

In blood vessels, NADA shares the ability of anandamide to activate both TRPV1 and the novel non-CB_1 endothelial receptor. O'Sullivan et al. [87] have identified that the compound effects relaxation of small mesenteric vessels via TRPV1 receptors and the novel endothelial CB receptor; in the superior mesenteric vessels the compound interacts with TRPV1 and CB_1 receptors. In the small mesenteric vessels, NADA is significantly more potent as a vasorelaxant than anandamide, the EC_{50} being 0.41 μM, as compared to 1.7 μM. In the superior mesenteric vessels, the compound is also more potent, but in the aorta the EC_{50} values of 1 μM were obtained for both anandamide and NADA.

In the isolated guinea pig bronchus, NADA has a contractile action that is antagonized by capsazepine [30]. However, the EC_{50} of approx. 13 μM, is considerably higher than observed in TRPV1 expressing HEK-293 cells (EC_{50} = 33 nM). The compound behaves as a partial agonist in the bronchus with an E_{max} value of 69%, which is significantly lower than that of capsaicin (93%). In isolated urinary bladder preparations from both the guinea pig and the rat, NADA behaves somewhat differently to expectations, having E_{max} values of 12 and 20%, respectively, as compared to those of approx. 90% for anandamide [30].

Premkumar et al. [88] further characterized NADA as potential novel endovanilloid. They found that in oocytes expressing TRPV1 receptors NADA has an EC_{50} of greater than 10 μM, a concentration that elicits currents that are <1% of those observed with 1 μM capsaicin. As has been previously demonstrated for anandamide, NADA possessed the ability to potentiate its own response with repeated application. This effect is not observed in cells expressing TRPV1 double mutants that are resistant to phosphorylation or in the presence of a PKC antagonist, indicating that the potentiation involves activation of PKC. There is evidence that by binding directly to the diacylglycerol site anandamide has a dual modulatory effect on PKC in rat brain *in vitro*: whereas anandamide increases phosphatidylserine-induced PKC activation, it also inhibits dioleylglycerol-induced potentiation of Ca^{2+}-induced PKC activation [89]. In DRG neurons the mean current amplitude of 10 μM NADA was 22% of that obtained with capsaicin (1 μM). In line with the hypothesis that NADA and anandamide need to cross the membrane to access the intracellular TRPV1 receptor-binding site, Premkumar et al. [88] report that the potency of NADA is increased to 41% of that obtained with capsaicin when it is applied intracellularly by inclusion in the pipette solution.

Oleamide and TRPV1

Oleoylethanolamide (oleamide) is an endocannabinoid that reduces feeding and has been proposed to be a peripherally acting satiety lipid. There is evidence that it may interact with CB_1 receptors (see [90]). Ahern [91] found that in oocytes expressing TRPV1, oleamide (up to 80 μM) did not activate TRPV1 currents but at a concentration of 20 μM the compound enhanced proton-activated currents by approx. 3-fold [91]. In contrast, the same concentration of oleamide significantly inhibited anandamide-evoked TRPV1 currents, whereas it had no effect on capsaicin-evoked currents. Following PKC phosphorylation of TRPV1, oleamide alone activated currents. Oleamide is a substrate for FAAH and as a consequence will inhibit metabolism of anandamide. In contrast to the findings of Ahern [91], Smart et al. [92] demonstrated that the compound enhances the effect of anandamide on Ca^{2+} influx in TRPV1-expressing HEK-239 cells, an effect the authors attributed to the inhibition of anandamide metabolism.

Conclusions

A large body of evidence now exists to substantiate that certain endocannabinoids activate TRPV1 receptors. Anandamide and NADA appear to be low-intrinsic-efficacy TRPV1 agonists behaving as partial agonists in tissues with low receptor reserve. However, in tissues with high receptor reserve and in circumstances associated with certain disease states they behave as full agonists. Recent findings indicate that the apparent low efficacy of anandamide and NADA in certain circumstances may be due to limited access to the intracellular environment. When released intracellularly, anandamide activates TRPV1 receptors in the same concentration range at which it is known to activate the CB_1 receptor. Furthermore, there is evidence that the endocannabinoid system may play a role in the modulation of TRPV1 receptor activation. The relative importance of endocannabinoids as physiological and/or pathophysiological TRPV1 receptor agonists – in comparison to other potential candidates, such as lipoxygenase products of arachidonic acid – has yet to be revealed.

The metabolism of anandamide is a divergent process, which differs in various cell types. In addition to metabolism by FAAH, it is also metabolized by COX-2 and lipoxygenase to form families of novel lipids that are pharmacologically active. Some enzymes display low activity towards anandamide; nevertheless, in the absence of other metabolic alternatives, such pathways may become significant. There is evidence that some of these metabolic products of anandamide may activate or sensitize the TRPV1 receptor; however, the physiological importance of these novel metabolic pathways remains to be established.

References

1 Pertwee RG, Ross RA (2002) Cannabinoid receptors and their ligands. *Prostaglandins Leukot Essent Fatty Acids* 66: 101–121

2 Di Marzo V, Blumberg PM, Szallasi A (2002) Endovanilloid signaling in pain. *Curr Opin Neurobiol* 12: 372–379

3 Parolaro D, Massi P, Rubino T, Monti E (2002) Endocannabinoids in the immune system and cancer. *Prostaglandins Leukot Essent Fatty Acids* 66: 319–332

4 Rice ASC, Farquhar-Smith WP, Nagy I (2002) Endocannabinoids and pain: spinal and peripheral analgesia in inflammation and neuropathy. *Prostaglandins Leukot Essent Fatty Acids* 66: 243–256

5 Walker JM, Huang SM (2002) Endocannabinoids in pain modulation. *Prostaglandins Leukot Essent Fatty Acids* 66: 235–242

6 Di Marzo V, De Petrocellis L, Fezza F, Ligresti A, Bisogno T (2002) Anandamide receptors. *Prostaglandins Leukot Essent Fatty Acids* 66: 377–391

7 Wiley JL, Martin BR (2002) Cannabinoid pharmacology: implications for additional cannabinoid receptor subtypes. *Chem Phys Lipids* 121: 57–63

8 Di Marzo V, Bisogno T, De Petrocellis L (2001) Anandamide: some like it hot. *Trends Pharmacol Sci* 22: 346–349

9 Di Marzo V, Blumberg PM, Szallasi A (2002) Endovanilloid signaling in pain. *Curr Opin Neurobiol* 12: 372–379

10 Huang SM, Bisogno T, Trevisani M, Al-Hayani A, De Petrocellis L, Fezza F, Tognetto M, Petros TJ, Krey JF, Chu CJ et al (2002) An endogenous capsaicin-like substance with high potency at recombinant and native vanilloid VR1 receptors. *Proc Natl Acad Sci USA* 99: 8400–8405

11 Evans R, Scott RH, Ross RA (2004) Multiple actions of anandamide on neonatal rat cultured sensory neurones. *Br J Pharmacol* 141: 1223–1233
12 Zygmunt PM, Petersson J, Andersson DA, Chuang H, Sorgard M, Di Marzo V, Julius D, Hogestatt ED (1999) Vanilloid receptors on sensory nerves mediate the vasodilator action of anandamide. *Nature* 400: 452–457
13 Smart D, Gunthorpe MJ, Jerman JC, Nasir S, Gray J, Muir AI, Chambers JK, Randall AD, Davis JB (2000) The endogenous lipid anandamide is a full agonist at the human vanilloid receptor (hVR1). *Br J Pharmacol* 129: 227–230
14 Ross RA, Gibson TM, Brockie HC, Leslie M, Pashmi G, Craib SJ, Di Marzo V, Pertwee RG (2001) Structure-activity relationship for the endogenous cannabinoid, anandamide, and certain of its analogues at vanilloid receptors in transfected cells and vas deferens. *Br J Pharmacol* 132: 631–640
15 Craib SJ, Ellington HC, Pertwee RG, Ross RA (2001) A possible role of lipoxygenase in the activation of vanilloid receptors by anandamide in the guinea-pig isolated bronchus. *Br J Pharmacol* 134: 30–37
16 Ross RA (2003) Minireview: anandamide and vanilloid TRP1 receptors. *Br J Pharmacol* 140: 790–801
17 Jennings EA, Vaughan CW, Roberts LA, Christie MJ (2003) The actions of anandamide on rat superficial medullary dorsal horn neurons *in vitro*. *J Physiol* 548: 121–129
18 Ralevic V, Kendall DA, Randall MD, Smart D (2002) Cannabinoid modulation of sensory neurotransmission via cannabinoid and vanilloid receptors: roles in regulation of cardiovascular function. *Life Sci* 71: 2577–2594
19 Sprague J, Harrison C, Rowbotham DJ, Smart D, Lambert DG (2001) Temperature-dependent activation of recombinant rat vanilloid VR1 receptors expressed in Hek293 cells by capsaicin and anandamide. *Eur J Pharmacol* 423: 121–125
20 Jerman JC, Gray J, Brough SJ, Ooi L, Owen D, Davis JB, Smart D (2002) Comparison of effects of anandamide at recombinant and endogenous rat vanilloid receptors. *Br J Anaesthesia* 89: 882–887
21 Andersson DA, Adner M, Hogestatt ED, Zygmunt PM (2002) Mechanisms underlying tissue selectivity of anandamide and other vanilloid receptor agonists. *Mol Pharmacol* 62: 705–713
22 Roberts LA, Christie MJ, Connor M (2002) Anandamide is a partial agonist at native vanilloid receptors in acutely isolated mouse trigeminal sensory neurons. *Br J Pharmacol* 137: 421–428
23 Jung J, Hwang SW, Kwak J, Lee SY, Kang CJ, Kim WB, Kim D, Oh U (1999) Capsaicin binds to the intracellular domain of the capsaicin-Activated Ion Channel. *J Neurosci* 19: 529–538
24 Jordt SE, Julius D (2002) Molecular basis for species specificity to 'hot' chili peppers. Cell 108: 421–430
25 Hillard CJ, Jarrahian A (2003) Cellular accumulation of anandamide: consensus and controversy. *Br J Pharmacol* 140: 802–808
26 Glaser S, Abumrad NA, Fatade F, Kaczocha M, Studholme KM, Deutsch D (2003) Evi-

dence against the presence of an anandamide transporter. *Proc Natl Acad Sci USA* 100: 4269–4274

27 De Petrocellis L, Bisogno T, Maccarrone M, Davis JB, Finazzi-Agro A, Di Marzo V (2001) The activity of anandamide at vanilloid VR1 receptors requires facilitated transport across the cell membrane and is limited by intracellular metabolism. *J Biol Chem* 276: 12856–12863

28 Harrison S, De Petrocellis L, Trevisani M, Benvenuti F, Bifulco M, Geppetti P, Di Marzo V (2003) Capsaicin-like effects of *N*-arachidonoyl dopamine in the isolated guinea pig bronchi and urinary bladder. *Eur J Pharmacol* 475: 107–114

29 Toth A, Kedei N, Wang Y, Blumberg PM (2003) Arachidonyl dopamine as a ligand for the vanilloid receptor VR1 of the rat. *Life Sci* 73: 487–498

30 Cravatt BF, Demarest K, Patricelli MP, Bracey MH, Giang DK, Martin BR, Lichtman AH (2001) Supersensitivity to anandamide and enhanced endogenous cannabinoid signaling in mice lacking fatty acid amide hydrolase. *Proc Natl Acad Sci USA* 98: 9371–9376

31 Orliac ML, Peroni R, Celuch SM, Adler-Grachinsky E (2002) Potentiation of anandamide effects in mesenteric beds isolated from endotoxemic rats. *J Pharmacol Exp* Therap 304: 179–184

32 Tucker RC, Kagaya M, Page CP, Spina D (2001) The endogenous cannabinoid agonist, anandamide stimulates sensory nerves in guinea-pig Airways. *Br J Pharmacol* 132: 1127–1135

33 Price TJ, Patwardhan A, Akopian AN, Hargreaves KM, Flores CM (2004) Modulation of trigeminal sensory neuron activity by the dual cannabinoid-vanilloid agonists anandamide, *N*-arachidonoyl-dopamine and arachidonyl-2-chloroethylamide. *Br J Pharmacol* 14: 1118–1130

34 Olah Z, Karai L, Iadarola MJ (2001) Anandamide activates vanilloid receptor 1 (VR1) at acidic Ph in dorsal root ganglia neurons and cells ectopically expressing VR1. *J Biol Chem* 276: 31163–31170

35 Hwang SW, Hawoon C, Kwak J, Lee S-Y, Kang C-J, Jung J, Cho S, Min KH, Suh Y-G, Kim G, Oh U (2000) Direct activation of capsaicin receptors by products of lipoxygenase: endogenous capsaicin-like substances. *Proc Natl Acad Sci USA* 97: 6155–6160

36 Ahluwalia J, Yaqoob M, Urban L, Bevan S, Nagy I (2003) Activation of capsaicin-sensitive primary sensory neurones induces anandamide production and release. *J Neurochem* 84: 585–591

37 Hermann H, De Petrocellis L, Bisogno T, Schiano Moriello A, Lutz B, Di Marzo V (2003) Dual effect of cannabinoid CB_1 receptor stimulation on a vanilloid VR1 receptor-mediated response. *Cell Mol Life Sci* 60: 607–616

38 Garcia DE, Brown S, Hille B, Mackie K (1998) Protein kinase C disrupts cannabinoid actions by phosphorylation of the CB_1 cannabinoid receptor. *J Neurosci* 18: 2834–2841

39 Chuang HH, Prescott ED, Kong HY, Shields S, Jordt SE, Basbaum AI, Chao MV, Julius D (2001) Bradykinin and nerve growth factor release the capsaicin receptor from Ptdins(4,5)P-2-mediated inhibition. *Nature* 411: 957–962

40 Netzeband JG, Conroy SM, Parsons KL, Gruol DL (1999) Cannabinoids enhance NMDA-elicited Ca^{2+} signals in cerebellar granule neurons in culture. *J Neurosci* 19: 8765–8777

41 Ho BY, Uezono Y, Takada S, Takase I, Izumi F (1999) Coupling of the expressed cannabinoid CB_1 and CB_2 receptors to phospholiase C and G-protein-coupled inwardly rectifying K^+ channels. *Receptors Channels* 6: 363–374

42 Gomez Del Pulgar T, De Ceballos ML, Guzman M, Velasco G (2002) Cannabinoids protect astrocytes from ceramide-induced apoptosis through the phosphatidylinositol 3-kinase/protein kinase B pathway. *J Biol Chem* 277: 36527

43 Ahluwalia J, Urban L, Capogna M, Bevan S, Nagy I (2000) Cannabinoid 1 receptors are expressed in nociceptive primary sensory neurons. *Neuroscience* 100: 685–688

44 Ahluwalia J, Urban L, Bevan S, Capogna M, Nagy I (2002) Cannabinoid 1 receptors are expressed by nerve growth factor – and glial cell-derived neurotrophic factor-responsive primary sensory neurones. *Neuroscience* 110: 747–753

45 Farquhar-Smith WP, Egertova M, Bradbury EJ, Mcmahon SB, Rice ASC, Elphick MR (2000) Cannabinoid CB1 receptor expression in rat spinal cord. *Mol Cell Neurosci* 15: 510–521

46 Khasabova IA, Simone DA, Seybold VS (2002) Cannabinoids attenuate depolarization-dependent Ca^{2+} influx in intermediate-size primary afferent neurons of adult rats. *Neuroscience* 115: 613–625

47 Bridges D, Rice ASC, Egertove M, Elphick MR, Winter J, Michael G (2003) Localisation of CB1 cannabinoid receptor in rat dorsal root ganglion using in situ hybridisation and immunohistochemistry. *Neuroscience* 119: 803–812

48 Price TJ, Helesic G, Parghi D, Hargreaves KM, Flores CM (2003) The neuronal distribution of cannabinoid receptor type one in the trigeminal ganglion of the rat. *Neuroscience* 120: 155–162

49 Khasabova IA, Harding-Rose C, Simone DA, Seybold VS (2004) Differential effects of CB_1 and opioid agonists on two populations of adult rat dorsal root ganglion neurons. *J Neurosci* 24: 1744–1753

50 Tognetto M, Amadesi S, Harrison S, Creminon C, Trevisani M, Carreras M, Matera M, Geppetti P, Bianchi A (2001) Anandamide excites central terminals of dorsal root ganglion neurons via vanilloid receptor-1 activation. *J Neurosci* 21 1104–1109

51 Ross RA, Coutts AA, Mcfarlane S, Irving AJ, Pertwee RG, Macewan DJ, Scott RH (2001) Actions of cannabinoid receptor ligands on rat cultured sensory neurones: implications for antinociception. *Neuropharmacology* 40: 221–232

52 Ahluwalia J, Urban L, Bevan S, Nagy I (2003). Anandamide regulates neuropeptide release from capsaicin-sensitive primary sensory neurons by activating both the cannabinoid 1 receptor and the vanilloid receptor 1 *in vitro*. *Neuroscience* 17: 2611–2618

53 Richardson JD, Kilo S, Hargreaves KM (1998) Cannabinoids reduce hyperalgesia and inflammation via interaction with peripheral CB_1 receptors. *Pain* 75: 111–119

54 Mang CF, Erbelding D, Kilbinger H (2001) Differential effects of anandamide on acetyl-

choline release in the guinea-pig ileum mediated via vanilloid and non-Cb_1 cannabinoid receptors. *Br J Pharmacol* 134: 161–167
55 Maccarrone M, Lorenzon T, Bari M, Melino G, Finazzi-Agro A (2000) Anandamide induces apoptosis in human cells via vanilloid receptors – evidence for a protective role of cannabinoid receptors. *J Biol Chem* 275: 31938–31945
56 Lever IJ, Malcangio M (2002) CB_1 receptor antagonist SR141716A increases capsaicin-evoked release of substance P from the adult mouse spinal cord. *Br J Pharmacol* 135: 21–24
57 Pertwee RG (2005) Inverse agonism and neutral antagonism at CB_1 receptors. *Life Sci* 76: 1307–1324
58 Ellington HC, Cotter MA, Cameron NE, Ross RA (2002) The effect of cannabinoids on capsaicin-evoked calcitonin gene-related peptide (CGRP) release from the isolated paw skin of diabetic and non-diabetic rats. *Neuropharmacology* 42: 966–975
59 Jia YL, Mcleod RL, Wang X, Parra LE, Egan RW, Hey JA (2002) Anandamide induces cough in conscious guinea-pigs through VR1 receptors. *Br J Pharmacol* 137: 831–836
60 Breivogel CS, Griffin G, Di Marzo V, Martin BR (2001) Evidence for a new G-protein coupled cannabinoid receptor in mouse brain. *Mol Pharmacol* 60: 155–163
61 Sagan S, Venance L, Torrens Y, Cordier J, Glowinski J, Giaume C (1999) Anandamide and Win 55212-2 inhibit cyclic AMP formation through G-protein-coupled receptors distinct from CB_1 cannabinoid receptors in cultured astrocytes. *Eur J Neurosci* 11: 691–699
62 Offertaler L, Mo FM, Batkai S, Liu J, Begg M, Razdan RK, Martin BR, Bukoski RD, Kunos G (2003) Selective ligands and cellular effectors of a G protein-coupled endothelial cannabinoid receptor. *Mol Pharmacol* 63: 699–705
63 Pertwee RG (2004) Novel targets for cannabinoids. *Curr Neuropharm* 2: 29
64 Zygmunt PM, Andersson DA, Hogestatt ED (2002) Δ9-tetrahydrocannabinol and cannabinol activate capsaicin-sensitive sensory nerves via a CB_1 and CB_2 cannabinoid receptor-independent mechanism. *J Neurosci* 22: 4720–4727
65 Jordt SE, Bautista DM, Chuang HH, Mckemy DD, Zygmunt PM, Hogestatt ED, Meng ID, Julius D (2004) Mustard oils and cannabinoids excite sensory nerve fibres through the TRP channel ANKTM1. *Nature* 427: 260–265
66 Kozak KR, Marnett LJ (2002) Oxidative metabolism of endocannabinoids. *Prostaglandins Leukot Essent Fatty Acids* 66: 211–220
67 Compton DR, Martin BR (1997) The effect of the enzyme inhibitor phenylmethylsulfonyl fluoride on the pharmacological effect on anandamide in the mouse model of cannabimimetic activity. *J Pharmacol Exp Ther* 283: 1138–1143
68 Willoughby KA, Moore SF, Martin BR, Ellis EF (1997) The biodisposition and metabolism of anandamide in mice. *J Pharmacol Exp Ther* 282: 243–247
69 Nilius B, Vriens J, Prenen J, Droogmans G, Voets T (2003) TRPV4 calcium entry channel: a paradigm for gating diversity. *Am J Cell Physiol* 286: C195–C205
70 Watanabe H, Vriens J, Prenen J, Droogmans G, Voets T, Nilius B (2003) Anandamide

and arachidonic acid use epoxyeicosatrienoic acids to activate TRPV4 channels. *Nature* 424: 434–439
71 Yu M, Ives D, Ramesha CS (1997) Synthesis of prostaglandin E2 ethanolamide from anandamide by cyclooxygenase-2. *J Biol Chem* 272: 21181–21186
72 Burstein SH, Rossetti RG, Yagen B, Zurier RB (2000) Oxidative metabolism of anandamide. *Prostaglandins Other Lipid Mediat* 61: 29–41
73 Kozak KR, Crews BC, Morrow JD, Wang LH, Ma YH, Weinander R, Jakobsson PJ, Marnett LJ (2002) Metabolism of the endocannabinoids, 2-arachidonylglycerol and anandamide, into prostaglandin, thromboxane, and prostacyclin glycerol esters and ethanolamides. *J Biol Chem* 277: 44877–44885
74 Matias I, Chen J, De Petrocellis L, Bisogno T, Ligresti A, Fezza F, Krauss AHP, Shi L, Protzman Ce, Li C et al (2004) Prostaglandin ethanolamides (prostamides): in vitro pharmacology and metabolism. *J Pharmacol Exp Ther* 309: 745–757
75 Smith JAM, Davis CL, Burgess GM (2000) Prostaglandin E_2-induced sensitization of bradykinin-evoked responses in rat dorsal root ganglion neurons is mediated by cAMP-dependent protein kinase A. *Eur J Neurosci* 12: 3250–3258
76 Ross RA, Craib SJ, Stevenson LA, Pertwee RG, Henderson A, Toole J, Ellington HC (2002) Pharmacological characterization of the anandamide cyclooxygenase metabolite: prostaglandin E_2 ethanolamide. *J Pharmacol Exp Ther* 301: 900–907
77 Ueda N, Yamamoto K, Yamamoto S, Tokunaga T, Shirakawa E, Shinkai H, Ogawa M, Sato T, Kudo I, Inoue K et al (1995) Lipoxygenase-catalyzed oxygenation of arachidonylethanolamide, a cannabinoid receptor agonist. *Biochim Biophys Acta* 1254: 127–134
78 Edgemond WS, Hillard CJ, Falck JR, Kearn CS, Campbell WB (1998) Human platelets and polymorphonuclear leukocytes synthesize oxygenated derivatives of arachidonylethanolamide (anandamide): their affinities for cannabinoid receptors and pathways of inactivation. *Mol Pharmacol* 54: 180–188
79 Hwang SW, Hawoon C, Kwak J, Lee S-Y, Kang C-J, Jung J, Cho S, Min KH, Suh Y-G, Kim G, Oh U (2000) Direct activation of capsaicin receptors by products of lipoxygenase: endogenous capsaicin-like substances. *Proc Natl Acad Sci USA* 97: 6155–6160
80 Piomelli D (2001) The ligand that came from within. *Trends Pharmacol Sci* 22: 17–19
81 Shin J, Cho H, Hwang SW, Jung J, Shin CY, Lee S-Y, Kim SH, Lee MG, Choi YH, Kim J et al (2002) Bradykinin-12-lipoxygenase-VR1 signaling pathway for inflammatory hyperalgesia. *Proc Natl Acad Sci USA* 99: 10150–10155
82 Carr MJ, Kollarik M, Meeker SN, Undem BJ (2002) A role for Trpv1 in bradykinin-induced excitation of vagal airway afferent nerve terminals. *J Pharmacol Exp Ther* 304: 1275–1279
83 Veldhuis WB, Van Der Stelt M, Wadman MW, Van Zadelhoff G, Maccarrone M, Fezza F, Veldink GA, Vliegenthart JFG, Bar PR, Nicolay K, Di Marzo V (2003) Neuroprotection by the endogenous cannabinoid anandamide and arvanil against *in vivo* excitotoxicity in the rat: role of vanilloid receptors and lipoxygenases. *J Neurosci* 23: 4127–4133
84 Kagaya M, Lamb J, Robbins J, Page CP, Spina D (2002) Characterization of the anan-

damide induced depolarization of guinea-pig isolated vagus nerve. *Br J Pharmacol* 137: 39–48

85 Boger DL, Sato H, Lerner AE, Hedrick MP, Fecik RA, Miyauchi H, Wilkie GD, Austin BJ, Patricelli MP, Cravatt BF (2000) Exceptionally potent inhibitors of fatty acid amide hydrolase: the enzyme responsible for degradation of endogenous oleamide and anandamide. *Proc Natl Acad Sci USA* 97: 5044–5049

86 Van Der Stelt M, Van Kuik JA, Bari M, Van Zadelhoff G, Leeflang BR, Veldink GA, Finazzi-Agro A, Vliegenthart JFG, Maccarrone M (2002) Oxygenated metabolites of anandamide and 2-arachidonoylglycerol: conformational analysis and interaction with cannabinoid receptors, membrane transporter, and fatty acid amide hydrolase. *J Med Chem* 45: 3709–3720

87 O'Sullivan SE, Kendall DA, Randall MD (2004) Characterisation of the vasorelaxant properties of the novel endocannabinoid N-arachidonoyl-dopamine (Nada). *Br J Pharmacol* 141: 803–812

88 Premkumar LS, Qi Zh, Van Buren J, Raisinghani M (2004) Enhancement of potency and efficacy of NADA by PKC-mediated phosphorylation of vanilloid receptor. *J Neurophysiol* 91: 1442–1449

89 De Petrocellis L, Orlando P, Di Marzo V (1995) Anandamide, an endogenous cannabinomimetic substance, modulates rat brain protein kinase C *in vitro*. *Biochem Mol Biol Int* 36: 1127–1133

90 Fowler CJ (2004) Oleamide: a member of the endocannabinoid family? *Br J Pharmacol* 141: 195–196

91 Ahern GP (2003) Activation of TRPV1 by the satiety factor oleoylethanolamide. *J Biol Chem* 278: 30429–30434

92 Smart D, Jonsson KO, Vandevoorde S, Lambert DM, Fowler CJ (2002) 'Entourage' effects of N-acyl ethanolamines at human vanilloid receptors. Comparison of effects upon anandamide-induced vanilloid receptor activation and upon anandamide metabolism. *Br J Pharmacol* 136: 452–458

Vanilloid receptor-mediated hyperalgesia and desensitization

Zoltán Sándor[1] *and Arpad Szallasi*[2]

[1]Department of Pharmacology and Pharmacotherapy, Faculty of Medicine, University of Pécs, H-7624, Pécs, Hungary, and [2]Department of Pathology, Monmouth Medical Center, Long Branch, NJ 07740, USA

Historical introduction

"There are compounds that can selectively desensitize sensory nerve endings to noxious chemical stimuli without causing local anesthesia.... Capsaicin is the paradigm of such desensitizing agents. Capsaicin effects last for days and protect against various chemicals...." With these words, the late Nicholas (Miklós) Jancsó opened a new chapter in sensory pharmacology in 1949 [1]. It is a misfortune of the scientific world that Jancsó chose to publish his seminal observation in his native Hungarian language in the journal *Kísérletes Orvostudomány* (*Experimental Medicine*) in the turbulent years following the descent of the Iron Curtain on post-World War Two Europe, and thus this new concept remained ignored for the next two decades.

By the late 1980s, it became generally accepted that sensitivity to capsaicin serves as a functional signature of a subset of primary sensory neurons [2, 3]. These neurons are unique in that their initial excitation by capsaicin (for structure, see Fig. 1) is followed by a lasting refractory state, referred to as desensitization by Jancsó [1]. This characteristic provides capsaicin with a clear therapeutic potential in disease states in which abnormal afferent sensory information conveyed by capsaicin-sensitive nerves is a major factor in the etiology [4]. An admittedly incomplete list of these diseases includes chronic neuropathic pain, migraine, pruritus, and overactive urinary bladder [5, 6].

The concept of a specific capsaicin receptor was formulated based on specific structure-activity relations in the 1970s [7, 8]. In 1990, specific binding of [^{3}H]resiniferatoxin ([^{3}H]RTX; Fig. 1), an ultrapotent capsaicin analog, furnished the first biochemical proof for the existence of this receptor [9], and the term vanilloid receptor was coined to stress the unifying chemical feature of capsaicin and RTX [4]. In 1997, the accelerated advances in vanilloid research culminated in the molecular cloning of the first (and, as of today, only) vanilloid (capsaicin) receptor [10], now known by the somewhat cumbersome name transient receptor potential channel, vanilloid subfamily member 1 (TRPV1; Fig. 2). However, in the brain of

Turning up the Heat on Pain: TRPV1 Receptors in Pain and Inflammation, edited by
Annika B. Malmberg and Keith R. Bley

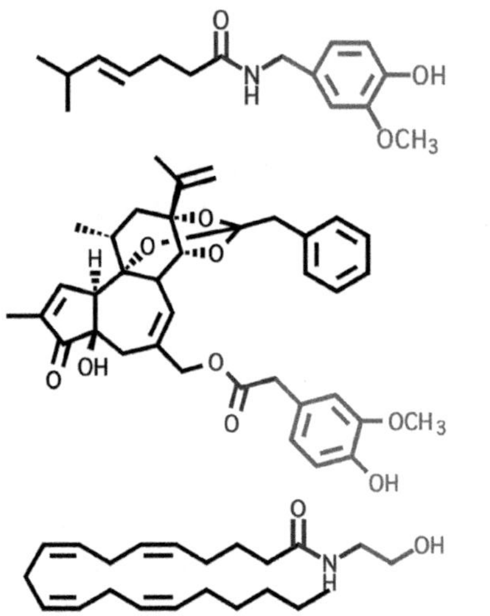

Figure 1
Vanilloid receptor agonist structures: capsaicin, resiniferatoxin, and anandamide.

TRPV1$^{-/-}$ mice, [^{3}H]RTX binding is markedly reduced but not eliminated, implying the existence of other, as-yet-unidentified, RTX-binding sites [11].

The cloning of TRPV1 has revolutionized the ways that we think about this receptor. It is now clear that instead of being merely a membrane recognition site for vanilloids and related compounds, in the periphery TRPV1 functions as a molecular integrator of noxious stimuli (Fig. 3), including heat, acids, pollutants with negative electric charge, and endogenous pro-inflammatory substances [12, 13]. In the brain, TRPV1 is believed to be a target for the endocannabinoid anandamide (for structure, see Fig. 1; reviewed in [14]) and related compounds such as *N*-arachidonoyl-dopamine (NADA) [15] and *N*-oleoyldopamine (OLDA) [16, 17], along with such substances of broad human use and abuse like ethanol [18] and nicotine [19].

Unexpectedly, TRPV1 (once believed to be a sensory neuron marker) turned out to have a rather widespread tissue distribution. The growing list of tissues with functional TRPV1 expression now includes keratinocytes [20, 21], urothelium [22], gastric epithelium [23], mast cells [24], sweat glands [24], smooth muscle [22], liver [25], macrophages [26], and polymorphonuclear neutrophils [27]. Recently, the recognition of disease-related changes is TRPV1 expression has been a much-anticipated breakthrough in the field (reviewed in [5]).

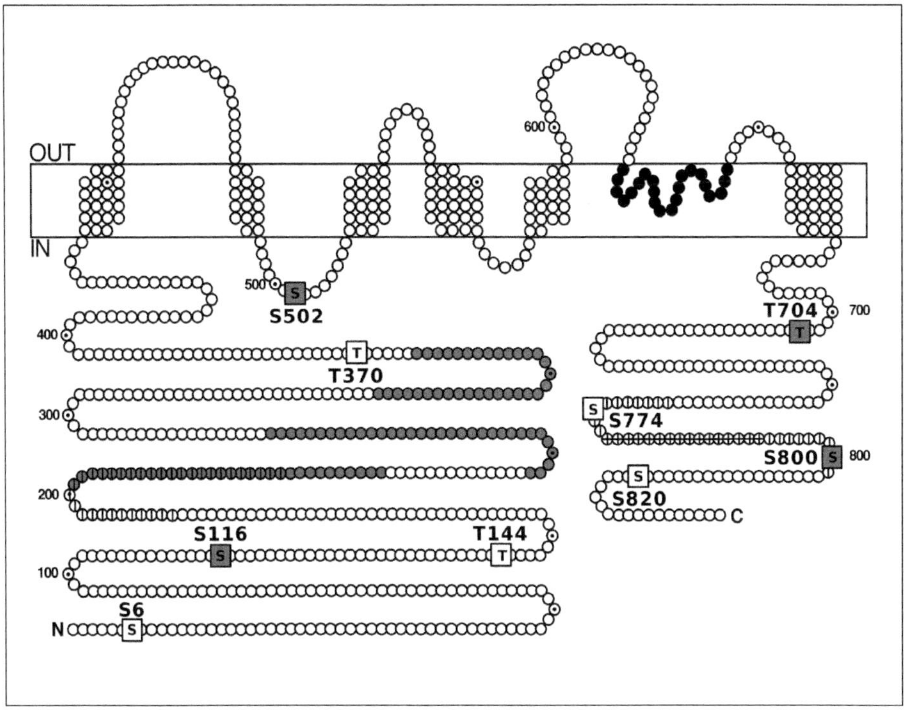

Figure 2
Schematic representation of amino acid residues involved in the modulation of the TRPV1 receptor. The six putative transmembrane domains, the loops, and the N- and C-terminal cytoplasmic domains are indicated by the arrangement of the small circles representing the amino acids. Residues corresponding to the three ankyrin repeats (200–232, 247–279, 332–364) are indicated with gray circles, whereas the putative pore loop (624–645) is black. Calmodulin-binding sites (189–222, 767–801) are represented by circles with a vertical stripe. The phosphatidylinositol 4,5-bisphosphate (PIP_2)-binding domain contained in the C-terminal calmodulin-binding site is represented by circles with a cross. Residues found to be phosphorylated are marked with big squares and the type and position of these amino acids are also indicated. Phosphorylated residues with significant effects on receptor function are marked with gray squares.

However, in order to avoid redundancy with other chapters in this volume, the above findings will not be discussed here. Instead, we will focus on the sensitization and desensitization mediated by vanilloid receptors, with an explicit recognition that the literature on vanilloid receptor-mediated hyperalgesia and desensitization is vast. Indeed, a Medline search has identified almost 1000 publications dealing with various aspects of these two topics.

Clearly, a comprehensive overview of this literature would be beyond the scope of this chapter. Admittedly incomplete and selective, this chapter will highlight only major shifts in our thinking about vanilloid sensitization and desensitization, with particular emphasis on data that have emerged in the past 2 years.

Sensitization and hyperalgesia by vanilloids

Overview

Connoisseurs of hot, spicy food know the enhanced burning sensation by heat of their tongue or lips from personal experience. No doubt, this is the best-known paradigm of capsaicin hyperalgesia. More scientifically, intradermal injection of capsaicin in humans results in primary hyperalgesia to heat and mechanical stimuli in the vicinity of the injection site [28, 29], followed by secondary hyperalgesia and allodynia in the area surrounding the site of primary hyperalgesia [30]. The observation that mice lacking TRPV1 show apparently normal heat sensation but exhibit reduced thermal hyperalgesia, compared to wild-type mice, supports the hypothesis that TRPV1 plays a pivotal role in the development of hyperalgesia to heat [31, 32]. As detailed below, primary hyperalgesia probably reflects sensitization of TRPV1 itself, whereas secondary hyperalgesia and allodynia may involve additional mechanisms such as the novel expression of TRPV1 on nerves that do not normally express this receptor and an architectural reorganization of the innervation ('sprouting') that may also reflect gross neurotoxicity by capsaicin.

In principle, sensitization of a membrane receptor might be a consequence of at least four distinct molecular mechanisms. First, existing receptors may acquire a higher apparent affinity via allosteric changes. Second, silent receptors may be unmasked by post-translational modifications. Third, receptors may be recycled from intracellular stores/compartments. And finally, increased transcriptional activity can yield higher receptor density. As discussed below, there is good evidence that vanilloid receptor-mediated sensitization utilizes all four of these cascades.

As already alluded to, TRPV1 acts as a molecular integrator of various noxious stimuli including, but not limited to, heat, protons, and capsaicin. Although the channel is independently activated by capsaicin (at submicromolar concentrations), heat (above 43 °C), and acidification ($pH < 6.5$) [10, 33], there is cross-sensitization among these agents with apparent synergism. For instance, the heat activation threshold of the channel is reduced to body temperature in the presence of mild (approx. pH 7.0) acidification [34].

There is an emerging notion that the recognition domains on TRPV1 for these stimuli are distinct. Indeed, a TRPV1 mutant in the region of transmembrane domain 6 showed disrupted capsaicin activation whereas it retained its ability to respond to protons (see Fig. 2) [35]. Of note, transmembrane domain 3 is also

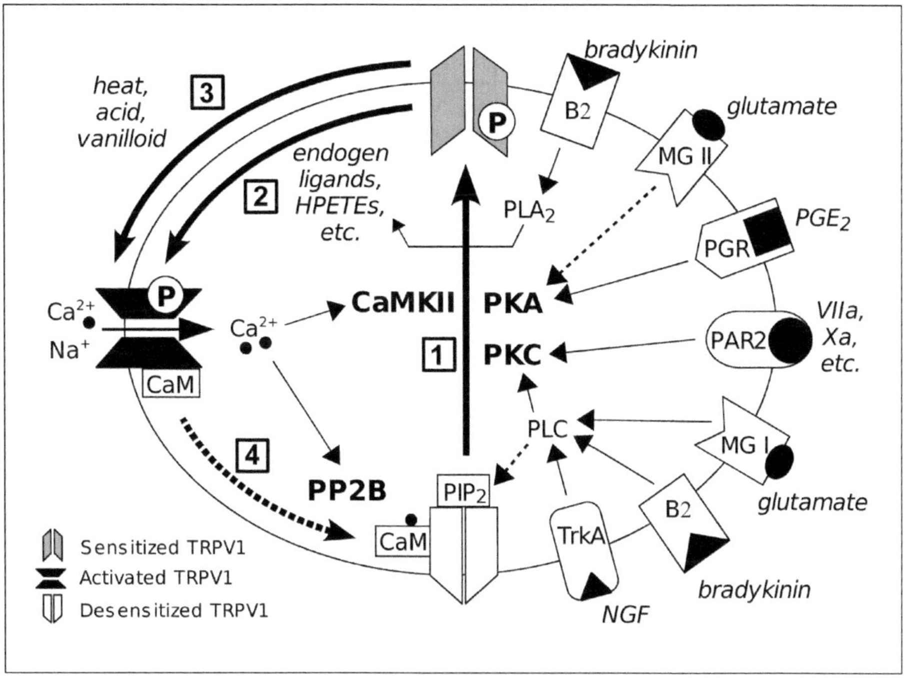

Figure 3
Schematic diagram of TRPV1 receptor modulation. (1) Sensitization of the receptor by phosphorylation (PKA, PKC, and CaMKII) and PLC-mediated PIP_2 removal. (2) Activation of the receptor by endogenous ligands, like hydroperoxyeicosatetraenoyl acids (HPETEs), N*-arachidonoyl-dopamine (NADA), etc. (3) Activation of the receptor by exogenous stimuli, like heat or vanilloids. (4) Calcium- and calmodulin-dependent dephosphorylation of the receptor by protein phosphatase 2B (PP2B; or calcineurin) leading to desensitization. On the right side of the diagram, the different signaling pathways affecting TRPV1 sensitization are depicted. Solid arrows represent activation; dashed arrows indicate inhibition. CaM, calmodulin; CaMKII, calmodulin dependent kinase II; PLA_2, phospholipase A_2; PLC, phospholipase C; PKA, protein kinase A; PKC, protein kinase C; B2, bradykinin receptor 2; MG I and MG II, group I and II metabotropic glutamate receptors; PAR2, protease-activated receptor 2; PGR, prostaglandin E_2 receptor; PIP_2, phosphatidylinositol 4,5-bisphosphate.*

important for attaining capsaicin sensitivity via hydrophobic interactions [36]. Lowering the pH was shown to enhance the apparent binding affinity for capsaicin, promote long openings, and stabilize the open conformation of the channel [37].

Regulation of TRPV1 sensitivity by protein kinases and phosphatases

The phosphorylation state of TRPV1, which reflects a dynamic balance between protein kinases and phosphatases, is now believed to be a key determinant of channel activity (Fig. 3). Of kinases, protein kinase C (PKC) has attracted the most attention [38]. For example, PKC links protease-activated receptor 2 (PAR2) to TRPV1 [39]. PAR2 is activated by factors like mast cell tryptase, coagulation factors VIIa and Xa, and trypsin, and is known to evoke hyperalgesia (reviewed in [40]). It has been speculated that capsaicin-evoked visceral pain and referred hyperalgesia is mediated via delayed sensitization by PAR2 of TRPV1 [41]. Furthermore, both bradykinin and acetylcholine augment TRPV1 activity via phosphorylation by PKC of residues Ser-502 and Ser-800 (Fig. 2) [42]. It is not known which of the nine PKC isozymes mediate(s) TRPV1 sensitization. Both PKCγ [43] and PKCε [44] have been implicated in pain transmission and thus represent obvious candidates for playing this role. Indeed, co-localization of TRPV1 and PKCε in capsaicin-sensitive neurons has been demonstrated [45].

The cAMP-dependent protein kinase (protein kinase A; PKA), which plays a major role in producing inflammatory hyperalgesia [46], was shown to sensitize TRPV1 (Fig. 3) [47]. For instance, prostaglandins utilize the PKA pathway that involves phosphorylation of three putative PKA consensus sites (Thr-144, Thr-370, and Ser-502) of which one (Ser-502) is shared with the PKC pathway [48]. Ser-502 is also a consensus site for Ca^{2+}/calmodulin-dependent kinase II (CaMKII). Indeed, a mutant TRPV1 at this site failed to exhibit capsaicin-evoked currents, with a concomitant loss of [^{3}H]RTX-binding sites [49]. Taken together, these findings suggest that Ser-502 is a key site for the regulation of TRPV1. Furthermore, phosphorylation by CaMKII of TRPV1 appears to be a prerequisite for vanilloid binding and subsequent channel activation. Conversely, as discussed below, TRPV1 dephosphorylation by calcineurin and other Ca^{2+}-dependent phosphatases promote tachyphylaxis. It is odd that Ca^{2+} can both sensitize and desensitize TRPV1, a non-selective cation channel that allows Ca^{2+} influx. One might visualize a complex pathway where increasing Ca^{2+} levels first stimulate CaMKII (resulting in a positive feedback) and then shut down the channel via calmodulin (negative feedback). Finally, CaMKII restores TRPV1 activity after intracellular Ca^{2+} has returned to normal. Somewhat unexpectedly, Ca^{2+}/calmodulin has been shown to bind to TRPV1 directly [50] and presumably block activity (Fig 3).

Importantly, CaMKII is not the only agent that has the ability to reverse capsaicin tachyphylaxis. Phorbol-12 myristate-13 acetate restores TRPV1 activity via PKC [51]. Nerve growth factor (NGF) can also sensitize TRPV1 in a phospholipase C (PLC)-dependent manner [52, 53]. In a related note, NGF was reported to increase TRPV1 expression [54]. Since NGF is produced during inflammation, both sensitization by NGF of TRPV1 and subsequent receptor upregulation may contribute to inflammatory hyperalgesia (Fig. 3).

A fourth class of kinases that has been implicated in vanilloid receptor-mediated sensitization is the extracellular signal-regulated kinases (ERKs) that are part of the mitogen-activated protein (MAP) kinase family. Phosphorylated ERKs (pERKs) appear to parallel sensitization in capsaicin-sensitive neurons [55] but the roles that pERKs might play in hyperalgesia remain to be elucidated.

Group I metabotropic glutamate receptors potentiate capsaicin responses via generation of diacylglycerol [56]. Interestingly, peripheral group II metabotropic glutamate receptors exert the opposite action: they can counteract the sensitization by PKA of TRPV1 via inhibition of adenylate cyclase, suggesting that these receptors might represent a therapeutic target for novel analgesic drugs [57].

Recruitment of TRPV1 from intracellular stores

TRPV1 exists in at least three intracellular compartments (for details, see [58]). Recently, Ferrer-Montiel and colleagues [59] used a yeast hybrid screen to identify proteins that may contribute to the potentiation of vanilloid receptor activity by PKC. They identified two proteins, snapin and synaptotagmin IX, involved in TRPV1 trafficking to the plasma membrane. It is speculated that a recruitment of TRPV1 from vesicles to the plasma membrane plays a major role in the development and maintenance of inflammatory hyperalgesia [59].

PLC liberates TRPV1 from the inhibitory control of phosphatidylinositol 4,5-bisphosphate (PIP_2)

TRPV1 is under the inhibitory control of PIP_2 [60, 61]. As depicted in Fig. 3, PLCβ cleaves PIP_2 and liberates TRPV1 from its inhibition, which is an important pathway utilized by both bradykinin and NGF. In addition, PLCβ generates diacylglycerol formation, which, in turn, activates PKC [38, 62]. Bradykinin can also activate TRPV1 through the generation of 12-lipoxygenase metabolites of arachidonic acid [63].

Clearly, potent algesic agents like bradykinin exert multifaceted actions of on TRPV1. They release TRPV1 from under the inhibitory action of PIP_2 via PLCβ [60]; they promote phosphorylation by PKC of TRPV1 via generation of diacylglycerol [38, 64]; and, finally, they stimulate the formation of pro-inflammatory substances that bind to and activate TRPV1 [63]. The resultant effect is a powerful sensitization of TRPV1 and subsequent hyperalgesia.

Upregulation of TRPV1 expression

Two days after intraplantar injection of Freund's complete adjuvant, a significant increase in TRPV1-like immunoreactivity was demonstrated [65], which was interpreted as a major mechanism of peripheral sensitization of nociceptors during inflammation. Of note, some of the new vanilloid receptors are apparently present on nerves that do not normally express TRPV1 [66]. In diabetic mice, anti-TRPV1 antiserum given intrathecally abolished thermal hyperalgesia and allodynia [67].

Vanilloid receptor agonist-induced desensitization

The initial excitation of sensory neurons by vanilloids is followed by a refractory state, traditionally termed desensitization [1–4, 68]. Although firmly entrenched in terminology, it would probably be better to avoid this term because desensitization is not a well-defined biochemical process but rather a complex cascade of events, ranging from short-term unresponsiveness to the irreversible damage of polymodal sensory neurons [69]. For didactic purposes, this cascade may be divided into three separate stages with distinct dose-response relationships. In real life, these stages are often present simultaneously without clear boundaries.

The first stage is often called specific or pharmacological desensitization [3] and it occurs at the level of the TRPV1 receptor. It can be defined as being restricted to subsequent vanilloid agonist challenges without preventing the neurons from responding to other types of stimulus [69]. In contrast, the second stage – nonspecific or functional desensitization [3] – implies a general impairment of neuronal function, during which the neuron becomes unresponsive not only to TRPV1 agonist but also to other types of noxious stimuli (e.g. heat, acids, and mechanical irritation). Specific desensitization usually lasts for minutes or hours, while functional desensitization can last for several weeks, but ultimately both are reversible [69]. This clearly distinguishes them from the third stage, termed vanilloid-induced neurotoxicity, which means the irreversible destruction of the sensitive neurons [69].

Specific desensitization of the TRPV1 receptor

Desensitization to capsaicin depends on both dose (topically, it should exceed 30 nM) and frequency of application and can be divided into tachyphylaxis and acute desensitization, with the former representing diminishing responses upon repeated short agonist administrations and the latter meaning decreasing receptor activity in continuous presence of the agonist [3, 69]. However, this division is artifactual, as both phenomena are quantitatively similar in terms of Ca^{2+} uptake [70, 71].

Both tachyphylaxis and acute desensitization depend on the presence of Ca^{2+} in the extracellular medium [70–73]. Tachyphylaxis requires the entry of extracellular Ca^{2+} through TRPV1, as raising the level of free intracellular Ca^{2+} by either depolarization or by using an ionophore is ineffective in this respect [72]. Of note, in isolated dorsal root ganglion (DRG) neurons the refractory period lasts up to 45 min, which is about 30–35 min longer than the elevation of the free intracellular Ca^{2+} level, indicating that although Ca^{2+} initiates the desensitization process other factors are also important in maintaining it [72].

Recently, the Ca^{2+}-binding protein calmodulin was implicated as the calcium-dependent modulator of TRPV1 receptor function. Calmodulin-binding sites were detected in both the N- [53] and C-termini [74] of TRPV1 (Fig. 2). Calmodulin is able to bind to these sites even in the absence of Ca^{2+}; therefore it may be positioned optimally to rapidly sense the Ca^{2+} influx through the open channel. Binding of the Ca^{2+}–calmodulin complex to the N-terminal tail of TRPV1 reduced the probability of channel opening [53]. Furthermore, the Ca^{2+}–calmodulin-dependent phosphatase, calcineurin, was shown to contribute to vanilloid desensitization by counteracting TRPV1 phosphorylation by various kinases [49]. The delicate balance between protein kinases and phosphatases was detailed above and is briefly summarized in Figure 3.

Functional desensitization of polymodal nociceptors

Functional desensitization is achieved when vanilloid agonists are applied at concentrations several-fold higher than that required for the threshold of stimulation. Capsaicin-sensitive polymodal nociceptors show decreased responses to various noxious stimuli detected by these neurons, like bradykinin, heat, and high-threshold mechanical stimuli [75–81]. Capsaicin-insensitive neurons reacting to cold and low-threshold mechanical stimuli are, however, not affected, indicating that the functional impairment is specific to the TRPV1-expressing neurons. This functional blockade affects the transduction process and action-potential generation in the nerve endings without blocking the conduction in the axon [82, 83]. Functional desensitization can last from hours to several weeks depending on the TRPV1 agonist dose and mode of administration, but it is reversible [69].

Various mechanisms can underlie this functional impairment, all of which seems to be secondary to the cation influx that accompanies the activation of the TRPV1 channel. For instance, an inhibition of voltage-gated ion channels was observed in capsaicin-treated trigeminal ganglial slice preparations [84] and isolated trigeminal ganglial neurons [85, 86]. Depolarization-induced slow inactivation of the tetrodotoxin-resistant Na^{+} channels appears to be responsible for the short-term adaptation of these neurons [87]. Long-term inhibition of action potential generation has also been observed by the inhibition of voltage-gated Na^{+} channels through the activation of the cAMP second-messenger system [85]. Other second-messenger systems involving Ca^{2+} and cGMP may also have a role as their levels are elevated during capsaicin receptor activation [88] and they are known modulators of voltage-gated Na^+ channels [89]. Capsaicin is also capable of blocking voltage-gated K^+ [90] and Ca^{2+} [91, 92] channels. The prolonged inhibition of voltage-gated Ca^{2+} channels in capsaicin-treated DRG neurons may be responsible for the nonspecific inhibition of sensory neuropeptide release [2–4].

From a therapeutic point of view, those compounds that can achieve desensitization without apparent excitation are the most interesting. These drugs are thought to open the TRPV1 channel slowly, without leading to action-potential generation and, at the same time, increase the intracellular Ca^{2+} level sufficiently to induce inhibition of voltage-gated ion channels and action-potential generation. Several non-pungent vanilloid analogs with reduced side effects, like olvanil [93, 94], nuvanil [95], and SDZ249-482 [96], show potential in this respect.

Vanilloid-induced complex changes in the level of various neurotransmitters, sensory neuropeptides, and their receptors can also contribute to the functional desensitization process (reviewed in [3, 4]). For example, high cumulative doses of capsaicin deplete neuropeptides including substance P, calcitonin-gene-related peptide (CGRP), somatostatin, and galanin [2–4]. The restoration of these neuropeptides can be impaired due to the blockage of axonal transport by capsaicin [97–99]. In contrast, a single moderate dose of RTX treatment induces only the depletion of substance P, while the levels of the other neuropeptides are not affected (or even increased), as in the case of galanin and vasoactive intestinal polypeptide (VIP) [100, 101]. The upregulation of the inhibitory neuropeptide galanin may play a role in the suppression of the C-fiber activation-induced spinal hyperexcitability (wind-up phenomenon) and contribute greatly to the prolonged analgesic action of RTX treatment [101].

The loss of specific RTX-binding sites in sensory ganglia, spinal cord, and urinary bladder of the rat following moderate systemic RTX treatment indicates the downregulation of TRPV1 [102–105] in sensory neurons. In the urinary bladder, the full recovery of the receptor to pre-treatment levels can take up to 8 weeks, which contributes to the long-lasting desensitization effect of the RTX treatment [105].

Following treatments with higher doses of vanilloid agonists, functional desensitization is accompanied by ultrastructural changes in capsaicin-sensitive neurons. After topical capsaicin treatment the nerve terminals of the rat cornea appears to be swollen [106], probably due to the osmotic effects of the cation influx. The loss of vesicles could also be observed reflecting the depletion of neuropeptides from the endings. The increased cytoplasmic Ca^{2+} levels lead to accumulation in the mitochondria which may result in the dissipation of the mitochondrial membrane potential [107], osmotic swelling, and re-organization of the inner structure of the mitochondria [108–110] in vanilloid agonist-sensitive neurons. The Ca^{2+} influx also activates calcium-dependent proteases which, in turn, can contribute to the ultrastructural changes and the functional impairment [111].

High-dose vanilloid agonist treatment can lead to the degeneration of nerve terminals in the skin [112–115], which after several weeks can be followed by re-innervation, as has been demonstrated in humans [114, 115]. These experiments demonstrate that the long time period sometimes required for the restoration of sensory neuronal function makes it difficult to draw a clear line between TRPV1 agonist-induced reversible functional desensitization and irreversible degeneration. However, it is clear that under certain conditions vanilloid agonist treatment can result in gross neurotoxicity meaning the irreversible damage and subsequent elimination of sensory neurons.

Vanilloid agonist-induced neurotoxicity

Vanilloid agonists can ablate capsaicin-sensitive neurons in culture [111, 116, 117] and in animals [118]; also, they can kill TRPV1 receptor-transfected heterologous cells [10, 119]. This excitotoxic effect is probably due to the uncontrolled influx of Ca^{2+}, which leads to mitochondrial damage [107] and protease activation [111], resulting in cell death. However, there are important differences between vanilloid-agonist induced neurotoxicity in cell culture and in animals.

Systemic capsaicin treatment of neonatal rats and mice results in the loss of small B-type neurons from the sensory ganglia of the animals [120, 121]. This phenomenon was originally described as an acute neurotoxic effect, leading to cell death within hours [120]. However, recent data suggest that, in fact, the cell death is delayed for several days and B-type neurons with swollen mitochondria are present in the sensory ganglia up to 20 weeks after the treatment [109, 110]. Furthermore, the death of these neurons could be prevented with daily NGF treatments of the animals [122]. It seems likely that capsaicin treatment kills polymodal sensory neurons in neonatal rats by depriving them of NGF, which is required for their survival [123, 124]. This can happen either by blocking axonal transport of NGF from the periphery to the cell bodies [97], or by completely destroying terminal parts of the neurons which are the most sensitive to systemic capsaicin treatment [112, 125]. Interestingly, adult rats that received neonatal capsaicin treatment still showed sensitivity to capsaicin challenges. In these animals the number of unmyelinated axons was decreased by about 40%, but there was no change in the proportion of the C-fiber receptor types [126]. These data indicate that these animals possess capsaicin-sensitive neurons that either survived the neonatal capsaicin treatment or appeared later. In agreement with this latter hypothesis, expression of TRPV1 receptor was observed in previously vanilloid agonist-insensitive neurons after axotomy [127, 128].

Peri-axonal capsaicin treatment (33 mM) of adult rats resulted in the degeneration of about 40% of the unmyelinated axons for 2–12 months after treatment [129–131], resulting in long-lasting anti-nociception. Systemic treatment of adult rats by subcutaneous administration of 35–300 mg/kg capsaicin results in swelling of the mitochondria in B-type sensory neurons when examined 1–60 days after treatment without the apparent destruction of the cell bodies [106, 132, 133]. On the other hand, severe degeneration of axons and axon terminals was observed in the periphery without the appreciable degeneration in the dorsal roots [112, 134]. This indicates that *in vivo* the central and peripheral endings of capsaicin-sensitive neurons are the most susceptible, to both the stimulatory [135] and degenerative [134] effects of capsaicin.

In certain aspects, TRPV1-knockout mice represent the ultimate model of vanilloid agonist-induced desensitization [31, 32]. The complete absence of vanilloid-sensitive DRG neurons and the abolished vanilloid sensitivity in these animals (as shown in behavioral tests) indicate that the TRPV1 gene encodes a single vanilloid-sensitive channel. The DRG neurons of these animals were also defective in responding to acids and moderate heat (<45 °C), while responses to high stimulus temperatures (<55 °C) remained intact, indicating the presence of other heat-sensing channels [136]. Paradoxically, TRPV1-knockout mice showed normal behavioral responses to moderate temperature challenges and only showed reduced pain behavior at higher temperatures [31]. These mice also completely lacked the ability to develop inflammation-induced thermal hyperalgesia, indicating the central role of the TRPV1 in inflammatory sensitization [32]. These results point to vanilloid-induced desensitization as a potentially useful therapy in inflammatory diseases.

Concluding remarks

Desensitization to vanilloids has a clear therapeutic potential (reviewed in [4–6]). In fact, capsaicin has been in clinical use for decades and its ultrapotent analog, RTX, has undergone advanced clinical evaluation. Following the cloning of the vanilloid receptor TRPV1 [10], the pharmaceutical industry has launched massive efforts to identify novel, potent, and orally active TRPV1 agonists and antagonists [137]. Whether these synthetic compounds can be developed into clinically useful drugs that surpass their natural cousins remains to be seen [137]. The past few years have brought us new insights into the molecular mechanisms that underlie vanilloid receptor-mediated desensitization. Mice deficient in TRPV1 confirmed that pivotal role that this receptor plays in the development and maintenance of thermal hyperalgesia during inflammation [31, 32]. Moreover, now there is good evidence to suggest disease-related upregulation of TRPV1 expression (reviewed in [5]). Apparently, biochemical mechanisms of TRPV1 sensitization and desensitization are complex, implying a benefit for the pharmacologic blockade of pathways that promote TRPV1 sensitization/re-activation. Potential targets include various protein kinases (PKC, PKA, CaMKII, PAR2, etc.), along with upstream receptors that activate these enzymes (e.g. bradykin receptors or metabotropic glutamate receptors). It is hoped that a multifaceted pharmacologic approach can be developed that simultaneously blocks active TRPV1 and prevents unmasking/recruitment of silent receptors.

References

1 Jancsó N, Jancsó A (1949) Desensitization of sensory nerve endings [in Hungarian]. *Kísérletes Orvostudomány* 2 (Suppl): 15

2 Buck SH, Burks TF (1986) The neuropharmacology of capsaicin: a review of some recent observations. *Pharmacol Rev* 38: 179–226

3 Holzer P (1991) Capsaicin: cellular targets, mechanisms of action, and selectivity for thin sensory neurons. *Pharmacol Rev* 43: 143–120

4 Szallasi A, Blumberg PM (1999) Vanilloid (capsaicin) receptors and mechanisms. *Pharmacol Rev* 51: 159–212

5 Szallasi A (2002) Vanilloid (capsaicin) receptors in health and disease. *Am J Clin Pathol* 118: 110–121

6 Szallasi A, Fowler CJ (2002) After a decade of intravesical vanilloid therapy: still more questions than answers. *Lancet Neurol* 1: 167–172

7 Szolcsányi J, Jancsó-Gábor A (1976) Sensory effects of capsaicin congeners. Part II: Importance of chemical structure and pungency in desensitizing activity of capsaicin-type compounds. *Arzneimittelforschung (Drug Res)* 26: 33–37

8 Walpole CSJ, Wrigglesworth R (1993) Structural requirements for capsaicin agonists and antagonists. In: JN Wood (ed): *Capsaicin in the Study of Pain*. Academic Press, San Diego, 63–82

9 Szallasi A, Blumberg PM (1990) Specific binding of resiniferatoxin, an ultrapotent capsaicin analog, by dorsal root ganglion membranes. *Brain Res* 524: 106–111

10 Caterina MJ, Schumacher MA, Tominaga M, Rosen TA, Levine JD, Julius D (1997) The capsaicin receptor: a heat-activated ion channel in the pain pathway. *Nature* 389: 816–824

11 Roberts JC, Davis JB, Benham CD (2004) [3H]Resiniferatoxin autoradiography in the CNS of wild-type and TRPV1 null mice defines TRPV1 (VR-1) protein distribution. *Brain Res* 995: 176–183

12 Caterina MJ, Julius D (2001) The vanilloid receptor: a molecular gateway to the pain pathway. *Annu Rev Neurosci* 24: 487–514

13 Jordt SE, McKemy DD, Julius D (2003) Lessons from peppers and peppermint: the molecular logic of thermosensation. *Curr Opin Neurobiol* 13: 487–492

14 Di Marzo V, Blumberg PM, Szallasi A (2002) Endovanilloid signaling in pain. *Curr Opin Neurobiol* 12: 372–380

15 Huang SM, Bisogno T, Trevisani M, Al-Hayani A, De Petrocellis L, Fezza M, Petros TJ, Krey FJ, Chu CI, Miller JD et al (2002) An endogenous capsaicin-like substance with high potency at recombinant and native vanilloid VR1 receptors. *Proc Natl Acad Sci USA* 99: 8400–8405

16 Chu CJ, Huang SM, De Petrocellis L, Bisogno T, Ewing SA, Miller JD, Zipkin RE, Daddario N, Appendino G, Di Marzo V et al (2003) *N*-oleoyldopamine, a novel endogenous capsaicin-like lipid that produces hyperalgesia. *J Biol Chem* 278: 13633–13639

17 Szolcsányi J, Sándor Z, Pethö G, Varga A, Bölcskei K, Almási R, Riedl Z, Hajos G, Czéh

G (2004) Direct evidence for activation and desensitization of the capsaicin receptor by *N*-oleoyldopamine on TRPV1-transfected cell line, in gene deleted mice, and in the rat. *Neurosci Lett* 361: 155–158

18 Trevisani M. Smart D, Gunthorpe MJ, Tognetto M, Barbieri M, Campi B, Amadesi S, Grey J, Jerman JC, Brough SJ et al (2002) Ethanol elicits and potentiates nociceptor responses via the vanilloid receptor-1. *Nat Neurosci* 5: 546–551

19 Liu L, Zhu W, Zhang ZS, Yang T, Grant A, Oxford G, Simon SA (2004) Nicotine inhibits voltage-dependent sodium channels and sensitizes vanilloid receptors. *J Neurophysiol* 91: 1482–1491

20 Inoue K, Koizumi S, Fiziwara S, Denda S, Inoue K, Denda M (2002) Functional vanilloid receptors in cultured normal human epidermal keratinocytes. *Biochem Biophys Res Commun* 291: 124–129

21 Southall MD, Li T, Gharibova LS, Pei Y, Nicol GD, Travers JB (2003) Activation of epidermal vanilloid receptor-1 induces release of proinflammatory mediators in human keratinocytes. *J Pharmacol Exp Ther* 304: 217–222

22 Birder LA, Kanai AJ, De Groat WC, Kiss S, Nealan ML, Burke NE, Dineley KE, Watkins S, Reynolds IJ, Caterina M (2001) Vanilloid receptor expression suggests a sensory role for urinary bladder epithelial cells. *Proc Natl Acad Sci USA* 98: 13396–13401

23 Kato S, Aihara E, Nakamura A, Xin H, Matsui H, Kohama K, Takeuchi I (2003) Expression of vanilloid receptors in rat gastric epithelial cells: role in cellular protection. *Biochem Pharmacol* 66: 1115–1121

24 Stander S, Moormann C, Schumacher M, Buddenkotte J, Artuc M, Shpakovitch V, Brzoska T, Lippert U, Henz BM, Luger TA et al (2004) Expression of the vanilloid receptor subtype 1 in cutaneous sensory nerve fibers, mast cells, and epithelial cells of appendage structures. *Exp Dermatol* 13: 129–139

25 Reilly CA, Taylor JL, Lanza DL, Carr BA, Crouch DJ, Yost GS (2003) Capsaicinoids cause inflammation and epithelial cell death through activation of vanilloid receptors. *Toxicol Sci* 73: 170–181

26 Chen CW, Lee ST, Wu WT, Fu WM, Ho FM, Lin WW (2003) Signal transduction for inhibition of iducible nitric oxide synthase and cyclooxygenase-2 induction by capsaicin and related analogs in macrophages. *Br J Pharmacol* 140: 1077–1087

27 Heiner I, Eisfeld J, Halaszovich CR, Wehage E, Jungling E, Zitt C, Luckhoff A (2003) Expression profile of the transient receptor potential (TRP) family in neutrophil granulocytes: evidence for currents through long TRP channel 2 induced by ADP-ribose and NAD. *Biochem J* 371: 1045–1053

28 Simone DA, Ngeow JYF, Putterman GJ, LaMotte RH (1987) Hyperalgesia to heat after intradermal injection of capsaicin. *Brain Res* 418: 201–203

29 Simone DA, Baumann TF, LaMotte RH (1989) Dose-dependent pain and mechanical hyperalgesia in humans after intradermal injection of capsaicin. *Pain* 38: 99–107

30 Torebjörk HE, Lundberg JM, LaMotte RH (1992) Central changes in processing of mechanoreceptive input in capsaicin-induced secondary hyperalgesia in humans. *J Physiol* 448: 765–780

31 Caterina MJ, Leffler A, Malmberg AB, Martin WJ, Trafton J, Petersen-Zeitz KR, Koltzenburg M, Basbaum AI, Julius D (2000) Impaired nociception and pain sensation in mice lacking the capsaicin receptor. *Science* 288: 306–313
32 Davis JB, Gray J, Gunthorpe MJ, Hatcher JP, Davey PT, Overend P, Harries MH, Latcham J, Clapham C, Atkinson K et al (2000) Vanilloid receptor-1 is essential for inflammatory thermal hyperalgesia. *Nature* 405:183-187
33 Jordt SE, Tominaga M, Julius D (2000) Acid potentiation of the capsaicin receptor determined by a key extracellular site. *Proc Natl Acad Sci USA* 97: 8134–8139
34 Reeh PW, Kress M (2001) Molecular physiology of proton transduction in nociceptors. *Curr Opin Pharmacol* 1: 45–51
35 Kuzhikandathil EV, Wang H, Szabo T, Morozowa N, Blumberg PM, Oxford GS (2001) Functional analysis of capsaicin receptor (vanilloid receptor subtype 1) multimerizationand agonist responseiveness using a dominant negative mutation. *J Neurosci* 21: 8697–8706
36 Jordt SE, Julius D (2002) Molecular basis for species-specific sensitivity to "hot" chili peppers. *Cell* 108: 421–430
37 Ryu S, Liu B, Qin F (2003) Low pH potentiates both capsaicin binding and channel gating of VR1 receptors. *J Gen Physiol* 122: 45–61
38 Vellani V, Mapplebeck S, Moriondo A, Davis JB, McNaughton PA (2001) Protein kinase C activation potentiates gating of the vanilloid receptor VR1 by capsaicin, protons, heat nand anandamide. *J Physiol* 534: 813–825
39 Amadesi S, Nie J, Vergnolle N, Cottrell GS, Grady EF, Trevisiani M, Manni C, Geppetti P, McRoberts JA, Ennes H et al (2004) Protease-activated receptor 2 sensitizes the capsaicin receptor transient receptor potential vanilloid receptor 1 to induced hyperalgesia. *J Neurosci* 24: 4300–4312
40 Ossovskaya VS, Bunnett NW (2003) Protease-activated receptors: contribution to physiology and disease. *J Physiol* 552: 589–601
41 Kawao N, Ikeda H, Kitano T, Kuroda R, Sekiguchi F, Kataoka K, Kamanaka Y, Kawataba A (2004) Modulation of capsaicin-evoked visceral pain and referred hyperalgesia by protease-activated receptors 1 and 2. *J Pharmacol Sci* 94: 277–285
42 Numazaki M, Tominaga T, Toyooka H, Tominaga M (2002) Direct phosphorilation of capsaicin receptor VR1 by protein kinase Ce and identification of two target serine residues. *J Biol Chem* 277: 13375–13378
43 Malmberg AB, Chen C, Tonegawa S, Basbaum AI (1997) Preserved acute pain and reduced neuropathic pain in mice lacking PKCγ. *Science* 278: 279–283
44 Khasar SG, Lin YH, Martin A, Dadgar J, McMahon T, Wang D, Hundle B, Aley KO, Isenberg W, McCarter G et al (1999) A novel nociceptor signaling pathway revealed in protein kinase C epsilon mutant mice. *Neuron* 24: 253–260
45 Zhou Y, Li GD, Zhao ZQ (2003) State-dependent phosphorylation of epsilon isozyme of prtein kinase C in adult rat dorsal root ganaglia after inflammation and nerve injury. *J Neurochem* 85: 571–580

46 Levine JD, Reichling DB (1999) Peripheral mechanisms of inflammatory pain. In: PD Wall, R Melzack (eds): *Textbook of pain*. Churchill Livingstone, Edinburgh, 59–84

47 Bhave G, Zhu W, Wang H, Brasier DJ, Oxford GS, Gereau RW (2002) cAMP-dependent protein kinase regulates desensitization of the capsaicin receptor (VR1) by direct phosphorylation. *Neuron* 35: 721–731

48 Rathee PK, Distler C, Obreja O, Neuhuber W, Wang GK, Wang SY, Nau C, Kress M (2002) PKA/AKAP/VR-1 module: A common link of Gs-mediated signaling to thermal hyperalgesia. *J Neurosci* 22: 4740–4745

49 Jung J, Shin JS, Lee SY, Hwang SW, Koo J, Cho H, Oh U (2004) Phosphorylation of vanilloid receptor 1 by Ca^{2+}/calmodulin-dependent kinase II regulates its vanilloid binding. *J Biol Chem* 279: 7048–7054

50 Rosenbaum T, Gordon-Shaag A, Munari M, Gordon SE (2004) Ca^{2+}/calmodulin modulates TRPV1 activation by capsaicin. *J Gen Physiol* 123: 53–62

51 Mandadi S, Numazaki M, Tominaga M, Bhat MB, Armati PJ, Roufogalis BD (2004) Activation of protein kinase C reverses capsaicin-induced calcium-dependent desensitization of TRPV1 ion channels. *Cell Calcium* 35: 471–478

52 Galoyan SM, Petruska JC, Mendell LM (2003) Mechanissm of sensitization of the response of single dorsal root ganglion cells from adult rat to noxious heat. *Eur J Neurosci* 18: 535–541

53 Lázár J, Szabó T, Marincsák R, Kovács L, Blumberg PM, Bíró T (2004) sensitization of recombinant vanilloid receptor-1 by various neurotrophic factors. *Life Sci* 75: 153–163

54 Winston J, Toma H, Shenoy M, Pasricha PJ (2001) Nerve growth factor regulates VR-1 mRNA in cultures of adult dorsal root ganglion neurons. *Pain* 89: 181–186

55 Dai Y, Iwata K, Fukuota T, Kondo E, Tokunaga A, Yamanaka H, Tachibana T, Noguchi, K (2002) Phosphorylation of extracellular signal-regulated kinase in primary afferent neurons by noxious stimuli and its involvement in peripheral sensitization. *J Neurosci* 22: 7737–7745

56 Hu HJ, Bhave G, Gereau RW (2002) Prostaglandin and protein kinase A-dependent modulation of vanilloid receptor function by metabotropic glutamate receptor 5: potential mechanism for thermal hyperalgesia. *J Neurosci* 22: 7444–7452

57 Yang D, Gereau RQ (2002) Peripheral group II metabotropic glutamate receptors (mGluR2/3) regulate prostaglandin E_2-mediated sensitization of capsaicin responses and thermal nociception. *J Neurosci* 22: 6388–6393

58 Cortright DN, Szallasi A (2004) Biochemical pharmacology of the vanilloid receptor TRPV1. *Eur J Biochem* 271: 1814–1819

59 Morenilla-Palao C, Planells-Cases R, Garcia-Sanz N, Ferrer-Montiel A (2004) Regulated exocytosis contributes to protein kinase C potentiation of vanilloid receptor activity. *J Biol Chem* 279: 25665–25672

60 Chuang HH, Prescott ED, Kong H, Shields S, Jordt SE, Basbaum AI, Chao MV, Julius D (2002) Bradykinin and nerve growth factor release the capsaicin receptor from PtdIns(4,5)P2-mediated inhibition. *Nature* 411: 957–962

61 Prescott ED, Julius D (2003) A modular PIP_2 binding site as a determinant of capsaicin receptor sensitivity. *Science* 300: 1284–1288

62 Crandall M, Kwash J, Yu W, White G (2002) Activation of protein kinase C sensitizes human VR1 to capsaicin and to moderate decreases in pH at physiological temperatures in *Xenopus* oocytes. *Pain* 98: 109–117

63 Shin J, Cho H, Hwang SW, Jung J, Shin CY, Lee SY, Kim SH, Lee MG, Choi YH, Kim J et al (2002) Bradykinin-12-lipoxygenase-VR1 signaling pathway for inflammatory hyperalgesia. *Proc Natl Acad Sci USA* 99: 10150–10155

64 Sugiura T, Tominaga M, Katsuya H, Mizumura K (2002) Bradykinin lowers the threshold temperature for heat activation of vanilloid receptor 1. *J Neurophysiol* 88: 544–548

65 Carlton SM, Coggeshall RE (2001) Peripheral capsaicin receptors increase in the inflamed rat hindpaw: a possible mechanism for peripheral sensitization. *Neurosci Lett* 310: 53–56

66 Rashid H, Inoue M, Kondo S, Kawashima T, Bakoshi S, Ueda H (2003) Novel expression of vanilloid receptor 1 on capsaicin-insensitive fibers acoounts for the analgesic effect of capsaicin cream in neuropathic pain. *J Pharmacol Exp Ther* 304: 940–948

67 Kamei J, Zushida K, Morita K, Sasaki M, Tanaka SI (2001) Role of vanilloid VR1 receptor in thermal allodynia and hyperalgesia in diabetic mice. *Eur J Pharmacol* 422: 83–86

68 Jancsó N (1955) *Speicherung. Stoffanreicherung in Retikuloendothelial und in der Niere.* Akadémiai kiadó, Budapest

69 Szolcsányi J (1993) Actions of capsaicin on sensory receptors. In: JN Woods (ed): *Capsaicin in the study of pain.* Academic Press, San Diego, 1–26

70 Docherty RJ, Yeats JC, Bevan S, Boddeke HW (1996) Inhibition of calcineurin inhibits the desensitization of capsaicin-evoked currents in cultured dorsal root ganglion neurons from adult rats. *Pflugers Arch* 431: 828–837

71 Piper AS, Yeats JC, Bevan S, Docherty RJ (1999) A study of the voltage dependence of capsaicin-activated membrane currents in rat sensory neurons before and after acute desensitization. *J Physiol* 518: 721–733

72 Cholewinski A, Burgess GM, Bevan S (1993) The role of calcium in capsaicin-induced desensitization in rat cultured dorsal root ganglion neurons. *Neuroscience* 55: 101

73 Koplas PA, Rosenberg RL, Oxford GS (1997) The role of calcium in the desensitization of capsaicin responses in rat dorsal root ganglion neurons. *J Neurosci* 17: 3525–3537

74 Numazaki M, Tominaga T, Takeuchi K, Murayama N, Toyooka H, Tominaga M (2003) Structural determinant of TRPV1 desensitization interacts with calmodulin. *Proc Natl Acad Sci USA* 100: 8002–8006

75 Kenins P (1982) Responses of single nerve fibres to capsaicin applied to the skin. *Neurosci Lett* 29: 83–88

76 He X, Schepelmann K, Schaible HG, Schmidt RF (1990) Capsaicin inhibits responses of fine afferents from the knee joint of the cat to mechanical and chemical stimuli. *Brain Res* 530: 147–150

77 Szolcsányi J (1977) A pharmacological approach to elucidation of the role of different nerve fibres and receptor endings in mediation of pain. *J Physiol (Paris)* 73: 251–259

78 Dray A, Hankins MW, Yeats JC (1989) Desensitization and capsaicin-induced release of substance P-like immunoreactivity from guinea-pig ureter *in vitro*. *Neuroscience* 31: 479–483

79 Dray A, Bettaney J, Forster P (1989) Capsaicin desensitization of peripheral nociceptive fibres does not impair sensitivity to other noxious stimuli. *Neurosci Lett* 99: 50–54

80 Dray A, Bettaney J, Forster P (1990) Actions of capsaicin on peripheral nociceptors of the neonatal rat spinal cord-tail in vitro: dependence of extracellular ions and independence of second messengers. *Br J Pharmacol* 101: 727–733

81 Lang E, Novak A, Reeh PW, Handwerker HO (1990) Chemosensitivity of fine afferents from rat skin *in vitro*. *J Neurophysiol* 63: 887–901

82 Szolcsányi J. (1987) Selective responsiveness of polymodal nociceptors of the rabbit ear to capsaicin, bradykinin and ultra-violet irradiation. *J Physiol* 388: 9–23

83 Belmonte C, Gallar J, Pozo MA, Rebollo I (1991) Excitation by irritant chemical substances of sensory afferent units in the cat's cornea. *J Physiol* 437: 709–725

84 Balla Z, Szöke E, Czéh G, Szolcsányi J (2001) Effect of capsaicin on voltage-gated currents of trigeminal neurons in cell culture and slice preparations. *Acta Physiol Hung* 88: 173–196

85 Liu L, Oortgiesen M, Li L, Simon SA (2001) Capsaicin inhibits activation of voltage-gated sodium currents in capsaicin-sensitive trigeminal ganglion neurons. *J Neurophysiol* 85: 745–758

86 Su X, Wachtel RE, Gebhart GF (1999) Capsaicin sensitivity and voltage-gated sodium currents in colon sensory neurons from rat dorsal root ganglia. *Am J Physiol Gastrointest Liver Physiol* 277: G1180–G1188

87 Blair NT, Bean BP (2003) Role of tetrodotoxin-resistant Na+ current slow inactivation in adaptation of action potential firing in small-diameter dorsal root ganglion neurons. *J Neurosci* 23: 10338–10350

88 Wood JN, Coote PR, Minhas A, McNiel M, Mullany I, Burgess G (1989) Capsaicin-induced ion fluxes increase cGMP but not cAMP levels in rat sensory neurons in culture. *J Neurosci* 53: 1203–1211

89 Marban E, Yamagishi T, Tomaselli GF (1998) Structure and function of voltage-gated sodium channels. *J Physiol* 508: 647–657

90 Kehl SJ (1994) Block by capsaicin of voltage-gated K+ currents in melanotrophs of the rat pituitary. *Br J Pharmacol* 112: 616–624

91 Bleakman D, Brorson JR, Miller RJ (1990) The effect of capsaicin on voltage-gated calcium currents and calcium signals in cultured dorsal root ganglion cells. *Br J Pharmacol* 101: 423–431

92 Docherty RJ, Robertson B, Bevan S (1991) Capsaicin causes prolonged inhibition of voltage-activated calcium currents in adult rat dorsal root ganglion neurons in culture. *Neuroscience* 40: 513–521

93 Brand L, Berman E, Schwen R, Loomans M, Janusz J, Bohne R, Maddin C, Gardner J, Lahann T, Farmer R et al (1987) NE-19550: a novel, orally active anti-inflammatory analgesic. *Drugs Exp Clin Res* 13: 259–265

94 Brand LM, Skare KL, Loomans ME, Reller HH, Schwen RJ, Lade DA, Bohne RL, Maddin CS, Moorehead DP, Fanelli R et al (1990) Anti-inflammatory pharmacology and mechanism of the orally active capsaicin analogs, NE-19550 and NE-28345. *Agents Actions* 31: 329–340
95 Dray A, Bettaney J, Rueff A, Walpole C, Wrigglesworth R (1990) NE-19550 and NE-21610, antinociceptive capsaicin analogues: studies on nociceptive fibres of the neonatal rat tail *in vitro*. *Eur J Pharmacol* 181: 289–293
96 Wrigglesworth R, Walpole CS, Bevan S, Campbell EA, Dray A, Hughes GA, James I, Masdin KJ, Winter J (1996) Analogues of capsaicin with agonist activity as novel analgesic agents: structure-activity studies. 4. Potent, orally active analgesics. *J Med Chem* 39: 4942–4951
97 Taylor DC, Pierau FK, Szolcsányi J (1985) Capsaicin-induced inhibition of axoplasmic transport is prevented by nerve growth factor. *Cell Tissue Res* 240: 569–573
98 Taylor DC, Pierau FK, Szolcsányi J (1984) Long lasting inhibition of horseradish peroxidase (HRP) transport in sensory nerves induced by capsaicin pretreatment of the receptive field. *Brain Res* 298: 45–49
99 Miller MS, Buck SH, Sipes IG, Yamamura HI, Burks TF (1982) Regulation of substance P by nerve growth factor: disruption by capsaicin. *Brain Res* 250: 193–196
100 Farkas-Szallasi T, Lundberg JM, Wiesenfeld-Hallin Z, Hökfelt T, Szallasi A (1995) Increased levels of GMAP, VIP and nitric oxide synthase, and their mRNAs, in lumbar dorsal root ganglia of the rat following systemic resiniferatoxin treatment. *Neuroreport* 6: 2230–2234
101 Xu XJ, Farkas-Szallasi T, Lundberg JM, Hökfelt T, Wiesenfeld-Hallin Z, Szallasi A (1997) Effects of the capsaicin analogue resiniferatoxin on spinal nociceptive mechanisms in the rat: behavioral, electrophysiological and in situ hybridization studies. *Brain Res* 752: 52–60
102 Szallasi A, Blumberg PM (1992) Vanilloid receptor loss in rat sensory ganglia associated with long term desensitization to resiniferatoxin. *Neurosci Lett* 140: 51–54
103 Szallasi A, Nilsson S, Farkas-Szallasi T, Blumberg PM, Hökfelt T, Lundberg JM (1995) Vanilloid (capsaicin) receptors in the rat: distribution in the brain, regional differences in the spinal cord, axonal transport to the periphery, and depletion by systemic vanilloid treatment. *Brain Res* 703: 175–183
104 Farkas-Szallasi T, Bennett GJ, Blumberg PM, Hökfelt T, Lundberg JM, Szallasi A. (1996) Vanilloid receptor loss is independent of the messenger plasticity that follows systemic resiniferatoxin administration. *Brain Res* 719: 213–218
105 Goso C, Piovacari G, Szallasi A (1993) Resiniferatoxin-induced loss of vanilloid receptors is reversible in the urinary bladder but not in the spinal cord of the rat. *Neurosci Lett* 162: 197–200
106 Szolcsányi J, Jancsó-Gábor A, Joó F (1975) Functional and fine structural characteristics of the sensory neuron blocking effect of capsaicin. *Naunyn Schmiedebergs Arch Pharmacol* 287: 157–169
107 Dedov VN, Mandadi S, Armati PJ, Verkhratsky A (2001) Capsaicin-induced depolari-

sation of mitochondria in dorsal root ganglion neurons is enhanced by vanilloid receptors. *Neuroscience* 103: 219–226
108 Szolcsányi J, Joó F, Jancsó-Gábor A (1971) Mitochondrial changes in preoptic neurons after capsaicin desensitization of the hypothalamic thermodetectors in rats. *Nature* 229: 116–117
109 Szöke E, Czéh G, Szolcsányi J, Seress L (2002) Neonatal anandamide treatment results in prolonged mitochondrial damage in the vanilloid receptor type 1-immunoreactive B-type neurons of the rat trigeminal ganglion. *Neuroscience* 115: 805–814
110 Szöke E, Seress L, Szolcsányi J. (2002) Neonatal capsaicin treatment results in prolonged mitochondrial damage and delayed cell death of B cells in the rat trigeminal ganglia. *Neuroscience* 113: 925–937
111 Chard PS, Bleakman D, Savidge JR, Miller RJ (1995) Capsaicin-induced neurotoxicity in cultured dorsal root ganglion neurons: involvement of calcium-activated proteases. *Neuroscience* 65: 1099–1108
112 Chung K, Klein CM, Coggeshall RE (1990) The receptive part of the primary afferent axon is most vulnerable to systemic capsaicin in adult rats. *Brain Res* 511: 222–226
113 Chung K, Schwen RJ, Coggeshall RE (1985) Ureteral axon damage following subcutaneous administration of capsaicin in adult rats. *Neurosci Lett* 53: 221–226
114 Simone DA, Nolano M, Johnson T, Wendelschafer-Crabb G, Kennedy WR (1998) Intradermal injection of capsaicin in humans produces degeneration and subsequent reinnervation of epidermal nerve fibers: correlation with sensory function. *J Neurosci* 18: 8947–8959
115 Nolano M, Simone DA, Wendelschafer-Crabb G, Johnson T, Hazen E, Kennedy WR (1999) Topical capsaicin in humans: parallel loss of epidermal nerve fibers and pain sensation. *Pain* 81: 135–145
116 Wood JN, Winter J, James IF, Rang HP, Yeats J, Bevan S (1988) Capsaicin-induced ion fluxes in dorsal root ganglion cells in culture. *J Neurosci* 8: 3208–3220
117 Jeftinija S, Liu F, Jeftinija K, Urban L (1992) Effect of capsaicin and resiniferatoxin on peptidergic neurons in cultured dorsal root ganglion. *Regul Pept* 39: 123–135
118 Karai L, Brown DC, Mannes AJ, Connelly ST, Brown J, Gandal M, Wellish OM, Neubert JK, Oláh Z, Iadarola MJ (2004) Deletion of vanilloid receptor-1 expressing primary afferent neurons for pain control. *J Clin Invest* 113: 1344–1352
119 Oláh Z, Szabó T, Karai L, Hough C, Fields RD, Caudle RM, Blumberg PM, Iadarola MJ (2001) Ligand-induced dynamic membrane changes and cell deletion conferred by vanilloid receptor 1. *J Biol Chem* 276: 11021–11030
120 Jancsó G, Király E, Jancsó-Gábor A (1977) Pharmacologically induced selective degeneration of chemosensitive primary sensory neurons. *Nature* 270: 741–743
121 Jancsó G, Király E (1981) Sensory neurotoxins: chemically induced selective destruction of primary sensory neurons. *Brain Res* 210: 83–89
122 Otten U, Lorez HP, Businger F (1983) Nerve growth factor antagonizes the neurotoxic action of capsaicin on primary sensory neurons. *Nature* 301: 515–517
123 Ruit KG, Elliott JL, Osborne PA, Yan Q, Snider WD (1992) Selective dependence of

mammalian dorsal root ganglion neurons on nerve growth factor during embryonic development. *Neuron* 8 :573–587
124 Mendell LM (1999) Neurotrophin action on sensory neurons in adults: an extension of the neurotrophic hypothesis. *Pain* 6: S127–S132
125 Winter J, Woolf C, Lynn B (1993) Degenerative and regenerative responses of sensory neurons to capsaicin-induced damage. In: JN Woods (ed): *Capsaicin in the Study of Pain*. Academic Press, San Diego, 139–160
126 Welk E, Fleischer E, Petsche U, Handwerker HO (1984) Afferent C-fibres in rats after neonatal capsaicin treatment. *Pflügers Arch* 400: 66–71
127 Fukuoka T, Tokunaga A, Tachibana T, Dai Y, Yamanaka H, Noguchi K (2002) VR1, but not P2X(3), increases in the spared L4 DRG in rats with L5 spinal nerve ligation. *Pain* 99: 111–120
128 HHudson LJ, Bevan S, Wotherspoon G, Gentry C, Fox A, Winter J (2001) VR1 protein expression increases in undamaged DRG neurons after partial nerve injury. *Eur J Neurosci* 13: 2105–2114
129 Pini A, Lynn B (1991) C-fibre function during the 6 weeks following brief application of capsaicin to a cutaneous nerve in the rat. *Eur J Neurosci* 3: 274–284
130 Pini A, Baranowski R, Lynn B (1990) Long-term reduction in the number of C-fibre nociceptors following capsaicin treatment of a cutaneous nerve in adult rats. *Eur J Neurosci* 2: 89–97
131 Jancsó G, Lawson SN (1990) Transganglionic degeneration of capsaicin-sensitive C-fiber primary afferent terminals. *Neuroscience* 39: 501–511
132 Joó F, Szolcsányi J, Jancsó-Gábor A (1969) Mitochondrial alterations in the spinal ganglion cells of the rat accompanying the long-lasting sensory disturbance induced by capsaicin. *Life Sci* 8: 621–626
133 Chiba T, Masuko S, Kawano H (1986) Correlation of mitochondrial swelling after capsaicin treatment and substance P and somatostatin immunoreactivity in small neurons of dorsal root ganglion in the rat. *Neurosci Lett* 64: 311–316
134 Hoyes AD, Barber P (1981) Degeneration of axons in the ureteric and duodenal nerve plexuses of the adult rat following in vivo treatment with capsaicin. *Neurosci Lett* 25: 19–24
135 Pethö G, Szolcsányi J (1996) Excitation of central and peripheral terminals of primary afferent neurons by capsaicin *in vivo*. *Life Sci* 58: 47–53
136 Caterina MJ, Rosen TA, Tominaga M, Brake AJ, Julius D (1999) A capsaicin receptor homologue with a high threshold for noxious heat. *Nature* 398: 436–441
137 Szallasi A, Appendino G (2004) Vanilloid receptor TRPV1 antagonists as the next generation of painkillers. Are we putting the cart before the horse? *J Med Chem* 47: 2717–2712

Capsaicin in human experimental pain models of skin, muscle and visceral sensitization

Lars Arendt-Nielsen and Ole K. Andersen

Center for Sensory-Motor Interaction, Laboratory for Experimental Pain Research, Aalborg University, Fredrik Bajers Vej 7, D3, DK-9220 Aalborg, Denmark

Introduction

It has always been the dream of pain researchers and clinicians to have objective measures of pain and phenomena such as hyperalgesia. Unfortunately, this is not yet possible and will most likely never be. Pain is a multidimensional, unpleasant sensory and emotional experience and cannot as such be represented or described by a single parameter or number. However, different possibilities exist to assess quantitatively various aspects of this complex sensory experience of pain and related phenomena such as hyperalgesia. By measuring different aspects of the pain experience and by a combination of the various measurements more can be learned about which dimensions are affected, which mechanisms/pathways are impaired and which are functioning normally [1]. The ultimate goal of pain-assessment procedures is to obtain a better understanding of mechanisms (e.g. hyperalgesia) involved in pain transduction, transmission and perception under normal and pathophysiological conditions, either in pain patients or after experimentally induced sensitization in normal volunteers.

Assessment of pain and evaluation of peripheral and central sensitization can be divided into four main categories (Fig. 1):
- assessment of ongoing clinical pain;
- assessment of experimentally evoked pain for diagnosis and monitoring of patients;
- assessment of experimentally evoked pain for basic studies in healthy volunteers under normal conditions;
- assessment of experimentally evoked pain for basic studies in healthy volunteers under conditions with experimentally induced hyperalgesia (for example, by capsaicin).

This chapter will focus on the fourth category and describe specifically how hyperalgesia can be induced in a standardized way by capsaicin, and the stimulation and

Turning up the Heat on Pain: TRPV1 Receptors in Pain and Inflammation, edited by Annika B. Malmberg and Keith R. Bley

assessment methods available to evaluate quantitatively the different aspects of capsaicin-induced hyperalgesia. The majority of studies have focused on topical and intradermal applications, whereas only a few experimental studies have applied capsaicin intramuscularly or to the viscera. Experimental investigations of peripheral and central aspects of capsaicin-induced hyperalgesia may increase the knowledge associated with diseases, such as neuropathic pain. In this chapter capsaicin induced cutaneous, muscular and visceral hyperalgesia are described separately.

Definitions

Hyperalgesia is defined as a leftward shift of the stimulus-response function that relates the magnitude of pain to stimulus intensity. From this definition it follows that the threshold for pain is lowered and pain to suprathreshold stimuli is enhanced. Due to central neuroplastic changes, normally non-nociceptive stimuli (warmth, cold, mechanical) can be perceived as pain (allodynia). The areas of hyperalgesia and allodynia produced by an injury to the skin include a zone that incorporates the injury site (primary hyperalgesia) and a much larger area extending well beyond the site of injury and into undamaged skin (secondary hyperalgesia) [2–4]. Primary hyperalgesia to heat can be explained by a peripheral sensitization of nociceptors [3, 4]. Primary hyperalgesia to mechanical stimuli most likely reflects central changes as no peripheral sensitization has been shown. Secondary hyperalgesia is normally characterized by enhanced pain to mechanical stimuli [5–8], whereas there are some controversies on the presence of secondary hyperalgesia to heat. Two different alterations are responsible for secondary mechanical hyperalgesia: (1) a change in the modality of the sensation evoked by low-threshold mechanoreceptors – this change produces touch-evoked pain, or allodynia; and (2) an increase of the pain sensitivity to mechanically sensitive nociceptors – this change produces hyperalgesia. Both of these changes are mediated by alterations in the central processing of the sensory input [9, 10]. For muscular and visceral pain, other definitions are normally used (see discussion below).

Induction of capsaicin-evoked cutaneous pain and hyperexcitability

Human experimental models of cutaneous hyperalgesia are based on a controlled stimulus to the skin that causes a strong selective afferent nociceptive barrage [11–13] leading to central changes. Cutaneous hyperalgesia can be induced by chemical, thermal, electrical or mechanical stimuli.

Capsaicin induces a 'chemical injury' via robust activation of nociceptive nerve endings, leading to peripheral release of inflammatory substances and central sensitization [4]. Capsaicin has been shown to activate nociceptors that are responsive to

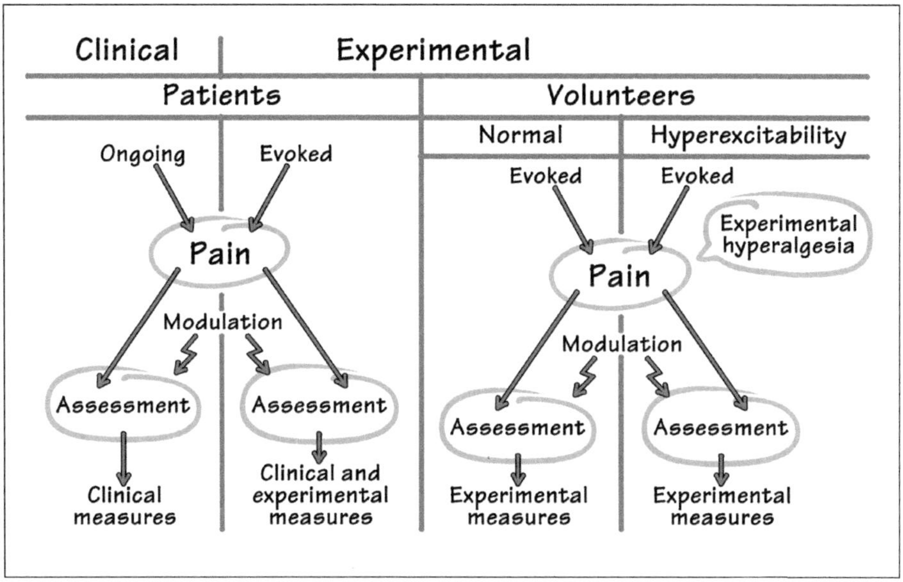

Figure 1
An illustration of how pain can be assessed in clinical and experimental settings. In the clinical situation, the various measures of the on-going pain can include pain intensity, pain quality, pain location and quality of life. Various experimental stimuli can also be applied to the patients and can be used, together with the clinical measures, to assess their responses to such experimental stimuli to determine the degree of hyperalgesia, allodynia, hyperpathia, after-sensations and wind-up-like pain. In healthy volunteers, responses to experimental stimuli can be evaluated under normal conditions and conditions with experimentally induced hyperalgesia (hyperexcitability). The experimental measures can include pain thresholds, stimulus-response function, temporal summation, reflexes and/or evoked potentials.

heat, primarily the polymodal A-fiber sensitive to mechanical and heat stimuli with rapid onset (AMH type II [13]) and the polymodal C-fiber responding to mechanical and heat stimuli (CMH [13, 14]). These two nociceptors express vanilloid receptor types, including TRPV1, while the A-fiber high-threshold mechanoreceptors [15] and type I AMHs [16] are not sensitive to capsaicin.

The models for induction of cutaneous hyperalgesia most commonly involve an injection or topical application of an algesic agent or a controlled heat injury [4, 13, 17, 18]. The controlled burn induces a true tissue injury; however, as blistering and skin hyper-pigmentation may occur following burn injuries [6, 18], it is therefore less optimal for experimental studies.

Topical application (e.g. 1.5 g of 1% capsaicin, cream applied to 4 cm^2 for 40 min) or intradermal injection (e.g. 100 μg of capsaicin in a 20 μl volume) of capsaicin are the most commonly used models (see Fig. 2). Topical application, causing burning pain, is better tolerated, but the manifestation and duration of hyperalgesia are less as compared with intradermal application. The effect of topical capsaicin varies with capsaicin concentration, and furthermore the stratum corneum of intact skin acts as a substantial diffusion barrier. Therefore it is difficult to estimate the amount of capsaicin that actually reaches the sensory receptors [13, 19].

Intradermal capsaicin elicits a severe stinging/burning pain lasting for few minutes, leaving well-characterized areas of primary and secondary hyperalgesia, which lasts up to 24 h [11, 13]. Capsaicin-evoked allodynia after intradermal injection lasts up to 6 h, and punctuate hyperalgesia can last up to 24 h [4]. It is generally agreed that allodynia disappears rapidly as the on-going pain induced by the capsaicin injection wanes, so in most experiments allodynia only lasts for approximately 30–60 min (see below). Either topical administration [20] or intradermal injection [21] can be applied repeatedly to study the progression of sensitization. Witting et al. [21] found no significant relation between capsaicin pain intensity and area of allodynia and punctuate hyperalgesia and concluded that the two phenomena are mediated by different central mechanisms. However, in other studies, a correlation between the area of allodynia and ongoing capsaicin-evoked [21] and ongoing neuropathic pain [22] has been described. Furthermore, Mohammadian et al. [20] described a correlation between the area of allodynia and the areas of punctuate hyperalgesia following repetitive applications of topical capsaicin.

The development of hyperalgesia depends on many factors. For example, reports have shown that capsaicin induced on-going pain and hyperalgesia are critically temperature-dependent. Mild cooling of the skin provided instant relief from on-going pain, and re-warming resulted in a re-appearance of on-going pain and hyperalgesia [17, 23]. Furthermore, we have recently shown that females develop larger areas of secondary hyperalgesia areas to intradermal capsaicin than males [24]. Capsaicin-induced mechanical hyperalgesia is also found to decline with age [25], so matched controls and standardized experimental conditions are essential in pharmacological screening studies.

Capsaicin as stimulus modality has also been used as a marker for central sensitization in chronic pain patients where the area of hyperalgesia is found to be larger in rheumatoid arthritis patients and compared with healthy matched controls [25]. Furthermore, a correlation between capsaicin-induced hyperalgesia and joint tenderness was assumed to be a result of central, and not exclusively peripheral, factors. The study suggests the value of capsaicin-based techniques to explore nociceptive mechanisms in clinical disorders characterized by chronic pain.

Given that heat conditioning of the primary hyperalgesic area rekindles the spontaneous afferent C-fiber input sustaining the central changes [26], Petersen and Rowbotham [27] proposed a model using both heat and capsaicin sensitization by

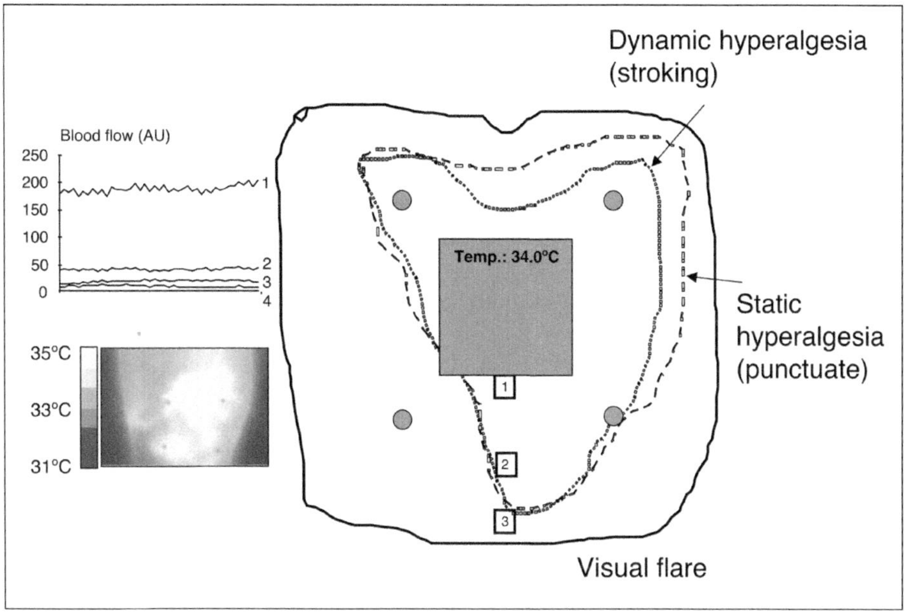

Figure 2
An example of laser Doppler flowmetry and thermographic recordings from a cutaneous area after topical application (30 min application under occlusion) and intradermal injection of capsaicin is illustrated. The pictures were taken 5 min after the occlusion was removed. In the top-left corner, the blood flow is depicted as measured at four different sites (three squares marked 1–3 while the fourth probe was placed 10 cm distal to the application area). After topical capsaicin application, primary heat hyperalgesia is seen where the capsaicin patch has been applied (marked by the four circles, which are also seen in the thermographic picture), surrounded by secondary hyperalgesia areas of pin-prick hyperalgesia (static hyperalgesia) and brush-evoked allodynia (dynamic hyperalgesia). After intradermal application, only the secondary hyperalgesic areas can be detected. AU, arbitrary units.

periodic rekindling of the on-going pain with heat. They concluded that combining low-intensity thermal stimulation and chemical nociceptor stimulation synergistically produce hyperalgesia with a low risk of tissue injury [27]. Dirks et al. [28] also found some indications that cutaneous sensitization induced by a combination of heat and topical capsaicin was more stable as compared with capsaicin alone. It was suggested that the stability and duration of the cutaneous sensitization were due to a synergistic effect between heat and capsaicin. However, these findings have not been reproduced in other studies using the topical application of capsaicin, the intradermal capsaicin model and the controlled-heat injury model [29].

Psychophysical methods to assess hyperalgesia

Once the hyperalgesia is induced, it is important to have methods available to assess quantitatively the excitability of the nociceptive and non-nociceptive pathways. Experimental pain-research tools developed for assessing hyperalgesia involve two separate topics: (1) standardized activation of the non-nociceptive and nociceptive pathways, and (2) quantitative assessment of the evoked responses. The main advantages of an experimental approach for assessing hyperalgesia are that:

- stimulus intensity, duration and modality are controlled and can be repeated over time;
- differentiated responses to activation of different structures (skin, muscle, viscera) can be assessed (multi-tissue sensory testing approach);
- differentiated responses to different stimulus modalities can be assessed (multi-modal sensory testing approach);
- the physiological and psychophysical responses can be assessed quantitatively and compared over time;
- hyperalgesia can be compared quantitatively between various normal/affected regions.

From a mechanistic point of view, the availability of an armamentarium of experimental tools is important for acquiring differentiated information about the different mechanisms/pathways involved/affected after induction of hyperalgesia. Stimuli can have phasic (short-lasting, milliseconds to a few seconds) or tonic (long-lasting, many seconds to minutes) properties. Most of the phasic stimuli can be applied either as a single stimulus or as a series of repeated stimuli (to evoke temporal summation), or as a stimulus activating a small or large/multiple area(s) for the study of spatial summation.

The psychophysical methods available to assess sensations evoked from hyperalgesic areas are developed on basis of the psychophysical laws and can be divided into response- and stimulus-dependent methods. The response-dependent methods rely on how the person evaluates the stimulus intensity or unpleasantness on a given scale (visual analog scales, verbal ratings scales, numerical ratings scales). The stimulus-dependent methods are based on adjustment of the stimulus intensity until a pre-defined response, typically a threshold, is reached. These methods are often used to assess hyperalgesia to specific stimulus modalities (e.g. heat hyperalgesia in the primary area of capsaicin-induced hyperalgesia) [29–31].

Stimulus-response functions are more informative than a threshold determination as suprathreshold response characteristics can be derived from the data. Stimulus-response functions are valuable to assess hyperpathia to various stimulus modalities in patients (e.g. those with neuropathic pain [32]).

Primary hyperalgesia

Primary hyperalgesia occurs in the skin underlying the actual site of stimulation (e.g. topical capsaicin application) and is mainly a result of peripheral sensitization of the nociceptors. Normally, thermal stimuli are used to quantify primary hyperalgesia, and a pain threshold can drop from approximately 44 to 33 °C and responses to suprathreshold stimulation increase. Hyperalgesia to tonic stimulation with a blunt probe (called pressure hyperalgesia), and impact hyperalgesia to shooting small bullets against the skin with controlled velocities, have also been described in the primary hyperalgesic zone [26] and are probably mediated by sensitized C-nociceptors [17].

Secondary mechanical hyperalgesia

Different forms of secondary mechanical hyperalgesia have been characterized. The secondary hyperalgesia is predominantly a central phenomenon reflecting neuroplastic changes in spinal cord neurons. Secondary hyperalgesia can be separated into an area of brush-evoked hyperalgesia (dynamic hyperalgesia or allodynia) where on-going pain seems essential and a slightly larger area of punctuate hyperalgesia (static hyperalgesia) [33, 34]. Allodynia is normally assessed by stroking the skin with a cotton swap in a standardized way, and the area marked where the sensation changes from touch to pain. Dynamic mechanical allodynia (stroking hyperalgesia) is mediated by Aβ-fibers. Static hyperalgesia (punctuate hyperalgesia) is usually determined by a nylon filament (e.g. with a von Frey hair with a bending force of 70 g) and is mediated by capsaicin-insensitive A-fiber nociceptors (high-threshold mechanoreceptors and type I AMH expressing the VRL1 cation channel [35]). Thus, secondary hyperalgesia to punctuate stimuli is induced by nociceptive C-fiber discharge but mediated by nociceptive A-fibers [36]. Conventional C-fiber nociceptors respond to heat stimuli, and yet heat hyperalgesia is in some studies absent in the region of secondary hyperalgesia. Fuchs et al. [37] therefore tried to desensitize heat-sensitive nociceptors by topical application of capsaicin. Interestingly, the intradermal capsaicin injection led to the development of a similar degree of secondary punctuate hyperalgesia before and after desensitization of heat-sensitive nociceptors, supporting the hypothesis that capsaicin-insensitive nociceptors mediate secondary punctuate hyperalgesia.

Different methods have been applied to evaluate the static and dynamic areas of secondary hyperalgesia, but normally the stimuli are applied along six vectors from outside and towards the center of the sensitized area in small steps [38, 39]. When the perception intensity/quality of the pin-prick or stroking stimuli change, the point is marked and the area of either hyperalgesia or allodynia is estimated.

Secondary heat hyperalgesia

Hardy et al. [3] studied secondary heat hyperalgesia by radiant heat stimuli elicited from a focused light bulb and reported that the pain threshold to thermal stimuli was unaltered in the secondary area while secondary heat hyperalgesia was detected for suprathreshold stimuli. Since then, their findings have been confirmed by the use of a contact thermode, although Hardy et al. [3] originally used a radiant heat stimulator. Secondary heat hyperalgesia has been successfully detected when high-temperature stimuli were applied [18, 30, 40–43], but not when low temperature stimuli were applied [4] or when heat pain threshold assessment was used for detection [6, 26]. It has been argued that suprathreshold contact heat stimuli are more sensitive to study secondary heat hyperalgesia than subthreshold contact heat stimuli. On the other hand, a few studies have investigated secondary heat hyperalgesia by radiant heat stimuli [5, 7, 40]. Arendt-Nielsen et al. [40] used brief argon laser heat pulses and demonstrated secondary heat hyperalgesia outside the area of topical application of capsaicin, while Raja et al. [5] and Ali et al. [7] failed to show secondary heat hyperalgesia to CO_2 laser stimuli after intradermal injection of capsaicin. From these studies, however, it is not clear if a difference between suprathreshold and subthreshold stimuli or different heat-stimulation techniques can explain the inconsistency.

Based on the study using brief high-intensity laser pulses [40], it was argued that Aδ-fibers are playing an important role in detecting secondary heat hyperalgesia as argon lasers predominantly activate Aδ-fibers. On the other hand, the study by Sumikura et al. [44] varied systematically the configuration of the xenon lamp-evoked heat stimuli between three intensity levels (0.8, 1.0 and 1.2×heat pain threshold) and four stimulus durations (200, 350, 500 and 750 ms). The results revealed secondary heat hyperalgesia to radiant heat stimuli and implied that not only suprathreshold stimuli, but also subthreshold stimuli, can be used to detect this. It should be noted that profiles of secondary heat hyperalgesia detected by radiant heat stimuli are different from those of secondary mechanical hyperalgesia [38]. Regarding the temporal profile, secondary heat hyperalgesia developed slowly, while secondary mechanical hyperalgesia developed quickly. Regarding the spatial profile, the border of secondary heat hyperalgesia was indistinct and the area of secondary heat hyperalgesia did not coincide with that of secondary mechanical hyperalgesia [38].

The possibility that secondary heat hyperalgesia can be detected by an activation of C-fibers is supported by recent study by Yucel et al. [30]. They studied heat hyperalgesia using a contact thermode, with different heating rates in the primary and secondary area induced by topical application of capsaicin. The study suggested that C-fibers are involved in the primary heat hyperalgesia, as the slow heating rate was perceived more painful than the fast heating rate in the primary area, and that both Aδ-fibers and C-fibers are involved in the secondary heat hyperalgesia, as there was no differential effect of the heating rates on the heat pain sensitivity in the secondary area using both sub- and suprathreshold thermal stimuli.

Temporal summation in secondary hyperalgesic areas

The phenomenon that a single painful or even a repeated sub-pain threshold stimulus causes exaggerated perceptions of pain in humans is called temporal summation (corresponding to the early phase of wind-up measured in animal dorsal horns) [45]. Temporal summation is a very potent mechanism, which is difficult to block pharmacologically [46] and is facilitated in patients suffering form neuropathic pain [47], chronic musculoskeletal pain [48] and chronic visceral pain syndromes [49].

The facilitation of temporal summation to either laser or electrical stimuli in secondary hyperalgesic areas seems controversial and depends on the stimulus modality and configuration [31, 40, 50, 51]. In the human topical capsaicin model, the summation of both electrical and radiant heat stimuli were facilitated in the secondary hyperalgesic area compared with normal skin [40]. Pedersen et al. [51] used the controlled-heat injury model for the induction of hyperalgesia and found no facilitation of temporal summation to painful electrical stimuli in the hyperalgesic areas. It is unclear if capsaicin and controlled-heat injury initiate different forms of central sensitization with different characteristics, which may result in different effects on temporal summation. Furthermore, it is not known if different stimulus modalities used to evoke temporal summation are of importance to explain the mechanisms of hyperalgesia and temporal summation. Therefore, Yucel et al. [31] performed a comprehensive study and showed facilitated temporal summation of pain in the secondary hyperalgesic area for punctuate and mechanical impact stimuli after the induction of hyperalgesia using both capsaicin injection and controlled-heat injury (see Fig. 3). There were no significant differences between the models.

In secondary hyperalgesic areas, summation of pain can be evoked by repetitive (2 Hz) von Frey hair stimulation (which is normally not painful) and the evoked after-sensations (the sensation after a series of repetitive stimuli) can be assessed. After-sensation is a phenomenon that is also seen when areas with neuropathic pain are stimulated [52], and interestingly animal experiments indicate that the pharmacology of the wind-up and the after-discharge are different [53].

Electrophysiological methods to assess hyperalgesia

Microneurography is the direct stimulation of or recording from single nerve fibers and has been used to characterize relationships between neuronal firing and perception intensity and quality. This method may identify activation patterns of nociceptive Aδ- and C-fibers and of different classes of nociceptors, although the vast

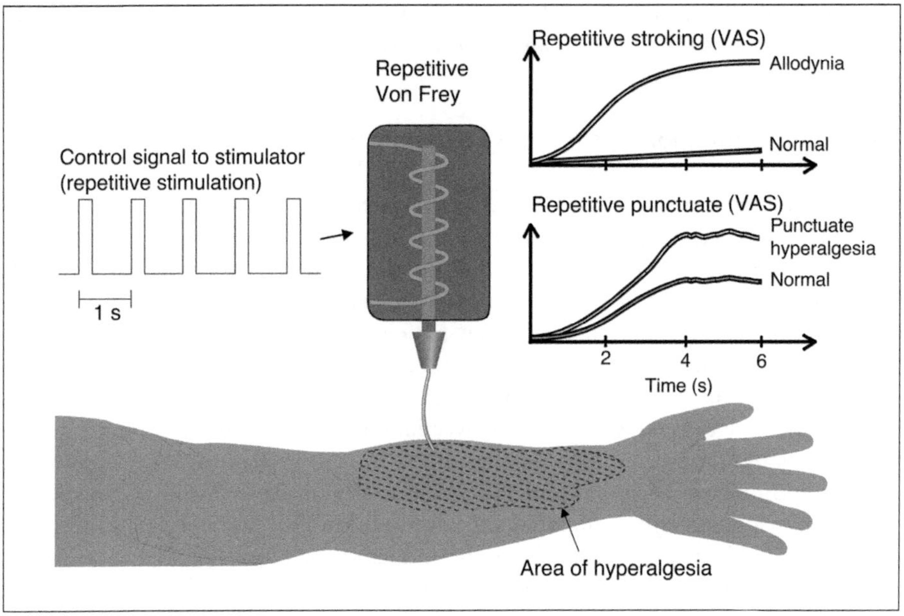

Figure 3
Repetitive punctuate stimulation with a von Frey hair within the area with hyperalgesia is illustrated. The device stimulates with a fixed time period in contact with the skin and with a fixed frequency. The thickness of the von Frey hair can be changed. After hyperalgesia has been induced, robust punctuate pain is induced with this repetitive paradigm, compared to normal skin. The temporal summation mechanism is facilitated for punctuate and mechanical impact stimuli within the secondary hyperalgesic area [31]. If the von Frey hair is replaced by a thinner filament, tactile allodynia can be assessed. Repetitive tactile stimulation causes pain after hyperalgesia has developed in contrast to normal skin. VAS, visual analog scale.

majority of studies have reported data from C-fibers. An elegant study [54] used micro-stimulation of a single tactile fiber (Aβ-fiber). The cutaneous receptive field of the stimulated Aβ-fiber was determined and the evoked tactile sensation characterized. Intradermal capsaicin was applied to the adjacent area so the receptive field became part of the secondary hyperalgesic area. The micro-stimulation was repeated and the tactile sensation was now perceived as pain, documenting that central components contribute significantly to secondary dynamic hyperalgesia.

Treede and Cole [55] studied secondary hyperalgesia in a subject who had a loss of myelinated afferent nerve fibers that spared the Aδ-group. Stroking with a cotton swab was not perceived anywhere on the affected skin either before or after intradermal injection of capsaicin. Thus, there was no hyperalgesia to light touch, while secondary hyperalgesia to punctuate stimuli was normal. These data support the finding of Torebjork et al. [54] that dynamic hyperalgesia to light touch (stroking hyperalgesia) is due to sensitization of central pain-signaling neurons to low-threshold mechanoreceptor input (Aβ-fibers) and static punctuate hyperalgesia is likely due to sensitization to nociceptor input (Aδ- or C-fibers).

The *nociceptive withdrawal reflex* is a spinal reflex of the lower limb that is elicited by painful somatic stimuli. The reflex can be used for assessment in two ways: either by a reflex threshold, or by the reflex amplitude to a fixed suprathreshold stimulus intensity. The reflex can be used to assess the response to both a single stimulus and a repeated stimulus (assessing temporal summation). The generation of a withdrawal reflex is initiated by the nociceptive input, but an extensive processing takes place within the spinal cord.

Under normal conditions, non-nociceptive (Aβ-afferents) afferents inhibit nociceptive activity at the spinal cord level, but together with central neuroplasicity activation of the Aβ-afferents causes pain (allodynia). The question is then whether summation between normal non-nociceptive mechanical stimuli and nociceptive stimuli may occur under pathologic conditions. In an experimental study, Andersen et al. [56] induced hyperalgesia by topical application of capsaicin and then used the nociceptive withdrawal reflex to prove summation between mechanical stimuli applied to the allodynic area and nociceptive stimuli applied outside. This summation between non-nociceptive and nociceptive afferents under experimental pathological conditions could be inhibited by the *N*-methyl-D-aspartate (NMDA) receptor antagonist ketamine [57]. These studies show evidently the plasticity of the human nociceptive system can be evoked experimentally by capsaicin and that the NMDA receptor system is involved.

Another reflex study utilized the inhibitory masseter reflex responses [58]. The inhibitory masseter reflex was evoked by non-painful electrical stimulation of the upper incisor while the subject was biting at low force. Capsaicin applied topically to the cheek produced a spontaneous burning pain sensation. During capsaicin treatment, the visual analog scale ratings for the sensation induced by tooth-pulp stimulation were significantly reduced, whereas no significant changes were found in the tooth-pulp-induced masseter reflex responses.

A new experimental approach is to map the reflex receptive fields associated with the organization of nociceptive withdrawal reflexes in individual muscles [59]. By stimulating several sites in random order, the spatial sensitivity of the reflex is determined, which indirectly depicts neuronal spatial encoding of the reflex. This non-invasive method is very interesting in relation to central sensitization as variation in neuronal receptive field size is often observed.

Evoked potentials to painful stimulation – and in particular the late-event-related vertex potential – have been widely used in pain research due to the relation between its amplitude and the pain intensity. The vertex potential can be elicited by any abrupt change in a sensory modality, including pain perception. The vertex potential elicited by an unspecific stimulus such as electrical stimulation may not correlate with the pain intensity, whereas a specific nociceptive stimulus (Aδ-fiber activation) produced by, for example, a laser, may generate potentials that are a result of traffic along the nociceptive pathways. As such the laser-evoked potential (LEP) has gained some clinical impact.

Most research has so far been focused on spinal changes after capsaicin-induced sensitization. However, one study [60], looked at the CO_2 LEPs elicited from topical capsaicin-generated areas of primary and secondary hyperalgesia. Source analysis revealed that a cingulate cortex generator was significantly decreased in strength during and after the capsaicin application. Moreover, the topography of the LEPs was modified by capsaicin, shifting from the central toward the parietal region. Dipolar modeling showed that the dipolar source in the anterior cingulate cortex moved backward after capsaicin. All these changes were not observed after stimulation of the hand, contralateral to the application of capsaicin, thus suggesting that functional changes are selective for the capsaicin-treated skin and the adjacent territories. The data suggest that acute cutaneous pain may inhibit the neural activity in the central nervous system processing nociceptive input. Further, the cortical representation of these inputs can rapidly be modified in presence of acute pain and sensitization.

A similar modification of CO_2 LEPs was found in a study by Beydoun et al. [61] where the subjects applied capsaicin cream to the dorsum of one hand and vehicle cream to the other hand three times daily for a period of 5 weeks. LEPs were recorded and psychophysical thresholds and magnitude estimation for several sensory modalities were determined. The study showed that topical capsaicin significantly and reversibly decreased the magnitude of the estimation of suprathreshold heat pain and laser pulses, and the amplitude of the LEPs indicated that topical capsaicin caused a definite functional and reversible inactivation of Aδ-nociceptive afferent transmission (receptor desensitization). This supports previous findings in healthy volunteers [62] and in patients with post-herpetic neuralgia [63].

LEPs and topical capsaicin-induced neurogenic inflammation and hyperalgesia have also been investigated in a pharmacological study where the effect of a potent tricyclic H_1-antagonist compound was evaluated [64]. The drug inhibited the N1/P2 peak-to-peak amplitude, where the main effect was on the N1 component (related to direct sensory input from the periphery) and it was concluded that the drug exerted its positive effect to reduce capsaicin-induced hyperalgesia primarily by a central mechanism.

One study [65] combined the orofacial laser-evoked brainstem reflex responses in the contracted jaw-closing muscle (laser-evoked silent period) and trigeminal LEPs after topical application of capsaicin to the cheek. The capsaicin-induced pain reduced the degree of reflex suppression and reduced the LEP by 42%. These effects were suggested to be mediated through the activation of segmental and suprasegmental inhibitory systems that may function interdependently.

A few studies have also looked at the magnetoencephalographic-evoked responses after capsaicin-evoked sensitization. Baron et al. [66] investigated the neural activation of the primary somatosensory cortex (SI) induced by intradermal capsaicin-evoked secondary hyperalgesia (dynamic mechanical hyperalgesia). The somatosensory-evoked magnetic fields induced by non-painful electrical stimulation of Aβ-afferents at the forearm skin were recorded. Capsaicin was injected adjacent to the stimulation site to induce secondary dynamic hyperalgesia. The electrical stimulus applied within the secondary hyperalgesic area was subsequently perceived as painful. Non-painful electrical stimulation of Aβ-afferents induced evoked responses in SI at latencies between 20 and 150 ms. Acute application of capsaicin produced an increase in the excitability of central neurons (magnetic dipole strengths) in SI. This might be due to sensitization of central neurons so that normally innocuous stimuli activate pain-signaling neurons, or alternatively, cortical neurons might increase their receptive fields.

The *magnethoencephalographic* parallel study [67] to the LEP study by Valeriani et al. [60] showed that intradermal injection of capsaicin to the thenar area of the hand caused an extension of the cortical hand representation and a decrease of the distance between the hand representation and the localization of the lip. A likely mechanism for this acute reorganization is that pain induced hyper-responsiveness of the left thenar to tactile input from neighboring body sites.

Since brain potentials evoked by painful stimuli mirror only the cortical electro- or magnetoencephalographic activity elicited by very brief painful stimuli, attempts have been made to use the spontaneous electroencephalographic (EEG) signals for assessing the cerebral activity accompanying experimental pain over long periods of time. Some of the spectral components of the EEG have shown a causal relationship to the presence of pain and to the intensity of pain. Furthermore, some of these spectral changes occur over specific cortical areas with relation to the somatic structure stimulated. However, studies on EEG and on-going clinical pain have not yet revealed a better understanding of the cortical structures involved in the processing of pain. To our best knowledge, only three studies have investigated the effect of intradermal and intramuscular capsaicin evoked pain on on-going EEG signals.

To examine the specific effects of cutaneous pain on EEG activity, tonic painful and non-painful sensations in the left forearm were induced by intradermal injection of capsaicin 100 μg/20 μl and the same volume of vehicle, respectively, in 15 healthy males [68]. The capsaicin injection evoked significant decreases of θ, α-1 and α-2 powers over the centro-parieto-occipital regions, compared with baseline. Yet, no significant difference in EEG activation between the non-painful vehicle injection and painful capsaicin injection was found. This implies that the observed topographic EEG activation was not specific for pain but probably related to the cutaneous stimulation.

In another study, the effect of as well skin as muscle capsaicin pain on the EEG was studied [69]. Painful and non-painful sensations were produced by intramuscular injections of capsaicin and vehicle solution in the left brachioradialis muscle in 15 male volunteers. Thirty-one-channel EEG data acquired before, during and after the two injections were analyzed in respect of the topography and the power spectrum. Comparing the EEG changes between baseline, non-painful and painful stimulations an increased β-2 activity during muscle pain was significant over extensive areas of the head, whereas a significant increase in α-2 activity took place at the posterior part of the head during the waning pain period. These results may imply that the painful and non-painful muscular stimulations evoke distinct EEG activation in different neural networks of the human brain and the intensity of nociceptive input from muscles may encode the variety of topographic EEG changes.

In the third study, the effect of intradermal and intramuscular capsaicin-evoked pain on the on-going EEG was compared directly and the data showed that intradermal capsaicin pain produced a similar but not identical EEG topographic pattern as muscle-evoked pain [70]. Muscle pain induced a significant increase of β-2 activity in the extensive frontal, parietal and occipital areas compared to skin pain. These results implicate that the nociceptive inputs from muscle and skin are processed differently in the brain.

Brain imaging methods to assess hyperalgesia

To disclose aspects of cortical processing of capsaicin-evoked nociceptive information, a change in the regional blood flow can be detected and represented by various brain-imaging techniques. For instance, *positron emission tomography* (PET) requires the injection of radioisotopes into the bloodstream towards the brain. These short half-life radioisotopes accumulate for a brief period of time in the active areas of brain and can then be localized by scanners sensitive to the transient increase in radioactivity. For instance, Coghill et al. [71] used PET to quantitatively examine pain-induced global cortical blood flow (CBF) changes after intradermal capsaicin injection. Capsaicin stimulation caused a 22.8% decrease in global CBF from resting levels. In addition, a number of studies have looked more specifically into which areas are activated following on-going capsaicin pain or after activation of sensitized areas (e.g. thermal- or brush-evoked allodynia).

To more specifically look into which structures are activated by sustained intradermal capsaicin-evoked pain, Andersson et al. [72] showed significant activations of the contralateral anterior cingulate gyrus, the ipsilateral anterior insular cortex and the ipsilateral prefrontal cortex. The results are consistent with an involvement of SI in the spatial discrimination of acute cutaneous pain.

PET scans have been obtained during heating of capsaicin-treated skin [73]. A comparison between scans evoked from non-treated and treated skin area reveals the specific activation of a medial thalamic pathway to the frontal lobe during thermal activation of the primary hyperalgesic area. The results suggest that different central pathways mediate the intensity and certain qualitative aspects of pain. In making this differentiation, the brain recognizes unique physiological features of different painful conditions, thus permitting adaptive responses to different pain states.

In another study [74], PET responses were measured before and after stroking-evoked pain produced by an innocuous brush applied within the area with dynamic hyperalgesia. Stroking-evoked pain, but not capsaicin pain alone, increased blood flow significantly in the contralateral right sensory association cortex (Brodmann area (BA) 5/7), and in bilateral prefrontal cortex (BA 9/10/47) and the insula. No significant activity was seen in thalamus or in the SI. Direct comparison between capsaicin pain and stroking-evoked pain revealed significant increase in the contralateral (BA 5/7) only. The specific activation of contralateral (BA 5/7) indicates that this brain region is important to the processing of stroking-evoked pain. The involvement of (BA 5/7) in stroking-evoked pain is claimed to reflect multi-sensory input to this region, its role in conscious pain perception, and its neuroplastic properties.

Iadarola et al. [75] performed a similar study and concluded that brush-evoked pain (allodynia) was characterized by bilateral activation of inferior prefrontal cortex, suggesting that prefrontal responses to pain are context dependent. Similarly, May et al. [76] performed an experimental study to explore mechanisms related to headache/migraine and administered capsaicin subcutaneously in the right forehead of healthy volunteers. Increases of regional CBF were found bilaterally in the insula, in the anterior cingulate cortex, the cavernous sinus and the cerebellum. The increase of activation in the region of the cavernous sinus, however, suggests that this structure is likely to be involved in trigeminal-transmitted pain.

Functional magnetic resonance imaging (fMRI) studies of capsaicin-evoked pain and sensitization are based on different magnetic properties of different tissues in the brain, which can be made visible in very strong magnetic fields. Red blood cells loaded with oxygen (oxygenated hemoglobin) present different magnetic properties than unloaded ones (deoxygenated hemoglobin). Active areas in the brain have higher levels of oxygenated hemoglobin. This signal dependent on blood oxygenation level allows the detection of active areas with good spatial and temporal resolution by magnetic field scanners.

The fMRI-detected neural network activated by the pain component of capsaicin-induced secondary mechanical hyperalgesia was studied by Baron et al. [77]. During hyperalgesia, significantly higher activation as compared to baseline was found in the contralateral prefrontal cortex: the middle (BAs 6, 8 and 9) and inferior frontal gyrus (BAs 44 and 45). No change was present within SI, secondary somatosensory cortex (SII) and the anterior cingulate cortex. The prefrontal activation was interpreted as a consequence of attention, cognitive evaluation and planning of motor behavior in response to pain. Further, a fMRI capsaicin study in animals has showed slightly different results with capsaicin-evoked changes in anterior cingulate (bilateral), frontal cortex (bilateral) and sensory motor cortex (contralateral) [78].

For PET, the temporal resolution and spatial resolution are not as good as with fMRI (see below), but PET and single photon-emission computed tomography (SPECT) allow, in addition, the assessment of the concentration of ligands and receptors with relevance for the pain transmission in the brain.

Vasomotor responses to assess hyperalgesia

The vasomotor manifestation of topical and intradermal capsaicin-induced cutaneous hyperexcitability is the flare reaction (also known as neurogenic inflammation; see Fig. 4). Flare reactions can be assessed by laser Doppler flowmetry, reflectance spectroscopy, thermography and visual inspection.

Not only can Aδ-nociceptive endings in the skin be inhibited by repetitive capsaicin application [62] but also by histamine-evoked neurogenic inflammation [79]. Topical application of capsaicin to the human skin induces a flare reaction, increased blood flow and elevated skin temperature 30–60 min after the application, and lasts approximately 3–4 h [12, 17, 20, 75]. Elevated temperature, visible flare and increased blood flow both in the area of application and in the surrounding area are most likely due to dilatation of local cutaneous arterioles because of activation of afferent C-fibers [34, 80]. Recent studies using thermography to assess the development of the flare have greatly improved our knowledge of flare mechanisms. Serra et al. [81] used thermography to demonstrate that a flare response occurs in a large scattered area. It was observed that the specific arterioles engaged in the flare reaction in response to a capsaicin injection are identical to the arterioles reacting in a post- ischemic period, suggesting there is no specific subgroup of arterioles for the flare reaction exclusively. Therefore it is conceivable that the flare response is not a single phenomenon, but may be due to several underlying mechanisms responsible for superficial warming and for deep arteriolar dilatation [82]. Yucel et al. [29] found an increased skin temperature (thermography) in the primary area for both topical capsaicin and controlled-heat injury. This result may be explained by increased C-fiber activity following heat conditioning resulting in increased dilatation via axon reflexes.

However, it still remains to be established how flare and hyperalgesia are linked, particularly when it is known that flare and hyperalgesia may occur independently [81]. Recently, Koppert et al. [82] have studied the effect of lidocaine on punctuate hyperalgesia and flare induced by intradermal capsaicin, and reported that the size of flare detected by laser Doppler flowmetry was larger than the size of punctuate hyperalgesia. This report seems to indicate a possible contribution of neurogenic inflammation to secondary hyperalgesia [82]. The study of Sumikura et al. [38] supports these findings, as they also suggested a contribution from neurogenic inflam-

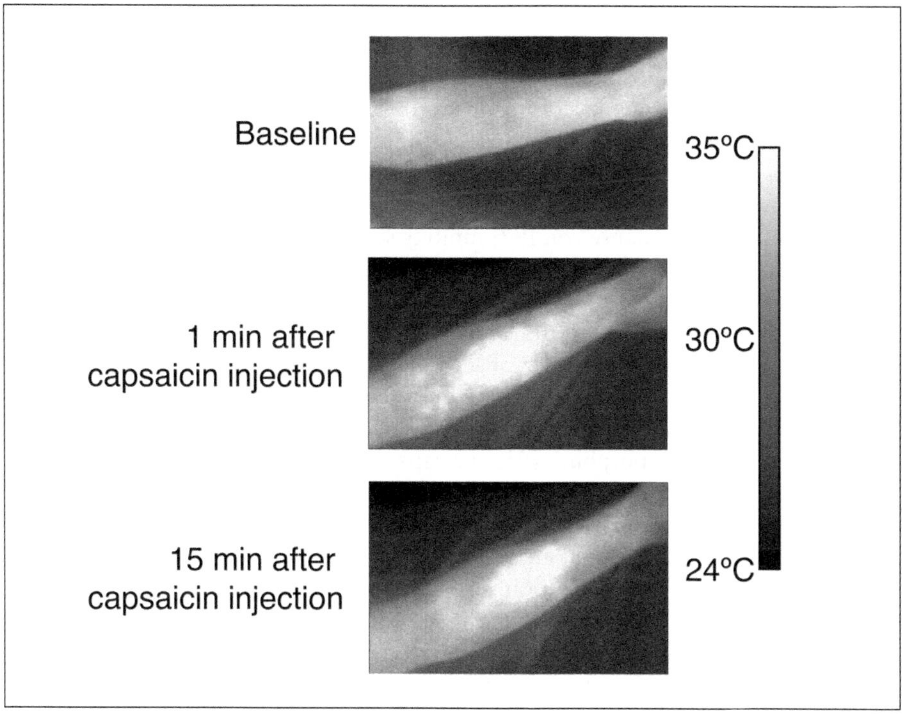

Figure 4
Following intradermal capsaicin injection, a pronounced increase in skin temperature is detected using thermography.

mation to secondary hyperalgesia to heat. The capsaicin-induced flux in the primary and secondary hyperalgesic areas after pretreating the capsaicin-injection site with local ketamine or lidocaine were examined by Gottrup et al. [83]. They found ketamine to reduce significantly the blood flow in both the primary and secondary hyperalgesic area, whereas lidocaine only reduced blood flow significantly in the primary hyperalgesic area.

Modulation of capsaicin-induced hyperalgesia

A number of studies have used topically applied drugs to modulate capsaicin-induced primary and secondary hyperalgesia. Topical administration of local analgesics (e.g. EMLA® [84], the cannabinoid receptor ligand HU210 [85] and iontophoresis of noradrenaline [86]) have been used to pretreat the skin prior to capsaicin application. Pretreatments significantly reduced the burning sensation from capsaicin and attenuated primary heat hyperalgesia. Interestingly EMLA® also inhibits capsaicin-evoked neurogenic inflammation [87].

Topical [88] and systemic [89] non-steroidal anti-inflammatory drugs are not efficient in inhibiting capsaicin- or burn- [90] induced secondary hyperalgesic areas. In contrast, topical acetylsalicylic acid seems to be more efficient in inhibiting secondary hyperalgesia [91], and topical administration of the 5-HT_3 (serotonin) receptor antagonist ondansetron [92] inhibits secondary mechanical hyperalgesia.

Different drugs have been applied intradermally in an attempt to modulate secondary hyperalgesia. The effect of intradermal fentanyl and ketamine on capsaicin-induced hyperalgesia has been studied by Koppert et al. [93]. Fentanyl and ketamine did not affect the area or intensity of secondary hyperalgesia, whereas larger doses of ketamine had some inhibitory effect on intensity of secondary hyperalgesia. In contrast, other studies showed no effect of subcutaneous infiltration of ketamine [94] but an effect of morphine [95] on capsaicin-induced pain and hyperalgesia. Subcutaneous infiltration of lidocaine [83, 94] reduces spontaneous pain and capsaicin-induced hyperalgesia. However, Koppert et al. [82] administered lidocaine as part of an intravenous regional anesthesia protocol to exclude possible central analgesic effects, and found no effect on secondary pin-prick hyperalgesia.

Although capsaicin-induced hyperalgesia is assumed to mimic some of the central manifestations (e.g. secondary hyperalgesia) seen in neuropathic pain syndromes, only limited pharmacological studies have supported that notion. Drugs used for management of neurogenic pain such as mexiletine [96], oral lamotrigine [97], intravenous magnesium [98], intravenous clonidine [99], systemic adenosine [100], systemic lidocaine [82, 101] and tricyclic antidepressants such as desipramine [102] or amitriptyline [103] do not have any effect on central manifestations evoked by capsaicin. However, a number of other drugs and administration regimes have shown to suppress secondary hyperalgesia induced by capsaicin; these include intrathecal adenosine [104, 105], intrathecal clonidine [99], blocking α-adrenoreceptors with intravenous phentolamine [106], intravenous remifentanil [97, 107], intravenous alfentanil [102, 103, 108], oral hydromorphone [97], intravenous lidocaine [43], the α-amino-3-hydroxy-5-methyl-4-isoxazolepropionic acid (AMPA)-kainate antagonist LY293558 [109], intravenous ketamine [56, 108] and the anticonvulsant gabapentin [110]. It is currently not clear which experimental model is the best to simulate clinically relevant conditions of central sensitization.

Induction and assessment of capsaicin-evoked muscular pain and hyperalgesia

There is only one human study which has investigated the peripheral neural activity of muscle pain evoked by intramuscular capsaicin [111]. Microneurography was used to excite and to record impulse activity in identified single afferent peroneal nerve fibers from skeletal muscles of human volunteers. Capsaicin (0.01%) injected into the receptive field of two slowly conducting muscle afferents (one Group III and one Group IV) produced spontaneous discharge of each fiber and caused intense cramping pain, suggesting that the units recorded were nociceptive. The results endorse the concept that the primary sensory apparatus that encodes the sensation of cramping muscle pain in humans is served by mechanical nociceptors with slowly conducting nerve fibers.

Intramuscular injection of capsaicin is known to cause pain and sensitize the muscle to cuff algometry [112] or standard pressure-pain stimulation [113, 114]. Capsaicin stimulation of muscle and skin producing the same pain-intensity levels resulted in different local and referred pains, indicating that the neurophysiological mechanisms underlying capsaicin-evoked skin and muscle pain differ [21].

Recently, an interaction between cutaneous itch and capsaicin-evoked muscle pain has been studied [115]. Capsaicin-induced muscle pain reduced the iontophoresed histamine-evoked cutaneous itch sensation. In contrast, capsaicin-induced muscle pain increased significantly after cutaneous histamine application. These novel data indicate that muscle pain inhibits itch and that histamine increases muscle pain. A bi-directional interaction between cutaneous histamine-sensitive afferents and nociceptive muscle afferents via central mechanisms is suggested.

Intramuscular injection of capsaicin has also been used to study the interaction between muscle pain and motor control. Muscle pain has been shown to inhibit the single motor-unit firing frequency [116] and potentiate the single motor-unit twitch force during low-level muscle contractions. Furthermore, intramuscular capsaicin facilitates the stretch reflex, possibly by increasing the sensitivity of the fusimotor system [117]. As chronic muscle pain and related central sensitization is an important clinical problem [118–121], more research should be done on assessing the manifestations and mechanisms evoked by intramuscular capsaicin. The manifestation of secondary hyperalgesia related to muscle sensitization has been suggested to be the size of the referred muscle pain area and the related somatosensory changes in these areas [122].

Induction and assessment of capsaicin-evoked visceral pain and hyperalgesia

Few human studies have been performed with capsaicin-evoked pain and hyperalgesia in visceral structures. Capsaicin has been applied to the gut of patients with an ileostoma [123]. Increasing volumes of capsaicin solution (0.25–3 ml, all containing 50 µg of capsaicin) were applied to the ileum via the stomal opening. Referred somatic pain developed around the stomal opening, with a correlation between the pain area and pain intensity. After application of capsaicin, significant hyperalgesia was found to distension of the gut (a 28% reduction in pressure-pain threshold). Specific activation of nociceptors in the gut mucosa provides new possibilities to study clinically relevant visceral pain mechanisms. Lee et al. [124] used healthy volunteers and found that administration of capsaicin decreases proximal gastric tone, inhibits phasic contractility of the proximal stomach and increases sensitivity to proximal gastric distension (i.e. lower discomfort threshold). Capsaicin perfused to human small intestine to study jejunal nociception caused crampy abdominal pain [125], which proves the existence of chemonociception in the human small intestine. The perfusion caused no sensitization to barostat balloon distention [126], which is in contrast to capsaicin application to the ileum.

The few data available on visceral sensitization by capsaicin generally support the hypothesis that capsaicin can cause pain in the human gastrointestinal tract. The manifestation of secondary hyperalgesia related to visceral pain has been suggested to be the size of the referred visceral pain area and the related somatosensory changes in these areas [122].

Conclusion

Capsaicin is a very useful experimental substance in human experimental and clinical pain research for the study of peripheral and central sensitization of skin, muscles and viscera. Sensitization related to deep tissues has been partly neglected in the past, as most research has been directed towards understanding the mechanisms underlying the cutaneous manifestation of sensitization related to neurogenic pain. Further research into the mechanisms behind capsaicin-induced hyperalgesia is needed in order to improve the experimental models and to facilitate pharmacological screening of new drugs for neuropathic pain using those experimental models. However, muscular and visceral pains have a much higher clinical prevalence than neurogenic pain; thus a better understanding of the central mechanisms of neuroplasticity related to these deep structures would have substantial socioeconomic impact.

References

1 Arendt-Nielsen L (1997) Induction and assessment of experimental pain from human skin, muscle and viscera. In: TS Jensen, JA Turner, Z Wiesenfeld-Hallin (eds): *Progress in pain research and management*. IASP Press, Seattle, 393–425
2 Lewis T (1942) *Pain*. The Macmillan Company, London
3 Hardy JD, Wolff HG, Goodel H (1950) Experimental evidence on the nature of cutaneous hyperalgesia. *J Clin Invest* 29: 115–140

4 LaMotte RH, Shain CN, Simone DA, Tsai EF (1991) Neurogenic hyperalgesia: psychophysical studies of underlying mechanisms. *J Neurophysiol* 66: 190–211
5 Raja SN, Campbell JN, Meyer RA (1984) Evidence for different mechanisms of primary and secondary hyperalgesia following heat injury to the glabrous skin. *Brain* 107: 1179–1188
6 Dahl JB, Brennum J, Arendt-Nielsen L, Jensen TS, Kehlet H (1993) The effect of preversus postinjury infiltration with lidocaine on thermal and mechanical hyperalgesia after heat injury to the skin. *Pain* 53: 43–51
7 Ali Z, Meyer RA, Campbell JN (1996) Secondary hyperalgesia to mechanical but not heat stimuli following a capsaicin injection in hairy skin. *Pain* 68: 401–411
8 Warncke T, Jørum E, Stubhaug A (1997) Local treatment with the N-methyl-D-aspartate receptor antagonist ketamine, inhibit development of secondary hyperalgesia in man by a peripheral action. *Neurosci Lett* 227: 1–4
9 Treede RD, Meyer RA, Raja SN, Campbell JN (1992) Peripheral and central mechanisms of cutaneous hyperalgesia. *Prog Neurobiol* 38: 397–421
10 Woolf CJ, Salter MW (2000) Neuronal plasticity: increasing the gain in pain. *Science* 288: 1765–1768
11 Simone DA, Baumann TK, LaMotte RH (1989) Dose-dependent pain and mechanical hyperalgesia in humans after intradermal injection of capsaicin. *Pain* 38: 99–107
12 Baumann TK, Simone DA, Shain CN, LaMotte RH (1991) Neurogenic hyperalgesia: the search for the primary cutaneous afferent fiber that contribute to capsaicin-induced pain and hyperalgesia. *J Neurophysiol* 66: 212–227
13 LaMotte RH, Lundberg LER, Torebjörk HE (1992) Pain, hyperalgesia and activity in nociceptive C units in humans after intradermal injection of capsaicin. *J Physiol (London)* 448: 749–764
14 Kenins P (1982) Responses of single nerve fibres to capsaicin applied to the skin. *Neurosci Lett* 29: 83–88
15 Szolcsanyi J, Anton F, Reeh PW, Handwerker HO (1988) Selective excitation by capsaicin of mechano-heat sensitive nociceptors in rat skin. *Brain Res* 446: 262–268
16 Ringkamp M, Peng YB, Wu G, Hartke TV, Campbell JN, Meyer RA (2001) Capsaicin responses in heat-sensitive and heat-insensitive A-fiber nociceptors. *J Neurosci* 21: 4460–4468
17 Koltzenburg M, Lundberg LER, Torebjörk HE (1992) Dynamic and static components of mechanical hyperalgesia in human hairy skin. *Pain* 51: 207–219
18 Pedersen JL, Kehlet H (1998) Hyperalgesia in a human model of acute inflammatory pain: a methodological study. *Pain* 74: 139–151
19 Magnusson BM, Koskinen LO (1996) Effects of topical application of capsaicin to human skin: a comparison of effects evaluated by visual assessment, sensation registration, skin blood flow and cutaneous impedance measurements. *Acta Derm Venereol* 76: 129–132
20 Mohammadian P, Andersen OK, Arendt-Nielsen L (1998) Correlation between local

vascular and sensory changes following tissue inflammation induced by repetitive application of topical capsaicin. *Brain Res* 792: 1–9

21 Witting N, Svensson P, Gottrup H, Arendt-Nielsen L, Jensen TS (2000) Intramuscular and intradermal injection of capsaicin: a comparison of local and referred pain. *Pain* 84: 407–412

22 Felsby S, Nielsen J, Arendt-Nielsen L, Jensen TS (1996) NMDA receptor blockade in chronic neuropathic pain: a comparison of ketamine and magnesium chloride. *Pain* 64: 283–291

23 Grönross M, Pertovaara A (1993) Capsaicin-induced central facilitation of a nociceptive flexion reflex in humans. *Neurosci Lett* 159: 215–218

24 Gazerani G, Bajaj P, Andersen OK, Arendt-Nielsen L (2004) *Intradermal capsaicin injection on the forehead: a model for trigeminal sensitization.* 7th European Headache Federation Congress, Rotterdam, The Netherlands, 16–19 June 2004, proceedings

25 Morris VH, Cruwys SC, Kidd BL (1997) Characterisation of capsaicin-induced mechanical hyperalgesia as a marker for altered nociceptive processing in patients with rheumatoid arthritis. *Pain* 71: 179–186

26 Kilo S, Schmelz M, Koltzenburg M, Handwerker HO (1994) Different patterns of hyperalgesia induced by experimental inflammation in human skin. *Brain* 117: 385–396

27 Petersen KL, Rowbotham MC (1999) A new human experimental pain model: the heat/capsaicin sensitization model. *Neuroreport* 10: 1511–1516

28 Dirks J, Petersen KL, Dahl JB (2003) The heat/capsaicin sensitization model: a methodologic study. *J Pain* 4: 122–128

29 Yucel A, Miyazawa A, Andersen OK, Arendt-Nielsen L (2001) The effect of heat conditioning of the primary area before and after the induction of hyperalgesia by topical/intradermal capsaicin or by controlled heat injury. Somatosens Mot Res 18: 295–302

30 Yucel A, Andersen OK, Nielsen J, Arendt-Nielsen L (2002) Heat hyperalgesia in humans: assessed by different stimulus temperature profiles. *Eur J Pain* 6: 357–364

31 Yucel A, Miyazawa A, Andersen OK, Arendt-Nielsen L (2004) Comparison of hyperalgesia induced by capsaicin injection and controlled heat injury: Effect on temporal summation. *Somatosens Mot Res* 21: 15–24

32 Nicolaisen L, Hansen CL, Nielsen J, Keller J, Arendt-Nielsen L, Jensen TS (1996) The effect of ketamine on phantom pain: a central neuropathic disorder maintained by peripheral input. *Pain* 67: 69–77

33 Bennett GJ (1994) Animal models of neuropathic pain. In: GB Gebhart, DL Hammond, TS Jensen (eds): *Progress in pain research and management,* vol. 2. IASP Press, Seattle, 495–510

34 Ochoa JL, Yarnitsky D, Marchettini P, Dotson R, Cline M (1993) Interactions between sympathetic vasoconstrictor outflow and C nociceptor-induced antidromic vasodilatation. *Pain* 54: 191–196

35 Magerl W, Fuchs PN, Meyer RA, Treede R-D (2001) Roles of capsaicin-insensitive nociceptors in cutaneous pain and secondary hyperalgesia. *Brain* 124 (Pt 9): 1754–1764

36 Ziegler EA, Magerl W, Meyer RA, Treede RD (1999) Secondary hyperalgesia to punc-

tate mechanical stimuli. Central sensitization to A-fibre nociceptor input. *Brain* 122: 2245–2257

37 Fuchs PN, Campbell JN, Meyer RA (2000) Secondary hyperalgesia persists in capsaicin desensitized skin. *Pain* 84: 141–149

38 Sumikura H, Andersen OK, Drewes AM, Arendt-Nielsen L (2003) Spatial and temporal profiles of flare and hyperalgesia after intradermal capsaicin. *Pain* 105: 285–291

39 Sumikura H, Andersen OK, Drewes AM, Arendt-Nielsen L (2003) A comparison of hyperalgesia and neurogenic inflammation induced by melittin and capsaicin in humans. *Neurosci Lett* 337: 147–150

40 Arendt-Nielsen L, Andersen OK, Jensen TS (1996) Brief, prolonged and repeated thermal stimuli applied to hyperalgesic skin areas: a psychophysical study. *Brain Res* 712: 165–167

41 Pedersen JL, Kehlet H (1998) Secondary hyperalgesia to heat stimuli after burn injury in man. *Pain* 76: 377–384

42 Serra J, Campero M, Ochoa J (1998) Flare and hyperalgesia after intradermal capsaicin injection in human skin. *J Neurophysiol* 80: 2801–2810

43 Wallace MS, Laitin S, Licht D, Yaksh TL (1997) Concentration-effect relations for intravenous lidocaine infusions in human volunteers: effects on acute sensory thresholds and capsaicin-evoked hyperpathia. *Anesthesiology* 86: 1262–1272

44 Sumikura H, Miyazawa A, Yusel A, Andersen OK, Arendt-Nielsen L (2004) Secondary heat hyperalgesia detected by radiant heat stimuli in humans: Evaluation of stimulus intensity and duration. *Somatosens Mot Res; in press*

45 Arendt-Nielsen L, Sonnenborg FA, Andersen OK (2000) Facilitation of the withdrawal reflex by repeated transcutaneous electrical stimulation: an experimental study on central integration in humans. *Eur J Appl Physiol* 81: 165–173

46 Petersen-Felix S, Arendt-Nielsen L, Bak P, Fischer M, Bjerring P, Zbinden AM (1995) The effects of isoflurane on repeated nociceptive stimuli (central temporal summation). *Pain* 64: 277–281

47 Baumgartner U, Magerl W, Klein T, Hopf HC, Treede RD (2002) Neurogenic hyperalgesia versus painful hypoalgesia: two distinct mechanisms of neuropathic pain. *Pain* 96: 141–151

48 Staud R, Robinson ME, Vierck CJ Jr, Cannon RC, Mauderli AP, Price DD (2003) Ratings of experimental pain and pain-related negative affect predict clinical pain in patients with fibromyalgia syndrome. *Pain* 105: 215–222

49 Rössel P, Drewes AM, Petersen P, Nielsen J, Arendt-Nielsen L (1999) Pain produced by electrical stimulation of the rectum in patients with irritable bowel syndrome. Further evidence for visceral hyperalgesia. *Scand J Gastroenterol* 34: 1001–1006

50 Magerl W, Wilk SH, Treede R-D (1998) Secondary hyperalgesia and perceptual wind-up following intradermal injection of capsaicin in humans. *Pain* 74: 257–268

51 Pedersen JL, Andersen OK, Arendt-Nielsen L, Kehlet H (1998) Hyperalgesia and temporal summation of pain after heat injury in man. *Pain* 74: 189–197

52 Gottrup H, Kristensen AD, Bach FW, Jensen TS (2003) Aftersensations in experimental and clinical hypersensitivity. *Pain* 103: 57–64
53 You H-J, Mørch CD, Chen J, Arendt-Nielsen L (2003) Differential antinociceptive effects induced by a selective cyclooxygenase-2 inhibitor (SC-236) on dorsal horn neurons and spinal withdrawal reflexes in anesthetized spinal rats. *Neuroscience* 121: 459–472
54 Torebjork HE, Lundberg LE, LaMotte RH (1992) Central changes in processing of mechanoreceptive input in capsaicin-induced secondary hyperalgesia in humans. *J Physiol* 448: 765–780
55 Treede RD, Cole JD (1993) Dissociated secondary hyperalgesia in a subject with a large-fibre sensory neuropathy. *Pain* 53: 169–174
56 Andersen OK, Gracely RH, Arendt-Nielsen L (1995) Facilitation of the human nociceptive reflex by stimulation of Aβ-fibers in a secondary hyperalgesic area sustained by nociceptive input from the primary hyperalgesic area. *Acta Physiol Scand* 155: 87–97
57 Andersen OK, Felsby S, Nicolaisen L, Bjerring P, Jensen TS, Arendt-Nielsen L (1996) The effect of ketamine on stimulation of primary and secondary hyperalgesic areas induced by capsaicin – a double blind, placebo-controlled, human experimental study. *Pain* 66: 51–62
58 Kemppainen P, Waltimo A, Waltimo T, Kononen M, Pertovaara A (1997) Differential effects of noxious conditioning stimulation of the cheek by capsaicin on human sensory and inhibitory masseter reflex responses evoked by tooth pulp stimulation. *J Dent Res* 76: 1561–1568
59 Andersen OK, Sonnenborg FA, Arendt-Nielsen L (1999) Modular organization of human leg withdrawal reflexes elicited by electrical stimulation of the foot sole. *Muscle Nerve* 22: 1520–1530
60 Valeriani M, Arendt-Nielsen L, Le Pera D, Restuccia D, Rosso T, De Armas L, Maiese T, Fiaschi A, Tonali P, Tinazzi M (2003) Short-term plastic changes of the human nociceptive system following acute pain induced by capsaicin. *Clin Neurophysiol* 114: 1879–1890
61 Beydoun A, Dyke DB, Morrow TJ, Casey KL (1996) Topical capsaicin selectively attenuates heat pain and A delta fiber-mediated laser-evoked potentials. *Pain* 65: 189–196
62 Bjerring P, Arendt-Nielsen L (1989) Use of a new argon laser technique to evaluate changes in sensory and pain thresholds in human skin following topical capsaicin treatment. *Skin Pharmacol* 2: 162–167
63 Bjerring P, Arendt-Nielsen L, Soderberg U (1990) Argon laser induced cutaneous sensory and pain thresholds in post-herpetic neuralgia. Quantitative modulation by topical capsaicin. *Acta Derm Venereol* 70: 121–125
64 Schaffler K, Seibel K, Thomsen M, Edwards M (2004) Effect of the new H1-antagonist ReN1869 on capsaicin-induced hyperalgesia in human skin/Human phase-I trial using somatosensory evoked potentials induced by a CO_2 laser. *Arzneimittelforschung* 54: 187–191

65 Romaniello A, Arendt-Nielsen L, Cruccu G, Svensson P (2002) Modulation of trigeminal laser evoked potentials and laser silent periods by homotopical experimental pain. *Pain* 98: 217–228
66 Baron R, Baron Y, Disbrow E, Roberts TP (2000) Activation of the somatosensory cortex during Abeta-fiber mediated hyperalgesia. A MSI study. *Brain Res* 871: 75–82
67 Soros P, Knecht S, Bantel C, Imai T, Wusten R, Pantev C, Lutkenhoner B, Burkle H, Henningsen H (2001) Functional reorganization of the human primary somatosensory cortex after acute pain demonstrated by magnetoencephalography. *Neurosci Lett* 298: 195–198
68 Chang PF, Arendt-Nielsen L, Graven-Nielsen T, Svensson P, Chen AC (2001) Topographic effects of tonic cutaneous nociceptive stimulation on human electroencephalograph. *Neurosci Lett* 305: 49–52
69 Chang PF, Arendt-Nielsen L, Graven-Nielsen T, Svensson P, Chen AC (2001) Different EEG topographic effects of painful and non-painful intramuscular stimulation in man. *Exp Brain Res* 141: 195–203
70 Chang PF, Arendt-Nielsen L, Graven-Nielsen T, Svensson P, Chen AC (2004) Comparative EEG activation to skin pain and muscle pain induced by capsaicin injection. *Int J Psychophysiol* 51: 117–126
71 Coghill RC, Sang CN, Berman KF, Bennett GJ, Iadarola MJ (1998) Global cerebral blood flow decreases during pain. *J Cereb Blood Flow Metab* 18: 141–147
72 Andersson JL, Lilja A, Hartvig P, Langstrom B, Gordh T, Handwerker H, Torebjork E (1997) Somatotopic organization along the central sulcus, for pain localization in humans, as revealed by positron emission tomography. *Exp Brain Res* 117: 192–199
73 Lorenz J, Cross D, Minoshima S, Morrow T, Paulson P, Casey K (2002) A unique representation of heat allodynia in the human brain. *Neuron* 35: 383–393
74 Witting N, Kupers RC, Svensson P, Arendt-Nielsen L, Gjedde A, Jensen TS (2001) Experimental brush-evoked allodynia activates posterior parietal cortex. *Neurology* 57: 1817–1824
75 Iadarola MJ, Berman KF, Zeffiro TA, Byas-Smith MG, Gracely RH, Max MB, Bennett GJ (1998) Neural activation during acute capsaicin-evoked pain and allodynia assessed with PET. *Brain* 121(Pt 5): 931–947
76 May A, Kaube H, Buchel C, Eichten C, Rijntjes M, Juptner M, Weiller C, Diener HC (1998) Experimental cranial pain elicited by capsaicin: a PET study. *Pain* 74: 61–66
77 Baron R, Baron Y, Disbrow E, Roberts TP (1999) Brain processing of capsaicin-induced secondary hyperalgesia: a functional MRI study. *Neurology* 53: 548–557
78 Malisza KL, Docherty JC (2001) Capsaicin as a source for painful stimulation in functional MRI. *J Magn Reson Imaging* 14: 341–347
79 Bjerring P, Arendt-Nielsen L (1990) Inhibition of histamine skin flare reaction following repeated topical applications of capsaicin. *Allergy* 45: 121–125
80 Wasner G, Baron R, Janig W (1999) Dynamic mechanical allodynia in humans is not mediated by a central presynaptic interaction of A beta-mechanoreceptive and nociceptive C-afferents. *Pain* 79: 113–119

81 Serra J, Campero M, Ochoa J (1998) Flare and hyperalgesia after intradermal capsaicin injection in human skin. *J Neurophysiol* 80: 2801–2810
82 Koppert W, Ostermeier N, Sittl R, Weidner C, Schmelz M (2000) Low-dose lidocaine reduces secondary hyperalgesia by a central mode of action. *Pain* 85: 217–224
83 Gottrup H, Bach FW, Jensen TS (2004) Differential effects of peripheral ketamine and lidocaine on skin flux and hyperalgesia induced by intradermal capsaicin in humans. *Clin Physiol Funct Imaging* 24: 103–108
84 Yosipovitch G, Maibach HI, Rowbotham MC (1999) Effect of EMLA pre-treatment on capsaicin-induced burning and hyperalgesia. *Acta Derm Venereol* 79: 118–121
85 Rukwied R, Watkinson A, McGlone F, Dvorak M (2003) Cannabinoid agonists attenuate capsaicin-induced responses in human skin. *Pain* 102: 283–288
86 Drummond PD, Lipnicki DM (2001) Repeated local administration of noradrenaline or saline inhibits thermal hyperalgesia in pain-sensitized human skin. *Br J Clin Pharmacol* 52: 289–295
87 Bjerring P, Arendt-Nielsen L (1990) A quantitative comparison of the effect of local analgesics on argon laser induced cutaneous pain and on histamine induced wheal, flare and itch. *Acta Dermato-Venereologica* 70: 126–131
88 Warncke T, Brennum J, Arendt-Nielsen L, Branebjerg PE (1996) Effects of local and systemic ibuprofen on primary and secondary hyperalgesia in man. *Curr Therap Res* 57: 937–949
89 Kilo S, Forster C, Geisslinger G, Brune K, Handwerker HO (1995) Inflammatory models of cutaneous hyperalgesia are sensitive to effects of ibuprofen in man. *Pain* 62: 187–193
90 Petersen KL, Brennum J, Dahl JB (1997) Experimental evaluation of the analgesic effect of ibuprofen on primary and secondary hyperalgesia. *Pain* 70: 167–174
91 Schmelz M, Kress M (1996) Topical acetylsalicylate attenuates capsaicin induced pain, flare and allodynia but not thermal hyperalgesia. *Neurosci Lett* 214: 72–74
92 Giordano J, Daleo C, Sacks SM (1998) Topical ondansetron attenuates nociceptive and inflammatory effects of intradermal capsaicin in humans. *Eur J Pharmacol* 354: R13–R14
93 Koppert W, Zeck S, Blunk JA, Schmelz M, Likar R, Sittl R (1999) The effects of intradermal fentanyl and ketamine on capsaicin-induced secondary hyperalgesia and flare reaction. *Anesth Analg* 89: 1521–1527
94 Gottrup H, Bach FW, Arendt-Nielsen L, Jensen TS (2000) Peripheral lidocaine but not ketamine inhibits capsaicin-induced hyperalgesia in humans. *Br J Anaesth* 85: 520–528
95 Kinnman E, Nygards EB, Hansson P (1997) Peripherally administrated morphine attenuates capsaicin-induced mechanical hypersensitivity in humans. *Anesth Analg* 84: 595–599
96 Ando K, Wallace MS, Braun J, Schulteis G (2000) Effect of oral mexiletine on capsaicin-induced allodynia and hyperalgesia: a double-blind, placebo-controlled, crossover study. *Reg Anesth Pain Med* 25: 468–474
97 Petersen KL, Maloney A, Hoke F, Dahl JB, Rowbotham MC (2003) A randomized study

of the effect of oral lamotrigine and hydromorphone on pain and hyperalgesia following heat/capsaicin sensitization. *J Pain* 4: 400–406
98 Mikkelsen S, Dirks J, Fabricius P, Petersen KL, Rowbotham MC, Dahl JB (2001) Effect of intravenous magnesium on pain and secondary hyperalgesia associated with the heat/capsaicin sensitization model in healthy volunteers. *Br J Anaesth* 86: 871–873
99 Eisenach JC, Hood DD, Curry R (1998) Intrathecal, but not intravenous, clonidine reduces experimental thermal or capsaicin-induced pain and hyperalgesia in normal volunteers. *Anesth Analg* 87: 591–596
100 Dirks J, Petersen KL, Rowbotham MC, Dahl JB (2001) Effect of systemic adenosine on pain and secondary hyperalgesia associated with the heat/capsaicin sensitization model in healthy volunteers. *Reg Anesth Pain Med* 26: 414–419
101 Dirks J, Fabricius P, Petersen KL, Rowbotham MC, Dahl JB (2000) The effect of systemic lidocaine on pain and secondary hyperalgesia associated with the heat/capsaicin sensitization model in healthy volunteers. *Anesth Analg* 91: 967–972
102 Wallace MS, Ridgeway B 3rd, Leung A, Schulteis G, Yaksh TL (2002) Concentration-effect relationships for intravenous alfentanil and ketamine infusions in human volunteers: effects on acute thresholds and capsaicin-evoked hyperpathia. *J Clin Pharmacol* 42: 70–80
103 Eisenach JC, Hood DD, Curry R, Tong C (1997) Alfentanil, but not amitriptyline, reduces pain, hyperalgesia, and allodynia from intradermal injection of capsaicin in humans. *Anesthesiology* 86: 1279–1287
104 Eisenach JC, Hood DD, Curry R (2002) Preliminary efficacy assessment of intrathecal injection of an American formulation of adenosine in humans. *Anesthesiology* 96: 29–34
105 Eisenach JC, Curry R, Hood DD (2002) Dose response of intrathecal adenosine in experimental pain and allodynia. *Anesthesiology* 97: 938–942
106 Liu M, Max MB, Parada S, Rowan JS, Bennett GJ (1996) The sympathetic nervous system contributes to capsaicin-evoked mechanical allodynia but not pinprick hyperalgesia in humans. *Neuroscience* 16: 7331–7335
107 Petersen KL, Jones B, Segredo V, Dahl JB, Rowbotham MC (2001) Effect of remifentanil on pain and secondary hyperalgesia associated with the heat--capsaicin sensitization model in healthy volunteers. *Anesthesiology* 94: 15–20
108 Park KM, Max MB, Robinovitz E, Gracely RH, Bennett GJ (1995) Effects of intravenous ketamine, alfentanil, or placebo on pain, pinprick hyperalgesia, and allodynia produced by intradermal capsaicin in human subjects. *Pain* 63: 163–172
109 Sang CN, Hostetter MP, Gracely RH, Chappell AS, Schoepp DD, Lee G, Whitcup S, Caruso R, Max MB (1998) AMPA/kainate antagonist LY293558 reduces capsaicin-evoked hyperalgesia but not pain in normal skin in humans. *Anesthesiology* 89: 1060–1067
110 Dirks J, Petersen KL, Rowbotham MC, Dahl JB (2002) Gabapentin suppresses cutaneous hyperalgesia following heat-capsaicin sensitization. *Anesthesiology* 97:102–107

111 Marchettini P, Simone DA, Caputi G, Ochoa JL (1996) Pain from excitation of identified muscle nociceptors in humans. *Brain Res* 740: 109–116
112 Polianskis R, Graven-Nielsen T, Arendt-Nielsen L (2002) Pressure-pain function in desensitized and hypersensitized muscle and skin assessed by cuff algometry. *J Pain* 3: 28–37
113 Graven-Nielsen T, Gibson SJ, Laursen RJ, Svensson P, Arendt-Nielsen L (2002) Opioid-insensitive hypoalgesia to mechanical stimuli at sites ipsilateral and contralateral to experimental muscle pain in human volunteers. *Exp Brain Res* 146: 213–222
114 Arima T, Svensson P, Arendt-Nielsen L (2000) Capsaicin-induced muscle hyperalgesia in the exercised and non-exercised human masseter muscle. *J Orofac Pain* 14: 213–223
115 Wasner G, Schwarz K, Schattschneider J, Binder A, Jensen TS, Baron R (2004) Interaction between histamine-induced itch and experimental muscle pain. *Eur J Pain* 8: 179–185
116 Sohn MK, Graven-Nielsen T, Arendt-Nielsen L, Svensson P (2004) Effects of experimental muscle pain on mechanical properties of single motor units in human masseter. *Clin Neurophysiol* 115: 76–84
117 Wang K, Arendt-Nielsen L, Svensson P (2002)Capsaicin-induced muscle pain alters the excitability of the human jaw-stretch reflex. *J Dent Res* 81: 650–654
118 Johansen MK, Graven-Nielsen T, Olesen AS, Arendt-Nielsen L (1999) Generalised muscular hyperalgesia in chronic whiplash syndrome. *Pain* 83: 229–234
119 Sörensen J, Graven-Nielsen T, Henriksson KG, Bengtsson M, Arendt-Nielsen L (1998) Hyperexcitability in fibromyalgia. *J Rheumatol* 25: 152–155
120 Banic B, Petersen-Felix S, Andersen OK, Radanov BP, Villiger PM, Arendt-Nielsen L, Curatolo M (2004) Evidence for spinal cord hypersensitivity in chronic pain after whiplash injury and in fibromyalgia. *Pain* 107: 7–15
121 Curatolo M, Petersen-Felix S, Arendt-Nielsen L, Giani C, Zbinden AM, Radanov BP (2001) Central hypersensitivity in chronic pain after whiplash injury. *Clin J Pain* 17: 306–315
122 Arendt-Nielsen L, Graven-Nielsen T (2002) Deep tissue hyperalgesia. *J Musculoskeletal Pain* 10: 97–119
123 Drewes AM, Schipper KP, Dimcevski G, Petersen P, Gregersen H, Funch-Jensen P, Arendt-Nielsen L (2003) Gut pain and hyperalgesia induced by capsaicin: a human experimental model. *Pain* 104: 333–341
124 Lee KJ, Vos R, Tack J (2004) Effects of capsaicin on the sensorimotor function of the proximal stomach in humans. *Aliment Pharmacol Ther* 19: 415–425
125 Hammer J, Hammer HF, Eherer AJ, Petritsch W, Holzer P, Krejs GJ (1998) Intraluminal capsaicin does not affect fluid and electrolyte absorption in the human jejunum but causes pain. *Gut* 43: 252–255
126 Schmidt B, Hammer J, Holzer P, Hammer HF (2004) Chemical nociception in the jejunum induced by capsaicin. *Gut* 53: 1109–1116

Part IV
Vanilloid receptor involvement in disease states

TRPV1 in gut function, abdominal pain and functional bowel disorders

Peter Holzer

Department of Experimental and Clinical Pharmacology, Medical University of Graz, Universitätsplatz 4, A- 8010 Graz, Austria

Capsaicin and the gut: not just a matter of taste

Few would deny that seasoning heightens the joy of a meal, and history tells us that wars have been fought to ensure an unbroken supply of spices. One of the most unusual seasonings is the vanilloid derivative capsaicin, the pungent ingredient in a wide variety of red peppers of the genus *Capsicum* including chilli and jalapeño, given that the sensory experience associated with its intake ranges from pleasant to painful. This is probably a reason why mankind is divided into people who love and those who abhor chilli and related peppers, with a distinct geographical distribution. However, the great divide between pleasant/indifferent and repellent sensations associated with capsaicin is rooted much deeper within the vertebrate phylum. Individuals that have experienced capsaicin powder in the air know that, for a few milliseconds, the attractive smell/taste of vanilla precedes the repellent experience of severe irritation in the eyes, nose and mouth that immediately follows. Birds, on the contrary, are not repelled by capsaicin at all, because the avian ortholog of TRPV1, which represents the so-called capsaicin receptor in mammals, lacks the vanilloid binding site [1]. It may be that capsaicin offers only a favorable vanilla flavor to birds, providing an incentive for these creatures to distribute the seeds of red pepper.

In addition to being a spice, red pepper has also been used as a remedy against various illnesses in traditional medicine. For instance, chilli has been recommended to stimulate appetite, cure indigestion and relieve ulcer pain and has been employed as a carminative in cholera. Now that the implications of TRPV1 in gastrointestinal (GI) function are gradually revealed, it turns out that some of the indications of red pepper in folklore medicine may in fact be related to TRPV1 stimulation or inactivation. This chapter reviews the existing knowledge of the presence and function of TRPV1 in the gut and highlights emerging avenues that identify TRPV1 as a therapeutic target in GI inflammation and functional bowel disorders, including functional dyspepsia and irritable bowel syndrome (IBS).

Turning up the Heat on Pain: TRPV1 Receptors in Pain and Inflammation, edited by Annika B. Malmberg and Keith R. Bley

TRPV1 and the gut: from a tasty receptor for a painful vanilloid to a key player in hyperalgesia

In the past years transient receptor potential (TRP) ion channels have become not only a hot spot in sensory neuroscience and pain research but also in neurogastroenterology. The digestive system is in particular need of sensory surveillance systems [2], and the TRP ion channel superfamily represents a pleiotropic sensory apparatus that reacts to both physical and chemical stimuli from outside and inside the cell [3]. As already alluded to, TRP ion channels are particularly relevant to taste and the taste-related aspects of food intake and digestion. Much as TRPV1 is the receptor for capsaicin, the spicy and irritating ingredient in plants of the genus *Capsicum* [4], TRPA1 (ANKTM1) functions as a receptor for isothiocyanate (mustard oil), the spicy and irritating ingredient in plants of the genus *Brassica* such as mustard, horseradish and wasabi [5]. The pleasant cooling sensation caused by menthol is mediated by TRPM8, which is in keeping with this TRP channel being a sensor for cool temperatures [6, 7]. Other TRP channels enable humans to appreciate sweet, bitter and *umami* tastes [8].

TRPV1 has attracted additional attention because it is activated by a variety of potentially noxious stimuli including heat (> 42 °C), acidosis (< pH 6), ethanol and lipid mediators such as anandamide, *N*-arachidonoyl-dopamine, *N*-oleoyldopamine and oleoylethanolamide, as well as 12-, 15- and 5-lipoxygenase products, including 12-(*S*)-hydroperoxyeicosatetraenoic acid (12-HPETE), 15-HPETE and leukotriene B_4 [9–16]. H^+ ions have long been suspected to be endogenous activators of the so-called capsaicin receptor [17] and, given the high acid concentrations that occur physiologically in the esophagogastroduodenal region, TRPV1 may be an important acid sensor, relevant to the regulation of acid secretion and mucosal homeostasis [18]. However, genetic deletion of TRPV1 fails to modify the excitatory effect of acidosis on nodose ganglion neurons [19], which is in keeping with the concept that the acid sensitivity of afferent neurons involves a variety of H^+-sensitive ion channels [18].

TRPV1 has been envisaged as a polymodal nociceptor because it can be sensitized by various pro-algesic stimuli. Thus, mild acidosis (pH 6–7), ethanol (0.3–3%) and proalgesic stimuli such as prostaglandin E_2, bradykinin, ATP and nerve growth factor can enhance the probability of channel opening in response to heat or capsaicin [14, 20–23]. From these properties it would appear that TRPV1 integrates many noxious stimuli and thus plays a significant role in setting the gain of nociceptors. As acidosis, prostaglandins, bradykinin, ATP and nerve growth factor are formed under conditions of injury and inflammation, they could bring about hyperalgesia by lowering the temperature threshold of TRPV1 so that the channel becomes active at normal body temperature [20, 24]. The relevance of TRPV1 to inflammatory hyperalgesia is borne out by the finding that TRPV1-knockout mice do not develop thermal hyperalgesia in response to experimental inflammation or purinoceptor stimulation [23, 25, 26].

Specific contributions of TRPV1 to GI function in health and disease

Capsaicin as a neuropharmacological tool: demonstration that sensory neurons are important for gut function

Red pepper has been used since ancient times as both a spice and a treatment for GI disease. However, the biological actions of its pungent ingredient were not understood before capsaicin was found to be a selective stimulant of primary afferent neurons. The immediate but transient excitation is followed by a long-lasting desensitization of sensory neurons to capsaicin and other stimuli [27, 28]. Following this discovery, capsaicin became an important tool to probe sensory neuron functions in many organs, including the gut [27–30]. The collective outcome of these studies attests to the important role which afferent neurons play in the control of GI blood flow, mucosal ion transport, mucosal inflammation, mucosal protection, mucosal repair, motor activity and nociception [29, 30]. There is now increasing awareness that hypersensitivity of sensory neurons and a disturbed gut-brain axis contribute to functional bowel disorders such as functional dyspepsia and IBS [2].

Although it is widely assumed that most biological actions of capsaicin are mediated through TRPV1 expressed on sensory neurons, this inference in its generalized form is not true. Firstly, capsaicin acts on a variety of cells other than sensory neurons, and these actions often display characteristics that differ from those of its sensory neuron-selective actions [27]. Secondly, desensitization of sensory neurons to capsaicin involves not only inactivation of TRPV1 but also defunctionalization of the entire afferent neuron [27]. Thus, functional deficits in animals treated with a dose of capsaicin that defunctionalizes afferent neurons cannot be directly linked to TRPV1 [31], and the implication of TRPV1 in physiological or pathological processes needs to be explored by specific inhibition of TRPV1 gene expression or TRPV1 channel activity. This chapter focuses on recent advances that directly point to a role of TRPV1 in GI function in health and disease.

Incomplete match of capsaicin-sensitive afferent neurons with neurons expressing TRPV1

Long before TRPV1 was identified, a subgroup of primary sensory neurons had been classified as 'capsaicin-sensitive afferent neurons' on the basis of their susceptibility to the excitatory and desensitizing action of capsaicin [27, 32]. Although they are heterogeneous in terms of morphology, chemical coding and function, capsaicin-sensitive afferents typically have small to medium-sized cell bodies, possess unmyelinated (C-) fibers and contain a variety of peptide transmitters, among which calcitonin-gene-related peptide and substance P are the most prominent [27]. Visualization of TRPV1 by immunohistochemistry and *in situ* hybridization techniques has confirmed its presumed localization to sensory neurons that originate from the trigeminal, nodose and spinal ganglia and give rise to unmyelinated axons [4, 33–41]. However, this work has also revealed that the population of capsaicin-sensitive afferent neurons, as defined in neurochemical, neurophysiological and neuropharmacological investigations, does not completely match with the population of neurons that express TRPV1-like immunoreactivity (TRPV1-LI) or TRPV1 mRNA.

One remarkable aspect of this mismatch is that TRPV1 is much more widely distributed than envisaged from the functional studies and, for instance, is expressed by many neurons in the brain [42]. Another example of mismatch concerns the chemical coding of TRPV1-expressing sensory neurons. Capsaicin-sensitive afferents have been characterized to be largely peptidergic, as they contain and release calcitonin-gene-related peptide and substance P [27, 28, 43, 44], whereas TRPV1-positive dorsal root ganglion neurons rather lack these neuropeptides [34]. There is, however, a substantial co-expression of calcitonin-gene-related peptide and TRPV1 in visceral sensory neurons [38]. Thus, calcitonin-gene-related peptide is found in 81–99% of the spinal afferent neurons innervating the gut and other viscera of mouse, rat and guinea pig [39, 43, 45], compared with 71–82% that express TRPV1 [39, 40].

A further instance of mismatch concerns the expression of TRPV1 in the GI tract. Functional studies have indicated that the population of capsaicin-sensitive afferent neurons constitutes extrinsic primary afferent neurons exclusively, whereas intrinsic enteric and extrinsic autonomic neurons do not directly respond to capsaicin [27, 29, 30]. Accordingly, Patterson et al. [37], Ward et al. [38] and Schicho et al. [40] have failed to detect TRPV1-LI in enteric neurons of the rat, guinea pig and mouse GI tract. Consequently, these authors conclude that the numerous TRPV1-positive nerve fibers that occur in the enteric nerve plexuses, musculature and mucosa represent processes of spinal afferents and, in the stomach, of some vagal afferents [37, 38, 40, 46]. This conjecture is strongly supported by the disappearance of TRPV1 mRNA from the rat gastric wall following total extrinsic denervation of the stomach [40]. However, other investigators have reported that TRPV1-LI is expressed by enteric neurons of the guinea pig, porcine and human intestine [47–51]. In addition, TRPV1 mRNA, TRPV1 protein and TRPV1-like binding sites have been found on rat gastric epithelial cells [52, 53], which is reminiscent of the presence of TRPV1 on epithelial cells of the urinary bladder (where this ion channel acts as a sensor relevant to bladder function) [54, 55]. The disparity in the findings related to the gut awaits to be elucidated, much as it will be important to find out the molecular identity of TRPV1-LI contained in enteric neurons and epithelial cells. Capsaicin has been reported to excite enteric neurons [56], and it appears worth re-investigating this effect, which has been thought to be due to transmitter release from extrinsic afferents.

Double-edged implications of TRPV1 in GI mucosal functions

Extensive studies involving capsaicin have demonstrated that capsaicin-sensitive afferent neurons participate in the regulation of GI circulation, secretion, mucosal homeostasis, motility and nociception [29, 30]. These tasks of sensory neurons are brought about by two different modes of operation: an afferent and an efferent-like function [57–59]. By conveying information from the gut to the spinal cord and brainstem, capsaicin-sensitive afferents contribute to GI sensation and constitute the afferent arm of autonomic and neuroendocrine reflex circuits relevant to digestion. Other sensory neurons subserve a local efferent-like function in the gut, which is brought about by the release of calcitonin-gene-related peptide, substance P and other mediators from their peripheral fibers, these transmitters in turn acting on GI effector systems [57, 58].

This efferent-like mode of operation is exemplified by the reactions of the rat gastric mucosa to luminal capsaicin exposure or acid back-diffusion [60, 61]. Both stimuli increase mucosal blood flow through a peripheral mechanism of action and initiate other mechanisms of defense, such as bicarbonate and mucus secretion (Fig. 1) [2, 30]. In addition, capsaicin-sensitive afferent nerve fibers participate in the feedback regulation of gastric acid secretion: as they are activated by accumulation of secreted acid, they release calcitonin-gene-related peptide which, via somatostatin release, halts further acid secretion [30, 62]. While the gastric hyperemic response to luminal acid back-diffusion is mediated by spinal afferents [63], the afferent signaling of gastric acid challenge to the brain is carried by vagal afferents [64, 65]. These observations indicate that the local efferent-like and afferent functions in response to gastric acid are subserved by different populations of sensory neurons [59].

The capsaicin-evoked gastric hyperemia [66], gastric mucosal protection [46, 53] and duodenal bicarbonate secretion [67] in the rat are likely to be mediated by TRPV1, given that they are antagonized by the TRPV1 blocker capsazepine. The gastric mucus secretion evoked by stimulation of protease-activated receptor-2 (PAR2) is likewise blunted by capsazepine [68], which suggests a link between PAR2 and TRPV1 (Tab. 1). In contrast, it is less clear whether the reactions of the rat gastroduodenal mucosa to luminal acid back-diffusion, which are mediated by capsaicin-sensitive afferent neurons, involve TRPV1. Whereas the gastric mucosal hyperemia evoked by acid back-diffusion remains unaltered by capsazepine [66], the duodenal vasodilatation caused by exposure to luminal acid is blunted by this TRPV1 blocker [69]. On the contrary, the acid-evoked secretion of duodenal bicarbonate is left unchanged by capsazepine [67]. It thus seems as if there are regional differences in the receptor mechanisms whereby acid challenge activates afferent neurons in the foregut, a conjecture that is in keeping with the multiplicity of acid-sensing ion channels expressed by sensory neurons [18]. An approach to solve this issue would be to examine acid-induced GI mucosal reactions in TRPV1-knockout

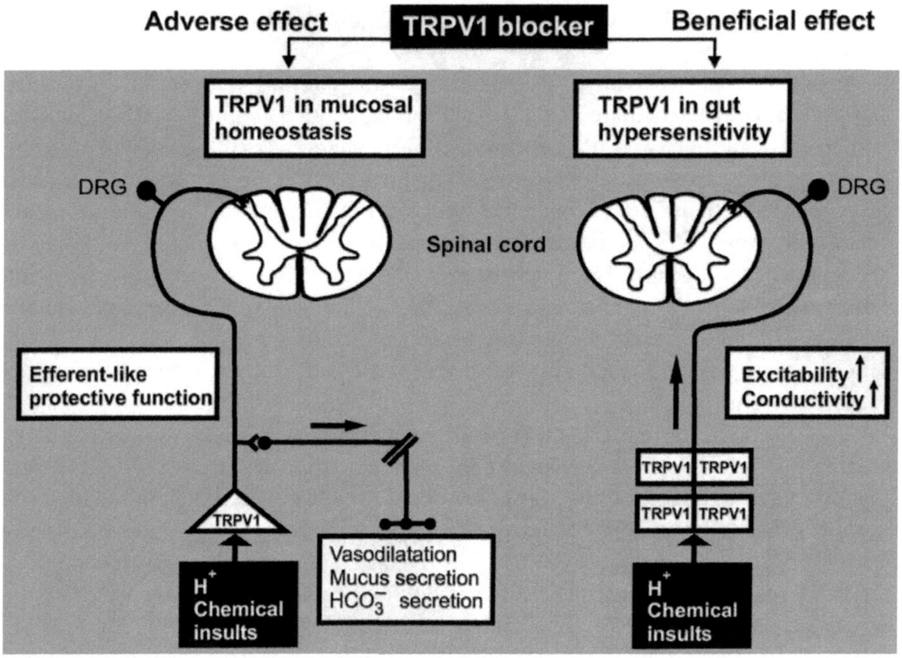

Figure 1
Double-edged role of TRPV1 in the maintenance of GI mucosal homeostasis and in the establishment of GI hyperalgesia. Being a polymodal detector of injurious chemicals, TRPV1 on sensory neurons with an efferent-like function is involved in the alarm of local protective reactions in the GI mucosa. The abdominal hypersensitivity associated with functional bowel disorders may in part be due to up-regulation and functional modification of TRPV1 on spinal afferents that signal to the central nervous system. TRPV1 blockers may thus have both beneficial and adverse effects on GI function, reducing hypersensitivity at the expense of increased mucosal vulnerability. DRG, dorsal root ganglion.

mice in which the stimulant action of capsaicin on vagal and spinal afferents is totally absent [19, 25, 26].

Whereas TRPV1 in the foregut mediates reactions that support mucosal homeostasis, there is evidence that TRPV1 in the mouse pancreas, rat ileum and rat colon facilitates processes of inflammation and tissue damage (Tab. 1). Thus, experimental pancreatitis induced by caerulein [70], ileitis induced by *Clostridium difficile* toxin A [71] and colitis induced by dextrane sulphate [72] are significantly ameliorated by capsazepine. Although it is yet to be disclosed how these inflammatory stimuli are linked to TRPV1, there is evidence that *C. difficile* toxin A enhances the formation of anandamide and 2-arachidonoylglycerol in the mucosa and that these

Table 1 - Implications of TRPV1 in gastrointestinal functions

GI function	Type of evidence	Ref.
Plasma protein extravasation in the rat esophagus evoked by ethanol	Inhibition by capsazepine	[14]
Mucus secretion in the rat stomach evoked by stimulation of PAR2	Inhibition by capsazepine	[68]
Vasodilatation in the rat duodenum caused by exposure to luminal acid	Inhibition by capsazepine	[69]
Inflammation of mouse pancreas induced by caerulein	Inhibition by capsazepine	[70]
Inflammation of rat ileum caused by *C. difficile* toxin A	Inhibition by capsazepine	[71]
Inflammation of rat ileum caused by anandamide	Inhibition by capsazepine	[73]
Inflammation of rat colon caused by dextrane sulphate sodium	Inhibition by capsazepine	[72]

PAR2, protease-activated receptor-2.

mediators, in turn, stimulate TRPV1 [73]. Activation of TRPV1 excites sensory nerve terminals in the mucosa, which leads to release of substance P, causes activation of enteric neurons and immune cells and ultimately results in hypersecretion, inflammation and mucosal damage [71, 73].

Contribution of TRPV1 to GI nociception

The molecular characteristics of TRPV1 as a polymodal nociceptor and its association with nociceptive afferent nerve fibers attribute this ion channel a particular role in pain and hyperalgesia. The gut is innervated by two populations of extrinsic afferents, vagal and spinal, both of which express TRPV1 [4, 33–41]. Of the nodose ganglion neurons that innervate the rat stomach, 42–80% stain for TRPV1, whereas 71–82% of the dorsal root ganglion neurons projecting to the rat stomach and mouse colon express TRPV1 [37, 39, 40]. However, the level of TRPV1-LI present in the peripheral nerve fibers differs considerably between vagal and spinal afferents. Thus, most TRPV1-positive nerve fibers in the gut appear to be processes of spinal afferents [37, 38, 40], whereas the level of TRPV1-LI in the peripheral fibers of vagal afferents is so low that it is difficult to detect [37]. This may explain why the proportion of capsaicin-sensitive fibers among vagal afferents supplying the esophagus and stomach is 30% [74–76].

There is ample evidence that capsaicin, most likely through gating of TRPV1, stimulates extrinsic afferents of the gut [75–78] and that administration of capsaicin into the alimentary canal gives rise to pain in humans [79–84] and mice [85, 86]. For instance, infusion of a red pepper sauce containing capsaicin (0.14 mM) into the human esophagus elicits heartburn and lowers the threshold for distension-induced perception and discomfort, apart from accelerating esophageal peristalsis and slowing gastric emptying [79]. Given that acid and ethanol can gate TRPV1 [14, 20], it has been speculated that heartburn in patients with gastroesophageal reflux disease may involve activation of TRPV1 by refluxing acid or swallowing of alcohol or hot food [31]. Indeed, when heartburn patients take capsaicin (0.016 mmol) in gelatin capsules before a test meal, the time to peak heartburn is significantly shortened, whereas dyspepsia, esophageal and gastric pH profiles as well as gastric emptying remain unaltered [81].

Infusion of capsaicin (0.13–0.65 mM) into the human jejunum induces pain whose abdominal localization and perceptional quality are similar to distension-induced pain [84]. Since it does not stimulate jejunal motility and does not alter jejunal mechanosensitivity, capsaicin is thought to evoke pain by stimulation of jejunal chemoreceptors, presumably TRPV1 [84]. Similarly, administration of a capsaicin solution (0.16 mM) into the ileum of patients with an ileostoma evokes abdominal pain, as well as referred somatic pain [83]. The effects of capsaicin-containing red pepper preparations have also been tested in IBS patients, given that spicy meals are said to exacerbate IBS symptoms. Acute ingestion of an intermediate chilli dose containing 46 mmol of capsaicin as single treatment does not alter the discomfort and pain thresholds for rectosigmoid distension in IBS patients, but lowers the pain threshold in healthy controls [87]. In contrast, intake of a low chilli dose containing 0.33 mmol of capsaicin with each meal for 1 week has been found to evoke upper abdominal discomfort and to decrease the rectal pain threshold for distension, a change that is not related to rectal compliance [88]. Ingestion of a high-chilli diet (15 g per day) for 3 days has inconsistent effects on IBS symptoms and does not alter rectosigmoid motility [89]. From these observations it would seem that repeated ingestion of a relatively low capsaicin dose enhances the rectosigmoid pain sensitivity of IBS patients, but a systematic investigation as to how IBS symptoms are related to dose, form and duration of the capsaicin treatment has not yet been carried out [88].

Direct evidence for a role of TRPV1 in the pain associated with GI disease has not yet been provided, but this has been increasingly implied in both experimental and clinical studies. While experimental paradigms of visceral hypersensitivity await to be tested in TRPV1-deficient mice, it has been shown that TRPV1-knockout mice, which fail to exhibit any vanilloid-evoked pain behavior, do not develop inflammation-induced cutaneous hypersensitivity to noxious heat [25, 26]. This observation, which demonstrates that TRPV1 is relevant to sensitization of dermal afferents, is matched by indirect evidence that TRPV1 contributes to sensitization of GI afferents. Thus, acute administration of capsaicin into the ileum of patients with an ileostoma has been reported to cause mechanical hypersensitivity [83], much as the ingestion of a low capsaicin dose for 1 week induces upper-abdominal discomfort and decreases rectal pain threshold for distension in IBS patients [88].

Furthermore, experimental studies suggest that stimulation of PAR2 in the GI tract, which brings about a delayed hypersensitivity to colorectal distension [90], is linked to activation of TRPV1. In particular, application of a PAR2 agonist into the pancreatic duct sensitizes spinal afferents to the excitatory effect of capsaicin [91], and nociception caused by intracolonic capsaicin is facilitated following intraperitoneal administration of a PAR2 agonist to mice [86]. The molecular mechanism behind the interaction between PAR2 and TRPV1 [68, 86, 91] involves protein kinase C, which phosphorylates TRPV1 and thereby sensitizes it to activation by other stimuli [92, 93]. Since trypsin and tryptase are released during tissue inflammation, it would appear that PAR2-mediated sensitization of TRPV1 could be an important mechanism underlying inflammatory hyperalgesia.

Further indirect evidence for a role of TRPV1 in abdominal hyperalgesia comes from reports that capsaicin desensitization is beneficial in patients with functional dyspepsia or an irritable (overactive) bladder. This approach is based on the ability of capsaicin to induce a state of sensory refractoriness [27] which, depending on the dose of capsaicin, may be due to desensitization (inactivation) of TRPV1 [94–98], down-regulation of TRPV1 [28], loss of sensory neuron excitability or Ca^{2+}-related neurotoxicity for sensory neurons [27, 28]. Such a state of functional desensitization can be achieved by systemic administration of high doses of capsaicin to experimental animals or by repeated topical administration of moderate doses of capsaicin to humans. Sensory refractoriness is usually reversible after several weeks to months, unless very high doses of capsaicin are administered systemically to cause permanent neurotoxicity. Capsaicin pretreatment of rats blocks the visceromotor response to gastric acid challenge [65], suppresses the cardiovascular pain response to noxious jejunal distension in the rat [99] and prevents the inflammation-induced hypersensitivity to colonic distension [100, 101]. Pretreatment of rats with SDZ 249-665, a vanilloid compound reproducing capsaicin desensitization, attenuates inflammatory bladder hyperreflexia, referred hyperalgesia [102] and behavioral pain responses to intraperitoneal acetic acid in rats [103].

Chronic administration of capsaicin is likewise beneficial in patients with abdominal pain. For instance, intravesicular administration of capsaicin or resiniferatoxin has been found to reverse urinary bladder hyperreflexia in humans [104]. Intractable idiopathic pruritus ani can be relieved by a 4-week treatment course with topical capsaicin [105], and daily intragastric administration of red pepper containing 5.7 mmol of capsaicin for 5 weeks significantly reduces epigastric pain and other symptoms of functional dyspepsia [82]. However, the initial phase of red pepper administration, when capsaicin is still stimulating afferent neurons, is often associated with an exacerbation of dyspeptic and IBS symptoms [82, 87, 88].

Alterations of TRPV1 expression and function in GI disease

Consistent with a role in GI pain and hyperalgesia is that TRPV1-LI on submucosal nerve fibers in the colon is enhanced in patients with painful inflammatory bowel disease [106]. The density of TRPV1-LI in extrinsic nerve bundles of the aganglionic bowel in Hirschsprung's disease is likewise increased [107]. Rectal hypersensitivity and fecal urgency are associated with a rise in the number of TRPV1-positive nerve fibers in the muscle, submucosa and mucosa of the rectum and of TRPV1-positive neurons in the myenteric and submucosal plexus [51].

There is experimental evidence that inflammation leads to up-regulation of TRPV1 expression and function (Fig. 1), a process in which nerve growth factor plays a particular role. *In vitro*, neurotrophic factors can augment the expression of TRPV1 mRNA in cultured sensory neurons and enhance their capsaicin sensitivity [108, 109], whereas *in vivo* nerve growth factor acts mostly by a post-transcriptional mechanism [110]. Thus, cutaneous inflammation induced by complete Freund's adjuvant leads, via nerve growth factor formation, to activation of the mitogen-activated protein kinase p38 in the somata of dorsal root ganglion neurons, which in turn increases the protein but not mRNA levels of TRPV1. The TRPV1 protein is then transported into the peripheral terminals of sensory neurons where it contributes to the maintenance of inflammatory heat hyperalgesia [110]. A similar process seems to take place in the rat gastric mucosa exposed to a noxious HCl concentration [40]. Six hours post-challenge, TRPV1 protein (but not TRPV1 mRNA) is enhanced in dorsal root ganglion neurons innervating the stomach, whereas TRPV1 mRNA and protein in the nodose ganglia stay unchanged [40].

Therapeutic options provided by TRPV1 channel blockers: a double-edged approach to abdominal pain?

The polymodal nociceptor properties of TRPV1 make these ion channel an intriguing target for novel therapies of abdominal pain and inflammation. From a therapeutic perspective it would appear, therefore, that TRPV1 channel blockers may be of substantial value in suppressing GI hyperalgesia related to inflammation and other circumstances where there is activation (sensitization) or up-regulation of TRPV1 (Fig. 1). That such an approach is worth pursuing is emphasized by the beneficial effect of chronic capsaicin desensitization in functional dyspepsia and overactive bladder [82, 104]. However, a disadvantage of this approach is the initial pungency that capsaicin and related TRPV1 agonists bring about. Early attempts to circumvent this problem led to the use and development of vanilloid-related compounds with reduced pungency but preserved ability to desensitize. Various capsaicin analogs differ in the relative ratio of initial excitation to subsequent refractoriness, with resiniferatoxin and particularly the capsaicin analog SDZ 249-665 being examples that hardly stimulate but effectively desensitize afferent neurons [28, 102, 103].

Clearly, selective TRPV1 channel blockers would lack the unwanted property of pungency, and the feasibility to antagonize capsaicin was proved by the discovery of the first TRPV1 blocker, capsazepine [111]. Whereas the selectivity of this compound for TRPV1 channels is limited [112], the new knowledge of the membrane topology of TRPV1 and its function-relevant domains provides a good basis for the design of highly specific channel blockers. Indeed, the search is on, and several compounds have recently been reported to behave as TRPV1 channel blockers, which are thought to provide novel drugs to fight persistent and chronic pain arising from inflammation or nerve injury [31, 113]. This indication is very likely to comprise visceral pain, although the utility of TRPV1 blockers has not yet been ascertained in established paradigms of GI hyperalgesia.

One important caveat, however, needs be considered. The physiological role of capsaicin-sensitive afferent neurons is to monitor the physical and chemical condition of the GI tract and to provide essential information for the local, autonomic and central regulation of digestive activity in health and disease [2, 30]. Owing to their sensory modalities, capsaicin-sensitive afferent neurons constitute a neural alarm system which helps to maintain GI mucosal homeostasis in the face of pending injury (Fig. 1), especially in the foregut [30]. As TRPV1 seems to be a relevant sensor involved in these tasks [46, 53, 66–69], blockade of TRPV1 may interfere with the physiological function of sensory neurons at the expense of an increased vulnerability of the GI mucosa. In addressing this issue it will be important to find out whether TRPV1 involved in GI mucosal protection can pharmacologically be differentiated from TRPV1 mediating GI hyperalgesia in inflammatory and functional bowel diseases [2].

Acknowledgements
The help of Evelin Painsipp in preparing the figure is greatly appreciated. Work in the author's laboratory is supported by the Austrian Research Funds, the Jubilee Funds of the Austrian National Bank, the Austrian Federal Ministry of Education, Science and Culture and the Zukunftsfonds Steiermark.

References

1 Jordt SE, Julius D (2002) Molecular basis for species-specific sensitivity to "hot" chili peppers. *Cell* 108: 421–430
2 Holzer P (2002) Sensory neurone responses to mucosal noxae in the upper gut: relevance to mucosal integrity and gastrointestinal pain. *Neurogastroenterol Motil* 14: 459–475

3 Clapham DE (2003) TRP channels as cellular sensors. *Nature* 426: 517–524
4 Caterina MJ, Schumacher MA, Tominaga M, Rosen TA, Levine JD, Julius D (1997) The capsaicin receptor: a heat-activated ion channel in the pain pathway. *Nature* 389: 816–824
5 Jordt SE, Bautista DM, Chuang HH, McKemy DD, Zygmunt PM, Högestätt ED, Meng ID, Julius D (2004) Mustard oils and cannabinoids excite sensory nerve fibers through the TRP channel ANKTM1. *Nature* 427: 260–265
6 McKemy DD, Neuhausser WM, Julius D (2002) Identification of a cold receptor reveals a general role for TRP channels in thermosensation. *Nature* 416: 52–58
7 Peier AM, Moqrich A, Hergarden AC, Reeve AJ, Andersson DA, Story GM, Earley TJ, Dragoni I, McIntyre P, Bevan S, Patapoutian A (2002) A TRP channel that senses cold stimuli and menthol. *Cell* 108: 705–715
8 Zhang Y, Hoon MA, Chandrashekar J, Mueller KL, Cook B, Wu D, Zuker CS, Ryba NJ (2003) Coding of sweet, bitter, and umami tastes: different receptor cells sharing similar signaling pathways. *Cell* 112: 293–301
9 Caterina MJ, Julius D (2001) The vanilloid receptor: a molecular gateway to the pain pathway. *Annu Rev Neurosci* 24: 487–517
10 Di Marzo V, Blumberg PM, Szallasi A (2002) Endovanilloid signaling in pain. *Curr Opin Neurobiol* 12: 372–379
11 Gunthorpe MJ, Benham CD, Randall A, Davis JB (2002) The diversity in the vanilloid (TRPV) receptor family of ion channels. *Trends Pharmacol Sci* 23: 183–191
12 Huang SM, Bisogno T, Trevisani M, Al-Hayani A, De Petrocellis L, Fezza F, Tognetto M, Petros TJ, Krey JF, Chu CJ et al (2002) An endogenous capsaicin-like substance with high potency at recombinant and native vanilloid VR1 receptors. *Proc Natl Acad Sci USA* 99: 8400–8405
13 Hwang SW, Oh U (2002) Hot channels in airways: pharmacology of the vanilloid receptor. *Curr Opin Pharmacol* 2: 235–242
14 Trevisani M, Smart D, Gunthorpe MJ, Tognetto M, Barbieri M, Campi B, Amadesi S, Gray J, Jerman JC, Brough SJ et al (2002) Ethanol elicits and potentiates nociceptor responses via the vanilloid receptor-1. *Nat Neurosci* 5: 546–551
15 Ahern GP (2003) Activation of TRPV1 by the satiety factor oleoylethanolamide. *J Biol Chem* 278: 30429–30434
16 Chu CJ, Huang SM, De Petrocellis L, Bisogno T, Ewing SA, Miller JD, Zipkin RE, Daddario N, Appendino G, Di Marzo V, Walker JM (2003) *N*-oleoyldopamine, a novel endogenous capsaicin-like lipid that produces hyperalgesia. *J Biol Chem* 278: 13633–13639
17 Bevan S, Geppetti P (1994) Protons: small stimulants of capsaicin-sensitive sensory nerves. *Trends Neurosci* 17: 509–512
18 Holzer P (2003) Acid-sensitive ion channels in gastrointestinal function. *Curr Opin Pharmacol* 3: 618–625
19 Kollarik M, Undem BJ (2004) Activation of bronchopulmonary vagal afferent nerves

with bradykinin, acid and vanilloid receptor agonists in wild-type and TRPV1$^{-/-}$ mice. *J Physiol (London)* 555: 115–123
20 Tominaga M, Caterina M, Malmberg AB, Rosen TA, Gilbert H, Skinner K, Raumann BE, Basbaum AI, Julius D (1998) The cloned capsaicin receptor integrates multiple painproducing stimuli. *Neuron* 21: 531–543
21 Premkumar LS, Ahern GP (2000) Induction of vanilloid receptor channel activity by protein kinase C. *Nature* 408: 985–990
22 Chuang HH, Prescott ED, Kong H, Shields S, Jordt SE, Basbaum AI, Chao MV, Julius D (2001) Bradykinin and nerve growth factor release the capsaicin receptor from PtdIns(4,5) P2-mediated inhibition. *Nature* 411: 957–962
23 Moriyama T, Iida T, Kobayashi K, Higashi T, Fukuoka T, Tsumura H, Leon C, Suzuki N, Inoue K, Gachet C et al (2003) Possible involvement of P2Y2 metabotropic receptors in ATP- induced transient receptor potential vanilloid receptor 1-mediated thermal hypersensitivity. *J Neurosci* 23: 6058–6062
24 Reeh PW, Pethö G (2000) Nociceptor excitation by thermal sensitization – a hypothesis. *Prog Brain Res* 129: 39–50
25 Caterina MJ, Leffler A, Malmberg AB, Martin WJ, Trafton J, Petersen-Zeitz KR, Koltzenburg M, Basbaum AI, Julius D (2000) Impaired nociception and pain sensation in mice lacking the capsaicin receptor. *Science* 288: 306–313
26 Davis JB, Gray J, Gunthorpe MJ, Hatcher JP, Davey PT, Overend P, Harries MH, Latcham J, Clapham C, Atkinson K et al (2000) Vanilloid receptor-1 is essential for inflammatory thermal hyperalgesia. *Nature* 405: 183–187
27 Holzer P (1991) Capsaicin: cellular targets, mechanisms of action, and selectivity for thin sensory neurons. *Pharmacol Rev* 43: 143–201
28 Szallasi A, Blumberg PM (1999) Vanilloid (capsaicin) receptors and mechanisms. *Pharmacol Rev* 51: 159–212
29 Holzer P, Barthó L (1996). Sensory neurons in the intestine. In: P Geppetti, P Holzer (eds): *Neurogenic inflammation.* CRC Press, Boca Raton, 153–167
30 Holzer P (1998) Neural emergency system in the stomach. *Gastroenterology* 114: 823–839
31 Geppetti P, Trevisani M (2004) Activation and sensitisation of the vanilloid receptor: role in gastrointestinal inflammation and function. *Br J Pharmacol* 141: 1313–1320
32 Szolcsányi J (1982) Capsaicin type pungent agents producing pyrexia. In: AS Milton (ed): *Pyretics and Antipyretics, Handbook of Experimental Pharmacology*, Volume 60. Springer, Berlin, 437–478
33 Helliwell RJ, McLatchie LM, Clarke M, Winter J, Bevan S, McIntyre P (1998) Capsaicin sensitivity is associated with the expression of the vanilloid (capsaicin) receptor (VR1) mRNA in adult rat sensory ganglia. *Neurosci Lett* 250: 177–180
34 Guo A, Vulchanova L, Wang J, Li X, Elde R (1999) Immunocytochemical localization of the vanilloid receptor 1 (VR1): relationship to neuropeptides, the P2X3 purinoceptor and IB_4 binding sites. *Eur J Neurosci* 11: 946–958

35 Michael GJ, Priestley JV (1999) Differential expression of the mRNA for the vanilloid receptor subtype 1 in cells of the adult rat dorsal root and nodose ganglia and its down-regulation by axotomy. *J Neurosci* 19: 1844–1854
36 Ichikawa H, Sugimoto T (2003) The co-expression of VR1 and VRL-1 in the rat vagal sensory ganglia. *Brain Res* 980: 293–296
37 Patterson LM, Zheng H, Ward SM, Berthoud HR (2003) Vanilloid receptor (VR1) expression in vagal afferent neurons innervating the gastrointestinal tract. *Cell Tissue Res* 311: 277–287
38 Ward SM, Bayguinov J, Won KJ, Grundy D, Berthoud HR (2003) Distribution of the vanilloid receptor (VR1) in the gastrointestinal tract. *J Comp Neurol* 465: 121–135
39 Robinson DR, McNaughton PA, Evans ML, Hicks GA (2004) Characterization of the primary spinal afferent innervation of the mouse colon using retrograde labelling. *Neurogastroenterol Motil* 16: 113–124
40 Schicho R, Florian W, Liebmann I, Holzer P, Lippe IT (2004) Increased expression of TRPV1 receptor in dorsal root ganglia by acid insult of the rat gastric mucosa. *Eur J Neurosci* 19: 1811–1818
41 Zhang L, Jones S, Brody K, Costa M, Brookes SJ (2004) Thermosensitive transient receptor potential channels in vagal afferent neurons of the mouse. *Am J Physiol Gastrointest Liver Physiol* 286: G983–G991
42 Mezey E, Toth ZE, Cortright DN, Arzubi MK, Krause JE, Elde R, Guo A, Blumberg PM, Szallasi A (2000) Distribution of mRNA for vanilloid receptor subtype 1 (VR1), and VR1-like immunoreactivity, in the central nervous system of the rat and human. *Proc Natl Acad Sci USA* 97: 3655–3660
43 Green T, Dockray GJ (1988) Characterization of the peptidergic afferent innervation of the stomach in the rat, mouse, and guinea-pig. *Neuroscience* 25: 181–193
44 Sternini C (1992) Enteric and visceral afferent CGRP neurons. Targets of innervation and differential expression patterns. *Ann NY Acad Sci* 657: 170–186
45 Perry MJ, Lawson SN (1998) Differences in expression of oligosaccharides, neuropeptides, carbonic anhydrase and neurofilament in rat primary afferent neurons retrogradely labelled via skin, muscle or visceral nerves. *Neuroscience* 85: 293–310
46 Horie S, Yamamoto H, Michael GJ, Uchida M, Belai A, Watanabe K, Priestley JV, Murayama T (2004) Protective role of vanilloid receptor type 1 in HCl- induced gastric mucosal lesions in rats. *Scand J Gastroenterol* 39: 303–312
47 Kulkarni-Narla A, Brown DR (2001) Opioid, cannabinoid and vanilloid receptor localization on porcine cultured myenteric neurons. *Neurosci Lett* 308: 153–156
48 Anavi-Goffer S, McKay NG, Ashford ML, Coutts AA (2002) Vanilloid receptor type 1-immunoreactivity is expressed by intrinsic afferent neurones in the guinea-pig myenteric plexus. *Neurosci Lett* 319: 53–57
49 Poonyachoti S, Kulkarni-Narla A, Brown DR (2002) Chemical coding of neurons expressing delta- and kappa-opioid receptor and type I vanilloid receptor immunoreactivities in the porcine ileum. *Cell Tissue Res* 307: 23–33

50 Anavi-Goffer S, Coutts AA (2003) Cellular distribution of vanilloid VR1 receptor immunoreactivity in the guinea-pig myenteric plexus. *Eur J Pharmacol* 458: 61–71
51 Chan CL, Facer P, Davis JB, Smith GD, Egerton J, Bountra C, Williams NS, Anand P (2003) Sensory fibers expressing capsaicin receptor TRPV1 in patients with rectal hypersensitivity and fecal urgency. *Lancet* 361: 385–391
52 Nozawa Y, Nishihara K, Yamamoto A, Nakano M, Ajioka H, Matsuura N (2001) Distribution and characterization of vanilloid receptors in the rat stomach. *Neurosci Lett* 309: 33–36
53 Kato S, Aihara E, Nakamura A, Xin H, Matsui H, Kohama K, Takeuchi K (2003) Expression of vanilloid receptors in rat gastric epithelial cells: role in cellular protection. *Biochem Pharmacol* 66: 1115–1121
54 Birder LA, Kanai AJ, de Groat WC, Kiss S, Nealen ML, Burke NE, Dineley KE, Watkins S, Reynolds IJ, Caterina MJ (2001) Vanilloid receptor expression suggests a sensory role for urinary bladder epithelial cells. *Proc Natl Acad Sci USA* 98: 13396–13401
55 Birder LA, Nakamura Y, Kiss S, Nealen ML, Barrick S, Kanai AJ, Wang E, Ruiz G, De Groat WC, Apodaca G et al (2002) Altered urinary bladder function in mice lacking the vanilloid receptor TRPV1. *Nat Neurosci* 5: 856–860
56 Takaki M, Nakayama S (1989) Effects of capsaicin on myenteric neurons of the guinea pig ileum. *Neurosci Lett* 105: 125–130
57 Holzer P (1988) Local effector functions of capsaicin-sensitive sensory nerve endings: involvement of tachykinins, calcitonin gene-related peptide and other neuropeptides. *Neuroscience* 24: 739–768
58 Maggi CA (1995) Tachykinins and calcitonin gene-related peptide (CGRP) as co- transmitters released from peripheral endings of sensory nerves. *Prog Neurobiol* 45: 1–98
59 Holzer P, Maggi CA (1998) Dissociation of dorsal root ganglion neurons into afferent and efferent-like neurons. *Neuroscience* 86: 389–398
60 Holzer P, Livingston EH, Guth PH (1991) Sensory neurons signal for an increase in rat gastric mucosal blood flow in the face of pending acid injury. *Gastroenterology* 101: 416–423
61 Holzer P, Livingston EH, Saria A, Guth PH (1991) Sensory neurons mediate protective vasodilatation in rat gastric mucosa. *Am J Physiol Gastrointest Liver Physiol* 260: G363–G370
62 Manela FD, Ren J, Gao J, McGuigan JE, Harty RF (1995) Calcitonin gene-related peptide modulates acid- mediated regulation of somatostatin and gastrin release from rat antrum. *Gastroenterology* 109: 701–706
63 Raybould HE, Sternini C, Eysselein VE, Yoneda M, Holzer P (1992) Selective ablation of spinal afferent neurons containing CGRP attenuates gastric hyperemic response to acid. *Peptides* 13: 249–254
64 Schuligoi R, Jocic M, Heinemann A, Schöninkle E, Pabst MA, Holzer P (1998) Gastric acid-evoked c-fos messenger RNA expression in rat brainstem is signaled by capsaicin-resistant vagal afferents. *Gastroenterology* 115: 649–660

65 Lamb K, Kang YM, Gebhart GF, Bielefeldt K (2003) Gastric inflammation triggers hypersensitivity to acid in awake rats. *Gastroenterology* 125: 1410–1418

66 Tashima K, Nakashima M, Kagawa S, Kato S, Takeuchi K (2002) Gastric hyperemic response induced by acid back-diffusion in rat stomachs following barrier disruption - relation to vanilloid type-1 receptors. *Med Sci Monit* 8: BR157–BR163

67 Kagawa S, Aoi M, Kubo Y, Kotani T, Takeuchi K (2003) Stimulation by capsaicin of duodenal HCO3-secretion via afferent neurons and vanilloid receptors in rats: comparison with acid-induced HCO3-response. *Dig Dis Sci* 48: 1850–1856

68 Kawabata A, Kinoshita M, Kuroda R, Kakehi K (2002) Capsazepine partially inhibits neurally mediated gastric mucus secretion following activation of protease-activated receptor. *Clin Exp Pharmacol Physiol* 29: 360–361

69 Akiba Y, Guth PH, Engel E, Nastaskin I, Kaunitz JD (1999) Acid-sensing pathways of rat duodenum. *Am J Physiol Gastrointest Liver Physiol* 277: G268–G274

70 Nathan JD, Patel AA, McVey DC, Thomas JE, Prpic V, Vigna SR, Liddle RA (2001) Capsaicin vanilloid receptor-1 mediates substance P release in experimental pancreatitis. *Am J Physiol Gastrointest Liver Physiol* 281: G1322–G1328

71 McVey DC, Vigna SR (2001) The capsaicin VR1 receptor mediates substance P release in toxin A-induced enteritis in rats. *Peptides* 22: 1439–1446

72 Kihara N, De La Fuente SG, Fujino K, Takahashi T, Pappas TN, Mantyh CR (2003) Vanilloid receptor-1 containing primary sensory neurones mediate dextran sulphate sodium-induced colitis in rats. *Gut* 52: 713–719

73 McVey DC, Schmid PC, Schmid HH, Vigna SR (2003) Endocannabinoids induce ileitis in rats via the capsaicin receptor (VR1). *J Pharmacol Exp Ther* 304: 713–722

74 Berthoud HR, Patterson LM, Willing AE, Mueller K, Neuhuber WL (1997) Capsaicin-resistant vagal afferent fibers in the rat gastrointestinal tract: anatomical identification and functional integrity. *Brain Res* 746: 195–206

75 Blackshaw LA, Page AJ, Partosoedarso ER (2000) Acute effects of capsaicin on gastrointestinal vagal afferents. *Neuroscience* 96: 407–416

76 Berthoud HR, Lynn PA, Blackshaw LA (2001) Vagal and spinal mechanosensors in the rat stomach and colon have multiple receptive fields. *Am J Physiol Regul Integr Comp Physiol* 280: R1371–R1381

77 Maubach KA, Grundy D (1999) The role of prostaglandins in the bradykinin-induced activation of serosal afferents of the rat jejunum *in vitro*. *J Physiol (London)* 515: 277–285

78 Su X, Wachtel RE, Gebhart GF (1999) Capsaicin sensitivity and voltage-gated sodium currents in colon sensory neurons from rat dorsal root ganglia. *Am J Physiol Gastrointest Liver Physiol* 277: G1180–G1188

79 Gonzalez R, Dunkel R, Koletzko B, Schusdziarra V, Allescher HD (1998) Effect of capsaicin-containing red pepper sauce suspension on upper gastrointestinal motility in healthy volunteers. *Dig Dis Sci* 43: 1165–1171

80 Hammer J, Hammer HF, Eherer AJ, Petritsch W, Holzer P, Krejs GJ (1998) Intraluminal

capsaicin does not affect fluid and electrolyte absorption in the human jejunum but does cause pain. *Gut* 43: 252–255
81 Rodriguez-Stanley S, Collings KL, Robinson M, Owen W, Miner PB (2000) The effects of capsaicin on reflux, gastric emptying and dyspepsia. *Aliment Pharmacol Ther* 14: 129–134
82 Bortolotti M, Coccia G, Grossi G, Miglioli M (2002) The treatment of functional dyspepsia with red pepper. *Aliment Pharmacol Ther* 16: 1075–1082
83 Drewes AM, Schipper KP, Dimcevski G, Petersen P, Gregersen H, Funch- Jensen P, Arendt-Nielsen L (2003) Gut pain and hyperalgesia induced by capsaicin: a human experimental model. *Pain* 104: 333–341
84 Schmidt B, Hammer J, Holzer P, Hammer HF (2004) Chemical nociception in the jejunum induced by capsaicin. *Gut* 53: 1109–1116
85 Laird JM, Martinez-Caro L, Garcia-Nicas E, Cervero F (2001) A new model of visceral pain and referred hyperalgesia in the mouse. *Pain* 92: 335–342
86 Kawao N, Ikeda H, Kitano T, Kuroda R, Sekiguchi F, Kataoka K, Kamanaka Y, Kawabata A (2004) Modulation of capsaicin-evoked visceral pain and referred hyperalgesia by protease- activated receptors 1 and 2. *J Pharmacol Sci* 94: 277–285
87 Agarwal MK, Bhatia SJ, Desai SA, Bhure U, Melgiri S (2002) Effect of red chillies on small bowel and colonic transit and rectal sensitivity in men with irritable bowel syndrome. *Indian J Gastroenterol* 21: 179–182
88 Schmulson MJ, Valdovinos MA, Milke P (2003) Chili pepper and rectal hyperalgesia in irritable bowel syndrome. *Am J Gastroenterol* 98: 1214–1215
89 Shah SK, Abraham P, Mistry FP (2000) Effect of cold pressor test and a high-chili diet on rectosigmoid motility in irritable bowel syndrome. *Indian J Gastroenterol* 19: 161–164
90 Coelho AM, Vergnolle N, Guiard B, Fioramonti J, Bueno L (2002) Proteinases and proteinase-activated receptor 2: a possible role to promote visceral hyperalgesia in rats. *Gastroenterology* 122: 1035–1047
91 Hoogerwerf WA, Zou L, Shenoy M, Sun D, Micci MA, Lee- Hellmich H, Xiao SY, Winston JH, Pasricha PJ (2001) The proteinase-activated receptor 2 is involved in nociception. *J Neurosci* 21: 9036–9042
92 Amadesi S, Nie J, Vergnolle N, Cottrell GS, Grady EF, Trevisani M, Manni C, Geppetti P, McRoberts JA, Ennes H et al (2004) Protease-activated receptor 2 sensitizes the capsaicin receptor transient receptor potential vanilloid receptor 1 to induce hyperalgesia. *J Neurosci* 24: 4300–4312
93 Dai Y, Moriyama T, Higashi T, Togashi K, Kobayashi K, Yamanaka H, Tominaga M, Noguchi K (2004) Proteinase-activated receptor 2-mediated potentiation of transient receptor potential vanilloid subfamily 1 activity reveals a mechanism for proteinase-induced inflammatory pain. *J Neurosci* 24: 4293–4299
94 Bhave G, Zhu W, Wang H, Brasier DJ, Oxford GS, Gereau RW (2002) cAMP-dependent protein kinase regulates desensitization of the capsaicin receptor (VR1) by direct phosphorylation. *Neuron* 35: 721–731

95 Mohapatra DP, Nau C (2003) Desensitization of capsaicin-activated currents in the vanilloid receptor TRPV1 is decreased by the cyclic AMP-dependent protein kinase pathway. *J Biol Chem* 278: 50080–50090
96 Numazaki M, Tominaga T, Takeuchi K, Murayama N, Toyooka H, Tominaga M (2003) Structural determinant of TRPV1 desensitization interacts with calmodulin. *Proc Natl Acad Sci USA* 100: 8002–8006
97 Jung J, Shin JS, Lee SY, Hwang SW, Koo J, Cho H, Oh U (2004) Phosphorylation of vanilloid receptor 1 by Ca2+/calmodulin- dependent kinase II regulates its vanilloid binding. *J Biol Chem* 279: 7048–7054
98 Rosenbaum T, Gordon-Shaag A, Munari M, Gordon SE (2004) Ca2+/calmodulin modulates TRPV1 activation by capsaicin. *J Gen Physiol* 123: 53–62
99 Lembeck F, Skofitsch G (1982) Visceral pain reflex after pretreatment with capsaicin and morphine. *Naunyn-Schmiedeberg's Arch Pharmacol* 321: 116–122
100 Plourde V, St.-Pierre S, Quirion R (1997) Calcitonin gene-related peptide in viscerosensitive response to colorectal distension in rats. *Am J Physiol Gastrointest Liver Physiol* 273: G191–G196
101 Delafoy L, Raymond F, Doherty AM, Eschalier A, Diop L (2003) Role of nerve growth factor in the trinitrobenzene sulfonic acid-induced colonic hypersensitivity. *Pain* 105: 489–497
102 Jaggar SI, Scott HCF, James IF, Rice ASC (2001) The capsaicin analogue SDZ 249-665 attenuates the hyperreflexia and referred hyperalgesia associated with inflammation of the rat urinary bladder. *Pain* 89: 229–235
103 Urban L, Campbell EA, Panesar M, Patel S, Chaudhry N, Kane S, Buchheit K, Sandells B, James IF (2000) In vivo pharmacology of SDZ 249-665, a novel, non-pungent capsaicin analogue. *Pain* 89: 65–74
104 Cruz F (2004) Mechanisms involved in new therapies for overactive bladder. *Urology* 63 (Suppl 1): 65–73
105 Lysy, J, Sistiery-Ittah M, Israelit Y, Shmueli A, Strauss-Liviatan N, Mindrul V, Keret D, Goldin E (2003) Topical capsaicin - a novel and effective treatment for idiopathic intractable pruritus ani: a randomised, placebo controlled, crossover study. *Gut* 52: 1323–1326
106 Yiangou Y, Facer P, Dyer NH, Chan CL, Knowles C, Williams NS, Anand P (2001) Vanilloid receptor 1 immunoreactivity in inflamed human bowel. *Lancet* 357: 1338–1339
107 Facer P, Knowles CH, Tam PK, Ford AP, Dyer N, Baecker PA, Anand P (2001) Novel capsaicin (VR1) and purinergic (P2X3) receptors in Hirschsprung's intestine. *J Pediatr Surg* 36: 1679–1684
108 Winter J (1998) Brain derived neurotrophic factor, but not nerve growth factor, regulates capsaicin sensitivity of rat vagal ganglion neurones. *Neurosci Lett* 241: 21–24
109 Winston J, Toma H, Shenoy M, Pasricha PJ (2001) Nerve growth factor regulates VR-1 mRNA levels in cultures of adult dorsal root ganglion neurons. *Pain* 89: 181–186
110 Ji RR, Samad TA, Jin SX, Schmoll R, Woolf CJ (2002) p38 MAPK activation by NGF

in primary sensory neurons after inflammation increases TRPV1 levels and maintains heat hyperalgesia. *Neuron* 36: 57–68

111 Bevan S, Hothi S, Hughes G, James IF, Rang HP, Shah K, Walpole CS, Yeats JC (1992) Capsazepine: a competitive antagonist of the sensory neurone excitant capsaicin. *Br J Pharmacol* 107: 544–552

112 Docherty RJ, Yeats JC, Piper AS (1997) Capsazepine block of voltage-activated calcium channels in adult rat dorsal root ganglion neurones in culture. *Br J Pharmacol* 121: 1461–1467

113 Holzer P (2004) Gastrointestinal pain in functional bowel disorders: sensory neurons as novel drug targets. *Expert Opin Ther Targets* 8: 107–123

TRPV1 in the airways

Maria G. Belvisi and Peter J. Barnes

Respiratory Pharmacology Group and Airway Disease Section, National Heart & Lung Institute, Imperial College, Dovehouse Street, London SW3 6LY, UK

Introduction

Sensory nerves in the airways regulate central and local reflex events such as bronchoconstriction, airway plasma leakage, mucus secretion and cough [1]. Sensory nerve activity may be enhanced during inflammation such that these protective reflexes become exacerbated and deleterious [1]. Sensory nerve reflexes are under the control of at least two different classes of sensory fibre: the myelinated, rapidly-adapting stretch receptors and non-myelinated, capsaicin-sensitive, C-fibres [2, 3]. In the airways, activation of rapidly-adapting stretch receptors and C-fibres elicits cough, bronchoconstriction and mucus secretion via an afferent central reflex pathway [1, 4–6]. Activation of C-fibres in the airways also mediates efferent excitatory non-adrenergic, non-cholinergic (e-NANC) responses such as bronchoconstriction, mucus secretion, plasma exudation and vasodilatation, via the peripheral release of neuropeptides, a phenomenon known as neurogenic inflammation [1] (Fig. 1). A characteristic feature of many nociceptive sensory fibres is their sensitivity to capsaicin [7, 8]. However, until recently the molecular mechanisms involved in activation of sensory nociceptive fibres were unknown. Pharmacological evidence for the presence of a capsaicin receptor in sensory nerves was provided by the use of two capsaicin analogues, resiniferatoxin (a potent agonist) and capsazepine (a selective antagonist). Firstly, specific binding sites for resiniferatoxin were demonstrated on dorsal root ganglion membranes [9] and, secondly, capsazepine has been found to inhibit numerous capsaicin-evoked neuronal responses [10, 11], including those in the airways [12]. The capsaicin receptor has recently been identified and has been named the type 1 vanilloid receptor (VR1; or TRPV1, for transient receptor potential vanilloid 1) [13].

Turning up the Heat on Pain: TRPV1 Receptors in Pain and Inflammation, edited by
Annika B. Malmberg and Keith R. Bley

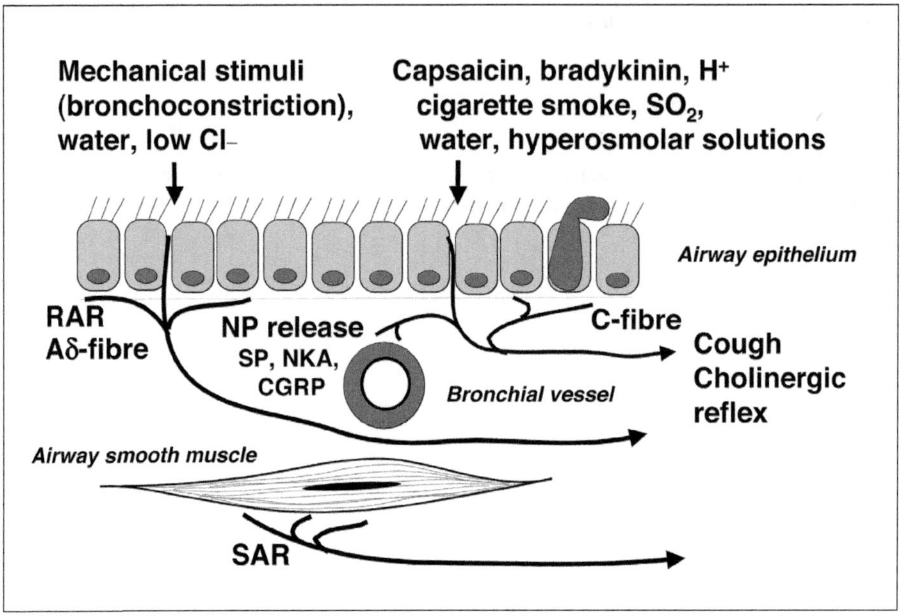

Figure 1
Sensory nerves in the airways. Myelinated Aδ-fibres are activated predominantly by mechanical stimuli (including bronchoconstriction), whereas unmyelinated C-fibres are activated by capsaicin, irritants and chemicals. TRPV1 is localized predominantly to C-fibres. CGRP, calcitonin-gene-related peptide; NKA, neurokinin A; NP, neuropeptide; RAR, rapidly-adapting stretch receptor; SAR, slowly adapting receptor; SP, substance P.

The type 1 vanilloid receptor (TRPV1)

TRPV1 is a membrane-associated vanilloid receptor. It is a ligand-gated ion channel expressed selectively on the neuronal plasma membrane of nociceptive C-fibres and is required for the activation of sensory nerves by vanilloids such as capsaicin, the pungent extract from plants in the *Capsicum* family (hot chilli peppers) [14, 15]. TRPV1 also mediates the response to painful heat, extracellular acidosis, protons and tissue injury [15, 16]. The convergence of these stimuli on TRPV1 channels, which are highly expressed in the sensory neurons of dorsal root and trigeminal ganglia, underlies the common perceptual experience of pain due to these stimuli. TRPV1 is an outwardly rectifying, cation-selective ion channel with a preference for calcium (P_{Ca}/P_{Na} ~10) and magnesium (P_{Mg}/P_{Na} ~5) [14], which depends on a single aspartic acid residue in the pore region of the protein. TRPV1 is activated by heat (>43 °C; although it is effectively dormant at normal body temperature) and low pH (<5.9), and may act as an integrator of chemical and physical pain-eliciting stimuli. When activated, TRPV1 produces depolarization through the influx of Na^+, but the high Ca^{2+} permeability of the channel is also important for mediating the response to stimuli associated with pain. Gating by heat is direct but the receptor can be opened by ligands or stimuli such as mild acidosis, which also reduces the threshold for temperature activation and potentiates the response to capsaicin. To summarize, TRPV1 mediates nociception and contributes to the detection and integration of diverse chemical and thermal stimuli [17].

The expected role for TRPV1 is in pain pathways and recent data from studies with TRPV1-knockout mice showed impaired inflammatory thermal hyperalgesia [17, 18]. This has led to a growing interest in developing small-molecule antagonists for this target. Recent pre-clinical data demonstrating efficacy in rodent models of both thermal and mechanical neuropathic or inflammatory pain has fuelled this enthusiasm [19, 20]. The established role of sensory nerve activation in the cough reflex and the role of TRPV1 in inflammatory pain has also alerted the respiratory community as to the therapeutic potential of TRPV1 antagonists as anti-tussives and as therapy for airway inflammatory diseases such as asthma and chronic obstructive pulmonary disease (COPD).

Ligands for TRPV1

Exogenous agonists of the TRPV1 include capsaicin and resiniferatoxin [15]. Endogenous agonists and modulators include the cannabinoid receptor agonist anandamide, *N*-arachidonoyl-dopamine (NADA), inflammatory mediators (bradykinin [21, 22], 5-hydroxytryptamine [22] and prostaglandin E2 [22]), hydrogen ions [14, 22], heat [14, 23], arachidonic acid [24, 25], lipoxin A_4 [26], prostacyclin [27], ethanol [28, 29] and several eicosanoid products of lipoxygenases, including 12-(*S*)- and 15-(*S*)-hydroperoxyeicosatetraenoic acids, 5-(*S*)-hydroxyeicosatetraenoic acid and leukotriene B_4 [30]. Interestingly, the effect of bradykinin may be indirect given that data exists suggesting that bradykinin appears to activate bradykinin B_2 receptors on afferent neurons leading to the generation of lipoxygenase metabolites that have agonist activity at TRPV1 [31]. Anandamide was first isolated from porcine brain [32] and is an endogenous lipid mediator, which is the ethanolamine amide of arachidonic acid. It is synthesized in the nervous system [33] as well as peripheral tissues including the lung [34]. Anandamide is a low-efficacy TRPV1 agonist that acts as a partial agonist in tissues with low receptor reserve; however, where a high receptor reserve exists, as in certain inflammatory disease states, then anandamide behaves as a full agonist [35, 36]. Interestingly, anandamide is also a known ligand of cannabinoid receptors [37]. NADA is a high-efficacy TRPV1 agonist that has recently been isolated from the central nervous system [38].

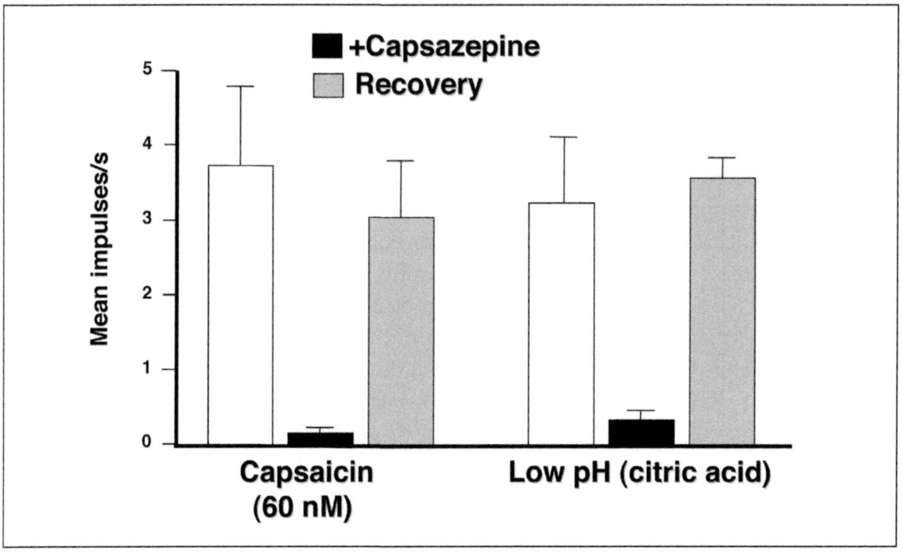

Figure 2
Effect of the TRPV1 antagonist capsazepine on single airway C-fibres of guinea pig activated by capsaicin or by low pH (citric acid). Adapted from [41].

The evidence supporting the hypothesis that TRPV1 can be gated by vanilloids such as capsaicin, but also by protons, heat, agonists at certain G-protein-coupled receptors, ethanol, cannabinoids and lipoxygenase products of arachidonic acid, has come from electrophysiological patch-clamp recording studies on the cell bodies of TRPV1-expressing cells [23, 28, 30, 39, 40]. Furthermore, pharmacological antagonism of TRPV1 has been shown to inhibit action potential discharge evoked by each of these stimuli [28, 30, 41, 42]. However, due to the non-selective effects of several of these tool compounds this pharmacological approach has its limitations. Furthermore, if the antagonist inhibits but does not completely block a given response it is not clear whether activation of the TRPV1 is obligatory for the response or merely contributes to the end response. In studies using TRPV1$^{-/-}$ mice it was concluded that whereas TRPV1 is required for action potential discharge of C-fibre terminals evoked by capsaicin and anandamide, it plays a contributory role in responses evoked by low pH or bradykinin [43]. These data suggest that TRPV1 is one of multiple ion channels responsible for the bradykinin-evoked generator potentials (i.e. membrane depolarization) in C-fibre terminals. Capsazepine has been demonstrated to abolish C-fibre responses to acid application in isolated guinea pig tracheal preparations, leading to the assumption that TRPV1 activation is an absolute requirement for C-fibre activation by acid [41, 44] (Fig. 2). However, it has been noted in recent studies that the airway C-fibre response to acid has transient and sustained components; the latter being inhibited, but not abolished, by capsazepine [44], giving weight to the data presented in the studies on TRPV1-knockout mice and suggesting that acid activates these fibres by both TRPV1-dependent and -independent pathways. The mechanism involved in the TRPV1-independent response to acid may involve the activation of other acid-sensitive channels, for example TRPV4 [45], members of the acid sensing ion channel (ASIC) family [46] and certain types of voltage-gated potassium channel [47].

It would appear that the pharmacology of TRPV1 is extremely complex. TRPV1 is a thermal sensor and as such is activated by temperatures exceeding 43 °C, but the heat threshold can be significantly lowered by mild acidification and by endogenous ligands such as bradykinin, ATP, ethanol, dithiothreitol and protein kinase C (PKC) activators [48]. Nicotine has recently been shown to sensitize TRPV1 channels and to accentuate responses to capsaicin. This effect is not dependent on nicotinic acetylcholine receptors. Nicotine has also been shown to inhibit tetrodotoxin (TTX)-resistant sodium channels, which are known to be important in mediating pain and inflammation. These effects occur at modest nicotine concentrations [49]. This lowering of the thermal activation threshold by inflammatory mediators may result in constant activation of TRPV1 even at normal body temperatures. In contrast, the endogenous lipid phosphatidylinositol (4,5)-bisphosphate (PIP_2) has an opposite effect on TRPV1, elevating its threshold for activation and maintaining the receptor in a closed (inactivated) state [50].

TRPV1 receptor antagonists

Antagonists of TRPV1 include ruthenium red, a dye that exhibits properties of non-competitive antagonism for the TRPV1 [51] and has a poorly defined mechanism of action and limited selectivity [16]. Capsazepine was another agent characterized as a weak but relatively selective, competitive TRPV1 receptor antagonist (over other TRPVs) [52]. However, at the concentrations required to antagonize the TRPV1 (approximately 10 μM), capsazepine has demonstrated non-specific effects such as inhibition of voltage-gated calcium channels and nicotinic receptors [53, 54]. Furthermore, this ligand has poor metabolic and pharmacokinetic properties in rodents, where it undergoes extensive first-pass metabolism when given orally [55].

Recently a number of antagonists with improved potency and/or selectivity have been described and these include antagonists that are structurally related to agonists such as iodo-resiniferatoxin, which is 100-fold more potent than capsazepine [56]. Previous evidence has indicated that iodo-resiniferatoxin is a potent antagonist at the TRPV1 both *in vitro* and *in vivo* [56, 57]. Since then several studies have used this antagonist as a pharmacological tool to explore the role for TRPV1 in various physiological settings. Recently, however, although the *in vitro* antagonistic activity of iodo-resiniferatoxin has been confirmed *in vivo* studies have demonstrated some agonist activity of this agent at high doses [58]. Therefore, there is a possibility that this agent retains some agonistic activity and that an agonist-dependent desensitization effect on sensory nerves may well contribute to some of the antagonistic activity observed. The GlaxoSmithKline compound SB-366791 [48], which has a good selectivity profile in a series of assays, is somewhat related to the inhibitor of anandamide transporter and potent vanilloid agonist AM-404, and should provide a useful tool for probing the physiology and pharmacology of TRPV1 [48].

Several TRPV1 receptor antagonists that share no obvious structural resemblance to any class of vanilloid agonist have also been developed. These compounds have been discovered by high-throughput, random screening projects. At present this group of antagonists includes (a) *N*-(haloanilino)carbonyl-*N*-alkyl-*N*-arylethylendiamines discovered at SmithKline Beecham, (b) *N*-diphenyl-*N*-napthylureas discovered at Bayer and (c) pyridol[2,3-*d*]-pyrimidin-4-ones discovered at Novartis [55, 59].

However, one potential issue in the development of TRPV1 antagonists is that since this receptor is targeted by heat, protons and a plethora of endogenous mediators that most likely interact at distinct recognition sites in a synergistic manner, a competitive receptor antagonist may not be able to block the activation of the receptor by blocking a single class of binding site.

Location of TRPV1 in airways

Airway sensory nerves

The capsaicin-sensitive vanilloid receptor is expressed mainly in sensory nerves including those emanating from the dorsal root ganglia and afferent fibres that innervate the airway, which originate from the vagal ganglia. In the dorsal root and trigeminal ganglion, TRPV1 is localized to small and medium-sized neurons [14, 60, 61]. Somatic sensory neurons of the vagus nerve are located in the jugular ganglion, whereas visceral sensory neurons of the nerve are located in the nodose ganglion. Previous studies have demonstrated that TRPV1 containing neurons are abundant in these ganglia [61].

There are classically two classes of sensory nerve in the airways; myelinated Aδ-fibres that are activated predominantly by mechanical stimuli, and unmyelinated C-fibres (Fig. 1). Physiological studies indicate that capsaicin-sensitive neurons are broadly defined as small cells with unmyelinated (C-) nerve fibres. Of these afferents, most capsaicin-sensitive neurons are polymodal nociceptors, chemonociceptors or heat-sensitive receptors. Aδ mechanoreceptors and cold receptors are not sensitive [62, 63]. More information on the distribution of TRPV1, and consequently the capsaicin sensitivities of subpopulations of primary sensory neurons, has been provided by combined *in situ* hybridization and histochemical techniques of TRPV1 mRNA. In these studies this receptor is almost exclusively expressed by neurofilament-negative small and medium-sized dorsal root ganglial cells [61]. In the same study the authors have shown that TRPV1 mRNA is expressed by most cells of the nodose ganglion in contrast to Caterina et al. [14], who did not find expression in these cells but in agreement with Helliwell et al. [64]. Furthermore, nodose cells that lack TRPV1 include large neurofilament-immunoreactive cells consistent with studies suggesting that low-threshold tracheal Aδ mechanoceptors originating in the nodose ganglia are neurofilament-positive and capsaicin-insensitive [65].

Non-neuronal location

The traditional view that TRPV1 is simply a marker of primary sensory nerves is now being challenged. The TRPV1 receptor has been detected in guinea pig and human airways by receptor-binding assays [66] and has been identified in non-neuronal cell types, such as mast cells, fibroblasts and smooth muscle [67].

Role of TRPV1 in airway responses

Cough

Cough is a dominant and persistent symptom of many inflammatory lung diseases, including asthma, COPD, viral infections, pulmonary fibrosis and bronchiectasis. Chronic cough can also be idiopathic in nature where no obvious causal mechanism is evident. Cough is the most common complaint for which medical attention is sought and although effective treatments for cough are not available, narcotic agents, such as the opioid codeine, are often used. However, such agents have only limited beneficial value due to the associated side effects such as constipation, nausea, vomiting and drowsiness. Therefore, the identification of novel therapies, devoid of central activity, for the treatment of chronic cough would be of significant therapeutic benefit and greatly enhance the quality of life of patients who suffer from this condition [68].

Inhalation challenge with capsaicin and low pH (e.g. citric acid) has been shown to evoke cough and has been used as a model for the last 50 years to investigate the action of potential anti-tussive therapies in clinical trials [69]. In fact, agonists at TRPV1 such as capsaicin and resiniferatoxin are the most potent stimulants of the cough reflex so far described in humans [70, 71]. Therefore, it has been suggested that TRPV1 activation may be one of the primary sensory mechanisms in cough [69]. In guinea pigs capsazepine inhibits coughing induced by inhaled capsaicin and citric acid (Fig. 3) [6]. However, if activation of TRPV1 is an important initiating

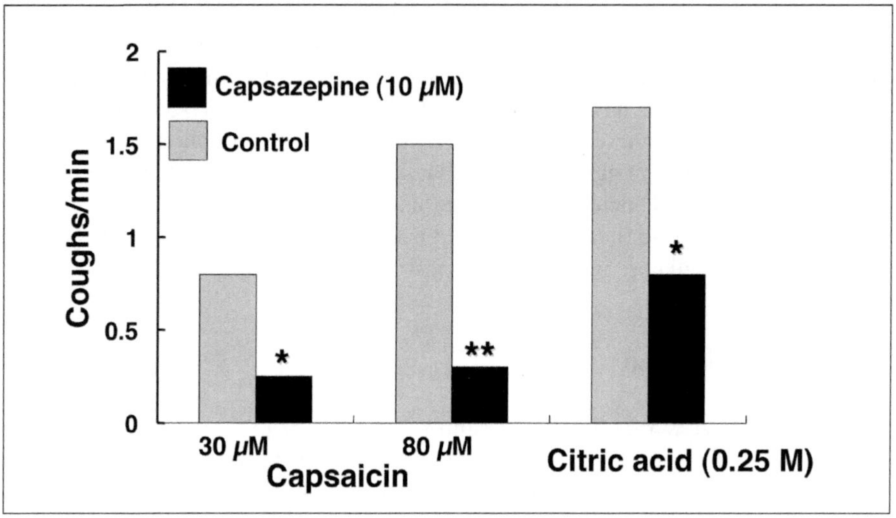

Figure 3
Effect of capsazepine on cough in conscious guinea pigs induced by inhaled capsaicin or citric acid. Mean ± S.E.M. values are shown; $^{*}P < 0.05$, $^{**}P < 0.01$ *compared to vehicle controls. Adapted from [6].*

factor for the cough reflex then an endogenous capsaicin-like ligand must be present. A number of putative endogenous ligands that are known to activate TRPV1 have been demonstrated to cause cough. TRPV1 has been shown to be sensitive to a fall in the extracellular pH, and H^+ ions not only stimulate the receptor directly [14, 23] but also increase the sensitivity of the receptor to capsaicin [14] and low-pH solutions are known to elicit cough in animals [6, 70] and in humans [72].

Anandamide has been shown to activate human and rat TRPV1 receptors in TRPV1-transfected cells and in rat dorsal root ganglial cells [36, 40]. Furthermore, anandamide has been shown to release intracellular Ca^{2+} (measured by the fluorometric imaging plate reader (FLIPR) technique) in guinea pig nodose ganglial cells and this increase was blocked by a TRPV1 antagonist [73]. These data together with the finding that anandamide activates rat pulmonary vagal C-fibres through the activation of TRPV1 [42] and depolarizes the guinea pig vagus nerve axon [74] provides a mechanistic explanation for the tussive effect observed with anandamide. Cough elicited by anandamide in the guinea pig can be inhibited by the TRPV1 inhibitor capsazepine, but not by CB_1 and CB_2 receptor antagonists [73]. However, when anandamide was inhaled by a single capsaicin-sensitive subject, it did not have any demonstrable activity [69]. The lack of tussive activity in humans could be due to the fact that although anandamide is an agonist at TRPV1 it also acts as a cannabinoid receptor agonist. Activation of cannabinoid receptors has been shown to inhibit the cough reflex either via a central effect on CB_1 receptors [34] or a peripheral inhibitory effect on CB_2 receptors [75]. Therefore, the pro-tussive activity via TRPV1 receptor activation could be negated by an opposing anti-tussive effect on the cannabinoid $CB_{1/2}$ receptors.

Interestingly, the TRPV1 antagonists capsazepine and iodo-resiniferatoxin have been shown to inhibit capsaicin- [6, 76], citric acid- [6, 76] and anandamide- [73] induced cough in conscious guinea pigs, suggesting that these agents are inducing cough via a common mechanism; the activation of TRPV1 [6]. However, it is unlikely that TRPV1 activation is the only stimulus for cough since some tussogenic agents, such as hypertonic saline, are not inhibited by TRPV1 antagonism [6, 76]. Recent data generated in anaesthetized guinea pigs have added to the confusion regarding the TRPV1-dependent nature of citric acid-induced cough. In these studies it has been suggested that citric acid-evoked cough (at least in anaesthetized guinea pigs) is mediated by capsaicin-insensitive nodose ganglion neurons and that capsaicin-sensitive nerve activation plays a permissive but non-essential role [77].

Neurogenic inflammation

Activation of TRPV1 in sensory nerves leads to the activation of nociceptive and protective reflex responses and the release of neurotransmitters from both peripheral and central nerve endings (Fig. 4). This latter effect causes a collection of inflammatory responses often referred to as neurogenic inflammation, which in the airways results in bronchoconstriction, plasma extravasation and mucus hypersecretion [1, 78, 79].

Bronchoconstriction

Early experiments using the relatively weak TRPV1 antagonist capsazepine provided pharmacological validation that capsaicin-induced bronchospasm in guinea pig bronchi did indeed involve the activation of TRPV1 [12, 80]. Contractile responses to resiniferatoxin and capsaicin were unaffected by the neurokinin-1 (NK_1) antagonist CP-96,345, partially inhibited by the NK_2 antagonist SR 48968 but nearly abolished by a combination of the antagonists. These data suggest that resiniferatoxin and capsaicin both release tachykinins that act on both NK_1 and NK_2 receptor subtypes in a TRPV1-dependent manner [80]. More recently a more potent and selective agent has been used to confirm these observations [81].

Interestingly, contractile and relaxant responses to capsaicin and resiniferatoxin have also been examined in human isolated bronchus (5–12 mm outer diameter). Bronchi isolated from 10 of 16 lungs contracted in response to capsaicin. The cap-

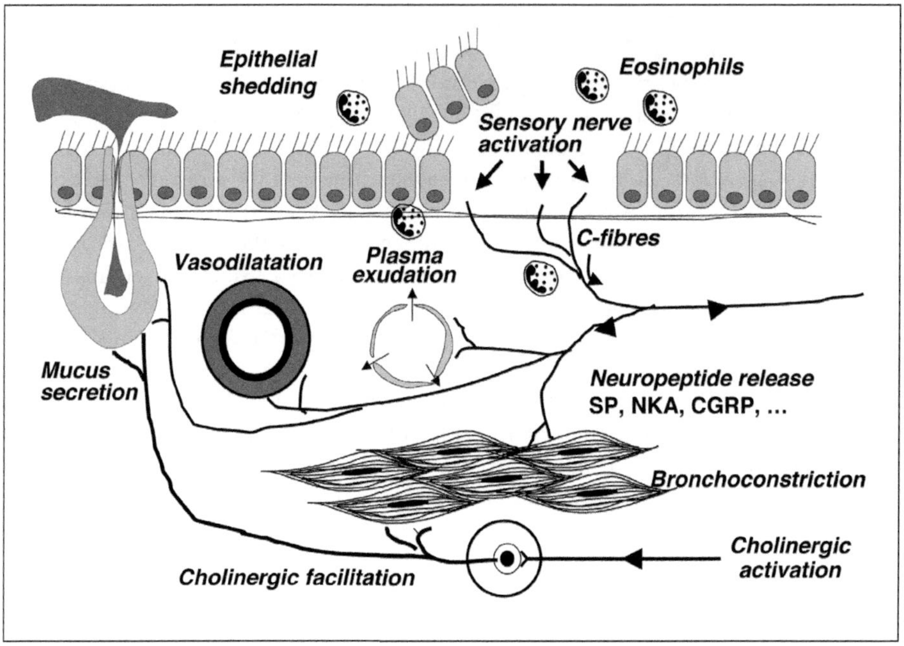

Figure 4
Neurogenic inflammation in asthmatic airways via retrograde release of peptides from sensory nerves via an axon reflex. C-fibres are activated via TRPV1 and cause reflex activation of cholinergic nerves. These nerves also release neuropeptides, including substance P (SP), which causes vasodilatation, plasma exudation and mucus secretion, neurokinin A (NKA), which causes bronchoconstriction and enhanced cholinergic reflexes, and calcitonin-gene-related peptide (CGRP), which causes vasodilatation. This axon reflex therefore amplifies the inflammation in asthmatic airways.

saicin-induced contractions were mimicked by resiniferatoxin and inhibited by capsazepine. The contractile response to capsaicin was not affected by the potent NK_2-selective antagonist SR 48968, whereas responses to concentrations of neurokinin A, neurokinin B, substance P (SP), neuropeptide γ and neuropeptide K, which produced similar-size contractions, were almost abolished by SR 48968. These results suggest that capsaicin and resiniferatoxin can alter smooth muscle tone in a TRPV1-dependent manner, but this response does not appear to involve SP or related neurokinins [82].

The endogenous cannabinoid agonist anandamide also produced a modest contractile response in guinea pig isolated bronchus compared with the vanilloid receptor agonist capsaicin. The contractile response to both anandamide and capsaicin was inhibited by the TRPV1 receptor antagonist, capsazepine. Furthermore, the NK_2-selective antagonist SR48968 but not the NK_1-selective antagonist SR140333 inhibited contractile responses to anandamide. The contractile response to anandamide was also abolished in tissues desensitized by capsaicin. The results demonstrate that anandamide induces a modest contractile response in guinea pig isolated bronchus that is dependent upon the activation of TRPV1 receptors on airway sensory nerves. However, cannabinoid receptors do not appear to play a role in this regard [83]. Recently, it has been demonstrated that ethanol is capable of stimulating airway sensory nerves via a TRPV1-dependent mechanism, thus causing the release of sensory neuropeptides and consequent bronchoconstriction [29]. Furthermore, this response was abolished by a combination of NK_1 and NK_2 receptor antagonists, indicating that neurokinin release from sensory nerves is also involved.

Airway hyperresponsiveness to bronchoconstrictor agents is recognized as a critical feature of bronchial asthma that correlates with clinical symptoms and sensory nerves in the airway have been implicated in hyperresponsiveness. Several studies have demonstrated that capsaicin pretreatment significantly inhibited the late bronchial response that was observed after ovalbumin inhalation, airway hyperresponsiveness and eosinophil accumulation in an allergic guinea pig model [84] and airway hyperresponsiveness to histamine in a rabbit model [85]. However, the role of capsaicin-sensitive nerves in the airway hyperresponsiveness of asthma patients is uncertain. It remains to be seen whether TRPV1 antagonists are effective in this regard.

Plasma extravasation

Activation of C-fibres in the airways also mediates efferent e-NANC responses, such as plasma exudation and vasodilatation, via the peripheral release of neuropeptides, a phenomenon known as neurogenic inflammation [1]. Plasma extravasation has been shown to be induced in rats or guinea pigs by intravenous injections of SP and capsaicin. The effect of intravenous capsaicin was absent in capsaicin-desensitized animals and in those pretreated with capsazepine [29, 86]. Capsaicin-induced plasma extravazation was also markedly inhibited by CP-96,345, a non-peptide antagonist of tachykinin NK1 receptors [87, 88]. These data suggest that capsaicin-induced plasma leakage is mediated by the release of neuropeptides and the activation of NK1 receptors via a TRPV1-dependent mechanism.

Endogenous TRPV1 ligands have also been shown to evoke plasma extravasation into the airways in animal models. Interestingly, it has been demonstrated that tachykinins, released in the airways by stimulating the oesophagus with hydrochloric acid, induce airway plasma extravasation. One normal HCl infusion into the oesophagus significantly increased plasma extravasation in the trachea which was inhibited in capsaicin-desensitized animals, suggesting a role for TRPV1-dependent mechanisms in this response [89]. In addition, it has recently been demonstrated that ethanol administered by the intravenous or intragastric route caused plasma extravasation in the guinea pig airways that was inhibited by capsazepine, and hence is a TRPV1-mediated effect [29].

Mucus secretion

Capsaicin-sensitive C-fibre afferents containing the neuropeptides SP, neurokinin A and calcitonin-gene-related peptide have also been shown, when stimulated, to evoke neurogenic secretion from airway mucus-secreting cells [90]. However, although capsaicin has been shown to elicit mucus secretion there are no data available describing the effect of TRPV1 antagonists on this response.

TRPV1 ligands and the airway inflammatory response

In addition to the stimulation of airway C-fibres, capsaicin has also been shown to have effects on the inflammatory response as assessed in human cell-based assay systems. Thus, capsaicin has been shown to interact with TRPV1 expressed by BEAS-2B and other airway epithelial cells to cause the calcium-dependent production of cytokines and, conversely, calcium-independent cell death. The mechanisms of these cellular responses to capsaicin appear to proceed via distinct cellular pathways, but both pathways are initiated by TRPV1. These results demonstrate that capsaicin and related compounds, which are contained in pepper-spray products, can produce airway inflammation and cause respiratory epithelial cell death [91].

Recent studies have suggested a role for the TRPV1 antagonists as a novel class of analgesic, anti-inflammatory agents. For example, studies utilizing animal models have shown that the thermal hyperalgesia response to bradykinin seen in wild-type mice was absent in mice lacking TRPV1 [92]. This study led to Ferreira et al. [93] investigating the role of TRPV1 channels in the nociceptive response induced by peripheral activation of kinin B_2 receptors. They found that bradykinin-induced nociception is induced by TRPV1 receptor stimulation via phospholipase C-pathway activation and the production of lipoxygenases.

However, the role of the TRPV1 in airway inflammatory disease would depend on the presence of endogenous activators of this channel under physiological and pathophysiological conditions. In fact, it is quite possible that this situation does in fact exist in the disease scenario given that TRPV1 activation can be initiated and responses to other activators potentiated in an acidic environment [94], which has been shown to be present in diseases such as asthma [95, 96] and COPD [96]. Interestingly, it has previously been established that the airway response to TRPV1 activation is enhanced in certain disease conditions. In particular, it has been reported that the cough response elicited in response to capsaicin is exaggerated in diseases such as asthma and COPD [97].

Anandamide is also an endogenous ligand known to be produced in central neurons [33]. However, more recent studies have also suggested that anandamide can be synthesized in lung tissue as a result of calcium stimulation, suggesting that this mediator could also be involved in the activation of TRPV1 under normal and disease conditions [34]. However, whether anandamide acts as a TRPV1 ligand (e.g. to elicit cough) or a cannabinoid receptor agonist (e.g. to inhibit cough) may be determined by the physiological or pathophysiological environment. Inflammatory agents (e.g. bradykinin, ATP, prostaglandin E_2 and nerve growth factor) can indirectly sensitize TRPV1 to cause hyperalgesia. Thus, bradykinin and nerve growth factor, which activate phospholipase Cβ, release TRPV1 from PIP_2-mediated inhibition [92]. However, phospholipase Cβ also regulates TRPV1 by diacylglycerol formation and subsequent activation of PKC, which in turn phosphorylates and sensitizes TRPV1 [98, 99]. Inflammatory stimuli – including prostaglandins, nerve growth factor and others – have also been shown to up-regulate the expression and function of TRPV1 via the activation of pathways dependent on p38 mitogen-activated protein kinase [100] and protein kinase A [101]. Alternatively, cannabinoid receptor activity may be down-regulated by PKC phosphorylation. Therefore, these data would seem to suggest that an inflammatory response as is present in asthma and COPD may favour anandamide acting at the TRPV1 receptor to elicit responses such as cough rather than acting at cannabinoid receptors to inhibit sensory nerve reflexes.

Previous data have suggested that there is an up-regulation of TRPV1 in inflammatory diseases. For example, TRPVR1 immunoreactivity is greatly increased in colonic nerve fibres of patients with active inflammatory bowel disease [102]. Furthermore, a recent study has found an increase in TRPV1 expression in sensory nerves found in the airway epithelial layer in biopsy specimens from patients with chronic cough compared to non-coughing, healthy volunteers. There was a significant correlation between the tussive response to capsaicin and the number of TRPV1-positive nerves within the patients with cough. Therefore it has been postulated that TRPV1 expression may be one of the determinants of the enhanced cough reflex found in patients with chronic cough and that the recently described TRPV1 antagonists could be effective in the treatment of chronic persistent cough due to diverse causes [103]. TRPV1 may therefore play an important role in enhanced sensitivity of the airways in asthma, COPD and chronic cough, with the implication that antagonism of TRPV1 may have therapeutic potential (Fig. 5).

Conclusions

Current data summarized in this chapter suggest that airway inflammatory diseases (e.g. asthma and COPD) may respond to treatment with an effective and selective

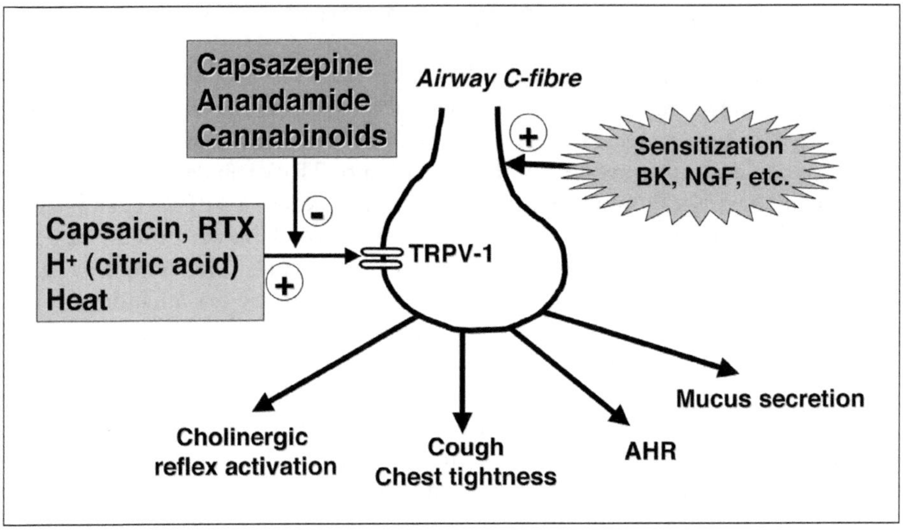

Figure 5
TRPV1 may play an important role in the enhanced sensitivity of the airways seen in asthma, COPD and chronic cough with the implication that antagonism of TRPV1 may have therapeutic potential. The enhanced sensitivity may be due to the release of inflammatory mediators such as BK and NGF upregulating the activity of TRPV1 leading to exaggerated functional responses. These effects can be elicited by agonists at the TRPV1 (e.g., capsaicin, H^+, heat, RTX) and inhibited by agents that inhibit sensory nerve activity (e.g., capsazepine, cannabinoid agonists, andandamide). RTX, resiniferatoxin; BK, bradykinin; NGF, nerve growth factor; AHR, airway hyperresponsiveness

inhibitor of TRPV1 and to this end much work is being carried out to develop novel inhibitors [104]. Interestingly, capsaicin-sensitive nerve stimulation in subjects with active allergic rhinitis produces reproducible and dose-dependent leucocyte influx, albumin leakage and glandular secretion. These results provide *in vivo* evidence for the occurrence of neurogenic inflammation in the human upper airway with active allergic disease and may therefore suggest the therapeutic utility of TRPV1 antagonists in the management of this disease [105]. In addition, the treatment of persistent cough is a facet of airway diseases that is sorely in need of effective treatments and a TRPV1 inhibitor may prove extremely effective against cough induced by, for example, gastroesophageal acid reflux [106], as well as that associated with asthma and other airways diseases as described.

References

1 Barnes PJ (2001) Neurogenic inflammation in the airways. *Respir Physiol* 125: 145–154
2 Coleridge HM, Coleridge JC (1994) Pulmonary reflexes: neural mechanisms of pulmonary defense. *Annu Rev Physiol* 56: 69–91
3 Sant'Ambrogio G, Remmers JE, de Groot WJ, Callas G, Mortola JP (1978) Localization of rapidly adapting receptors in the trachea and main stem bronchus of the dog. *Respir Physiol* 33: 359–366
4 Widdicombe JG (1954) Respiratory reflexes from the trachea and bronchi of the cat. *J Physiol* 123: 55–70
5 Karlsson JA (1993) A role for capsaicin sensitive, tachykinin containing nerves in chronic coughing and sneezing but not in asthma: a hypothesis. *Thorax* 48: 396–400
6 Lalloo UG, Fox AJ, Belvisi MG, Chung KF, Barnes PJ (1995) Capsazepine inhibits cough induced by capsaicin and citric acid but not by hypertonic saline in guinea pigs. *J Appl Physiol* 79: 1082–1087
7 Julius D, Basbaum AI (2001) Molecular mechanisms of nociception. *Nature* 413: 203–210
8 Maggi CA, Meli A (1988) The sensory-efferent function of capsaicin-sensitive sensory neurons. *Gen Pharmacol* 19: 1–43
9 Szallasi A, Blumberg PM (1990) Specific binding of resiniferatoxin, an ultrapotent capsaicin analog, by dorsal root ganglion membranes. *Brain Res* 524: 106–111
10 Bevan S, Hothi S, Hughes G, James IF, Rang HP, Shah K, Walpole CS, Yeats JC (1992) Capsazepine: a competitive antagonist of the sensory neurone excitant capsaicin. *Br J Pharmacol* 107: 544–552
11 Dickenson AH, Dray A (1991) Selective antagonism of capsaicin by capsazepine: evidence for a spinal receptor site in capsaicin-induced antinociception. *Br J Pharmacol* 104: 1045–1049
12 Belvisi MG, Miura M, Stretton D, Barnes PJ (1992) Capsazepine as a selective antagonist of capsaicin-induced activation of C-fibres in guinea-pig bronchi. *Eur J Pharmacol* 215: 341–344
13 Hwang SW, Oh U (2002) Hot channels in airways: pharmacology of the vanilloid receptor. *Curr Opin Pharmacol* 2: 235–242
14 Caterina MJ, Schumacher MA, Tominaga M, Rosen TA, Levine JD, Julius D (1997) The capsaicin receptor: a heat-activated ion channel in the pain pathway. *Nature* 389: 816–824
15 Caterina MJ, Julius D (2001) The vanilloid receptor: a molecular gateway to the pain pathway. *Annu Rev Neurosci* 24: 487–517
16 Szallasi A, Blumberg PM (1999) Vanilloid (Capsaicin) receptors and mechanisms. *Pharmacol Rev* 51: 159–212
17 Caterina MJ, Leffler A, Malmberg AB, Martin WJ, Trafton J, Petersen-Zeitz KR, Koltzenburg M, Basbaum AI, Julius D (2000) Impaired nociception and pain sensation in mice lacking the capsaicin receptor. *Science* 288: 306–313

18 Davis JB, Gray J, Gunthorpe MJ, Hatcher JP, Davey PT, Overend P, Harries MH, Latcham J, Clapham C, Atkinson K et al (2000) Vanilloid receptor-1 is essential for inflammatory thermal hyperalgesia. *Nature* 405: 183–187

19 Garcia-Martinez C, Humet M, Planells-Cases R, Gomis A, Caprini M, Viana F, De La PE, Sanchez-Baeza F, Carbonell T, De Felipe C et al (2002) Attenuation of thermal nociception and hyperalgesia by VR1 blockers. *Proc Natl Acad Sci USA* 99: 2374–2379

20 Walker KM, Urban L, Medhurst SJ, Patel S, Panesar M, Fox AJ, McIntyre P (2003) The VR1 antagonist capsazepine reverses mechanical hyperalgesia in models of inflammatory and neuropathic pain. *J Pharmacol Exp Ther* 304: 56–62

21 Stucky CL, Abrahams LG, Seybold VS (1998) Bradykinin increases the proportion of neonatal rat dorsal root ganglion neurons that respond to capsaicin and protons. *Neuroscience* 84: 1257–1265

22 Vyklicky L, Knotkova-Urbancova H, Vitaskova Z, Vlachova V, Kress M, Reeh PW (1998) Inflammatory mediators at acidic pH activate capsaicin receptors in cultured sensory neurons from newborn rats. *J Neurophysiol* 79: 670–676

23 Tominaga M, Caterina MJ, Malmberg AB, Rosen TA, Gilbert H, Skinner K, Raumann BE, Basbaum AI, Julius D (1998) The cloned capsaicin receptor integrates multiple pain-producing stimuli. *Neuron* 21: 531–543

24 Manzini S, Ballati L, Geppetti P, Rubini I, Meini S, Perretti F (1989) Arachidonic acid-induced bronchomotor responses are partially mediated by release of sensory neuropeptides from capsaicin-sensitive structures. *Br J Pharmacol* 98: 1077–1079

25 Manzini S, Meini S (1991) Involvement of capsaicin-sensitive nerves in the bronchomotor effects of arachidonic acid and melittin: a possible role for lipoxin A4. *Br J Pharmacol* 103: 1027–1032

26 Meini S, Evangelista S, Geppetti P, Szallasi A, Blumberg PM, Manzini S (1992) Pharmacologic and neurochemical evidence for the activation of capsaicin-sensitive sensory nerves by lipoxin A4 in guinea pig bronchus. *Am Rev Respir Dis* 146: 930–934

27 Mapp CE, Fabbri LM, Boniotti A, Maggi CA (1991) Prostacyclin activates tachykinin release from capsaicin-sensitive afferents in guinea-pig bronchi through a ruthenium red-sensitive pathway. *Br J Pharmacol* 104: 49–52

28 Trevisani M, Smart D, Gunthorpe MJ, Tognetto M, Barbieri M, Campi B, Amadesi S, Gray J, Jerman JC, Brough SJ et al (2002) Ethanol elicits and potentiates nociceptor responses via the vanilloid receptor-1. *Nat Neurosci* 5: 546–551

29 Trevisani M, Gazzieri D, Benvenuti F, Campi B, Dinh QT, Groneberg DA, Rigoni M, Emonds-Alt X, Creminon C, Fischer A et al (2004) Ethanol causes inflammation in the airways by a neurogenic and TRPV1-dependent mechanism. *J Pharmacol Exp Ther* 309: 1167–1173

30 Hwang SW, Cho H, Kwak J, Lee SY, Kang CJ, Jung J, Cho S, Min KH, Suh YG, Kim D, Oh U (2000) Direct activation of capsaicin receptors by products of lipoxygenases: endogenous capsaicin-like substances. *Proc Natl Acad Sci USA* 97: 6155–6160

31 Carr MJ, Kollarik M, Meeker SN, Undem BJ (2003) A role for TRPV1 in bradykinin-

induced excitation of vagal airway afferent nerve terminals. *J Pharmacol Exp Ther* 304: 1275–1279

32 Devane WA, Hanus L, Breuer A, Pertwee RG, Stevenson LA, Griffin G, Gibson D, Mandelbaum A, Etinger A, Mechoulam R (1992) Isolation and structure of a brain constituent that binds to the cannabinoid receptor. *Science* 258: 1946–1949

33 Di Marzo V, Fontana A, Cadas H, Schinelli S, Cimino G, Schwartz JC, Piomelli D (1994) Formation and inactivation of endogenous cannabinoid anandamide in central neurons. *Nature* 372: 686–691

34 Calignano A, Katona I, Desarnaud F, Giuffrida A, La Rana G, Mackie K, Freund TF, Piomelli D (2000) Bidirectional control of airway responsiveness by endogenous cannabinoids. *Nature* 408: 96–101

35 Ross RA (2003) Anandamide and vanilloid TRPV1 receptors. *Br J Pharmacol* 140: 790–801

36 Smart D, Gunthorpe MJ, Jerman JC, Nasir S, Gray J, Muir AI, Chambers JK, Randall AD, Davis JB (2000) The endogenous lipid anandamide is a full agonist at the human vanilloid receptor (hVR1). *Br J Pharmacol* 129: 227–230

37 Felder CC, Briley EM, Axelrod J, Simpson JT, Mackie K, Devane WA (1993) Anandamide, an endogenous cannabimimetic eicosanoid, binds to the cloned human cannabinoid receptor and stimulates receptor-mediated signal transduction. *Proc Natl Acad Sci USA* 90: 7656–7660

38 Huang SM, Bisogno T, Trevisani M, Al Hayani A, De Petrocellis L, Fezza F, Tognetto M, Petros TJ, Krey JF, Chu CJ et al (2002) An endogenous capsaicin-like substance with high potency at recombinant and native vanilloid VR1 receptors. *Proc Natl Acad Sci USA* 99: 8400–8405

39 Shin J, Cho H, Hwang SW, Jung J, Shin CY, Lee SY, Kim SH, Lee MG, Choi YH, Kim J et al (2002) Bradykinin-12-lipoxygenase-VR1 signaling pathway for inflammatory hyperalgesia. *Proc Natl Acad Sci USA* 99: 10150–10155

40 Zygmunt PM, Petersson J, Andersson DA, Chuang H, Sorgard M, Di Marzo V, Julius D, Hogestatt ED (1999) Vanilloid receptors on sensory nerves mediate the vasodilator action of anandamide. *Nature* 400: 452–457

41 Fox AJ, Urban L, Barnes PJ, Dray A (1995) Effects of capsazepine against capsaicin- and proton-evoked excitation of single airway C-fibres and vagus nerve from the guinea-pig. *Neuroscience* 67: 741–752

42 Lin YS, Lee LY (2002) Stimulation of pulmonary vagal C-fibres by anandamide in anaesthetized rats: role of vanilloid type 1 receptors. *J Physiol* 539: 947–955

43 Kollarik M, Undem BJ (2004) Activation of bronchopulmonary vagal afferent nerves with bradykinin, acid and vanilloid receptor agonists in wild-type and TRPV1$^{-/-}$ mice. *J Physiol* 555: 115–123

44 Kollarik M, Undem BJ (2002) Mechanisms of acid-induced activation of airway afferent nerve fibres in guinea-pig. *J Physiol* 543: 591–600

45 Suzuki M, Mizuno A, Kodaira K, Imai M (2003) Impaired pressure sensation in mice lacking TRPV4. *J Biol Chem* 278: 22664–22668

46 Kellenberger S, Schild L (2002) Epithelial sodium channel/degenerin family of ion channels: a variety of functions for a shared structure. *Physiol Rev* 82: 735–767
47 O'Connell AD, Morton MJ, Hunter M (2002) Two-pore domain K^+ channels-molecular sensors. *Biochim Biophys Acta* 1566: 152–161
48 Gunthorpe MJ, Rami HK, Jerman JC, Smart D, Gill CH, Soffin EM, Luis HS, Lappin SC, Egerton J, Smith GD et al (2004) Identification and characterisation of SB-366791, a potent and selective vanilloid receptor (VR1/TRPV1) antagonist. *Neuropharmacology* 46: 133–149
49 Liu L, Zhu W, Zhang ZS, Yang T, Grant A, Oxford G, Simon SA (2004) Nicotine inhibits voltage-dependent sodium channels and sensitizes vanilloid receptors. *J Neurophysiol* 91: 1482–1491
50 Prescott ED, Julius D (2003) A modular PIP2 binding site as a determinant of capsaicin receptor sensitivity. *Science* 300: 1284–1288
51 Amann R, Maggi CA (1991) Ruthenium red as a capsaicin antagonist. *Life Sci* 49: 849–856
52 Urban L, Dray A (1991) Capsazepine, a novel capsaicin antagonist, selectively antagonises the effects of capsaicin in the mouse spinal cord *in vitro*. *Neurosci Lett* 134: 9–11
53 Docherty RJ, Yeats JC, Piper AS (1997) Capsazepine block of voltage-activated calcium channels in adult rat dorsal root ganglion neurones in culture. *Br J Pharmacol* 121: 1461–1467
54 Liu L, Simon SA (1997) Capsazepine, a vanilloid receptor antagonist, inhibits nicotinic acetylcholine receptors in rat trigeminal ganglia. *Neurosci Lett* 228: 29–32
55 Szallasi A, Appendino G (2004) Vanilloid receptor TRPV1 antagonists as the next generation of painkillers. Are we putting the cart before the horse? *J Med Chem* 47: 2717–2723
56 Wahl P, Foged C, Tullin S, Thomsen C (2001) Iodo-resiniferatoxin, a new potent vanilloid receptor antagonist. *Mol Pharmacol* 59: 9–15
57 Rigoni M, Trevisani M, Gazzieri D, Nadaletto R, Tognetto M, Creminon C, Davis JB, Campi B, Amadesi S, Geppetti P, Harrison S (2003) Neurogenic responses mediated by vanilloid receptor-1 (TRPV1) are blocked by the high affinity antagonist, iodo-resiniferatoxin. *Br J Pharmacol* 138: 977–985
58 Seabrook GR, Sutton KG, Jarolimek W, Hollingworth GJ, Teague S, Webb J, Clark N, Boyce S, Kerby J, Ali Z et al (2002) Functional properties of the high-affinity TRPV1 (VR1) vanilloid receptor antagonist (4-hydroxy-5-iodo-3-methoxyphenylacetate ester) iodo-resiniferatoxin. *J Pharmacol Exp Ther* 303: 1052–1060
59 Wang Y, Szabo T, Welter JD, Toth A, Tran R, Lee J, Kang SU, Suh YG, Blumberg PM, Lee J (2002) High affinity antagonists of the vanilloid receptor. *Mol Pharmacol* 62: 947–956
60 Guo A, Vulchanova L, Wang J, Li X, Elde R (1999) Immunocytochemical localization of the vanilloid receptor 1 (VR1): relationship to neuropeptides, the P2X3 purinoceptor and IB4 binding sites. *Eur J Neurosci* 11: 946–958
61 Michael GJ, Priestley JV (1999) Differential expression of the mRNA for the vanilloid

receptor subtype 1 in cells of the adult rat dorsal root and nodose ganglia and its down-regulation by axotomy. *J Neurosci* 19: 1844–1854
62 Holzer P (1991) Capsaicin: cellular targets, mechanisms of action, and selectivity for thin sensory neurons. *Pharmacol Rev* 43: 143–201
63 Szolcsanyi J, Nagy J, Petho G (1993) Effect of CP-96,345 a non-peptide substance P antagonist, capsaicin, resiniferatoxin and ruthenium red on nociception. *Regul Pept* 46: 437–439
64 Helliwell RJ, McLatchie LM, Clarke M, Winter J, Bevan S, McIntyre P (1998) Capsaicin sensitivity is associated with the expression of the vanilloid (capsaicin) receptor (VR1) mRNA in adult rat sensory ganglia. *Neurosci Lett* 250: 177–180
65 Riccio MM, Myers AC, Undem BJ (1996) Immunomodulation of afferent neurons in guinea-pig isolated airway. *J Physiol* 491: 499–509
66 Szallasi A, Goso C, Manzini S (1995) Resiniferatoxin binding to vanilloid receptors in guinea pig and human airways. *Am J Respir Crit Care Med* 152: 59–63
67 Szallasi A (2002) Vanilloid (capsaicin) receptors in health and disease. *Am J Clin Pathol* 118: 110–121
68 Belvisi MG, Geppetti P (2004) Cough. 7: Current and future drugs for the treatment of chronic cough. *Thorax* 59: 438-440
69 Morice AH, Geppetti P (2004) Cough. 5: The type 1 vanilloid receptor: a sensory receptor for cough. *Thorax* 59: 257–258
70 Laude EA, Higgins KS, Morice AH (1993) A comparative study of the effects of citric acid, capsaicin and resiniferatoxin on the cough challenge in guinea-pig and man. *Pulm Pharmacol* 6: 171–175
71 Collier JG, Fuller RW (1984) Capsaicin inhalation in man and the effects of sodium cromoglycate. *Br J Pharmacol* 81: 113–117
72 Bickerman HA, Barach AL (1954) The experimental production of cough in human subjects induced by citric acid aerosols; preliminary studies on the evaluation of antitussive agents. *Am J Med Sci* 228: 156–163
73 Jia Y, McLeod RL, Wang X, Parra LE, Egan RW, Hey JA (2002) Anandamide induces cough in conscious guinea-pigs through VR1 receptors. *Br J Pharmacol* 137: 831–836
74 Kagaya M, Lamb J, Robbins J, Page CP, Spina D (2002) Characterization of the anandamide induced depolarization of guinea-pig isolated vagus nerve. *Br J Pharmacol* 137: 39–48
75 Patel HJ, Birrell MA, Crispino N, Hele DJ, Venkatesan P, Barnes PJ, Yacoub MH, Belvisi MG (2003) Inhibition of guinea-pig and human sensory nerve activity and the cough reflex in guinea-pigs by cannabinoid (CB2) receptor activation. *Br J Pharmacol* 140: 261–268
76 Trevisani M, Milan A, Gatti R, Zanasi A, Harrison S, Fontana G, Morice AH, Geppetti P (2004) Antitussive activity of iodo-resiniferatoxin in guinea pigs. *Thorax* 59: 769–772
77 Canning BJ, Mazzone SB, Meeker SN, Mori N, Reynolds SM, Undem BJ (2004) Identi-

fication of the tracheal and laryngeal afferent neurones mediating cough in anaesthetized guinea-pigs. *J Physiol* 557: 543–558
78 Belvisi MG (2003) Airway sensory innervation as a target for novel therapies: an outdated concept? *Curr Opin Pharmacol* 3: 239–243
79 Belvisi MG (2003) Sensory nerves and airway inflammation: role of A delta and C-fibres. *Pulm Pharmacol Ther* 16: 1–7
80 Ellis JL, Undem BJ (1994) Inhibition by capsazepine of resiniferatoxin- and capsaicin-induced contractions of guinea pig trachea. *J Pharmacol Exp Ther* 268: 85–89
81 Undem BJ, Kollarik M (2002) Characterization of the vanilloid receptor 1 antagonist iodo-resiniferatoxin on the afferent and efferent function of vagal sensory C-fibers. *J Pharmacol Exp Ther* 303: 716–722
82 Ellis JL, Sham JS, Undem BJ (1997) Tachykinin-independent effects of capsaicin on smooth muscle in human isolated bronchi. *Am J Respir Crit Care Med* 155: 751–755
83 Tucker RC, Kagaya M, Page CP, Spina D (2001) The endogenous cannabinoid agonist, anandamide stimulates sensory nerves in guinea-pig airways. *Br J Pharmacol* 132: 1127–1135
84 Mukaiyama O, Morimoto K, Nosaka E, Takahashi S, Yamashita M (2004) Greater involvement of neurokinins found in Guinea pig models of severe asthma compared with mild asthma. *Int Arch Allergy Immunol* 134: 263–272
85 Herd CM, Gozzard N, Page CP (1995) Capsaicin pre-treatment prevents the development of antigen-induced airway hyperresponsiveness in neonatally immunised rabbits. *Eur J Pharmacol* 282: 111–119
86 Saria A, Lundberg JM, Skofitsch G, Lembeck F (1983) Vascular protein linkage in various tissue induced by substance P, capsaicin, bradykinin, serotonin, histamine and by antigen challenge. *Naunyn-Schmiedeberg's Arch Pharmacol* 324: 212–218
87 Eglezos A, Giuliani S, Viti G, Maggi CA (1991) Direct evidence that capsaicin-induced plasma protein extravasation is mediated through tachykinin NK1 receptors. *Eur J Pharmacol* 209: 277–279
88 Lei YH, Barnes PJ, Rogers DF (1992) Inhibition of neurogenic plasma exudation in guinea-pig airways by CP-96,345, a new non-peptide NK1 receptor antagonist. *Br J Pharmacol* 105: 261–262
89 Kohrogi H, Hamamoto J, Kawano O, Iwagoe H, Fujii K, Hirata N, Ando M (2001) The role of substance P release in the lung with esophageal acid. *Am J Med* 111 (Suppl 8A): 25S–30S
90 Rogers DF (2002) Pharmacological regulation of the neuronal control of airway mucus secretion. *Curr Opin Pharmacol* 2: 249–255
91 Reilly CA, Taylor JL, Lanza DL, Carr BA, Crouch DJ, Yost GS (2003) Capsaicinoids cause inflammation and epithelial cell death through activation of vanilloid receptors. *Toxicol Sci* 73: 170–181
92 Chuang HH, Prescott ED, Kong H, Shields S, Jordt SE, Basbaum AI, Chao MV, Julius D (2001) Bradykinin and nerve growth factor release the capsaicin receptor from PtdIns(4,5)P2-mediated inhibition. *Nature* 411: 957–962

93 Ferreira J, da Silva GL, Calixto JB (2004) Contribution of vanilloid receptors to the overt nociception induced by B2 kinin receptor activation in mice. *Br J Pharmacol* 141: 787–794
94 McLatchie LM, Bevan S (2001) The effects of pH on the interaction between capsaicin and the vanilloid receptor in rat dorsal root ganglia neurons. *Br J Pharmacol* 132: 899–908
95 Hunt JF, Fang K, Malik R, Snyder A, Malhotra N, Platts-Mills TA, Gaston B (2000) Endogenous airway acidification. Implications for asthma pathophysiology. *Am J Respir Crit Care Med* 161: 694–699
96 Kostikas K, Papatheodorou G, Ganas K, Psathakis K, Panagou P, Loukides S (2002) pH in expired breath condensate of patients with inflammatory airway diseases. *Am J Respir Crit Care Med* 165: 1364–1370
97 Doherty MJ, Mister R, Pearson MG, Calverley PM (2000) Capsaicin responsiveness and cough in asthma and chronic obstructive pulmonary disease. *Thorax* 55: 643–649
98 Crandall M, Kwash J, Yu W, White G (2002) Activation of protein kinase C sensitizes human VR1 to capsaicin and to moderate decreases in pH at physiological temperatures in *Xenopus* oocytes. *Pain* 98: 109–117
99 Vellani V, Mapplebeck S, Moriondo A, Davis JB, McNaughton PA (2001) Protein kinase C activation potentiates gating of the vanilloid receptor VR1 by capsaicin, protons, heat and anandamide. *J Physiol* 534: 813–825
100 Ji RR, Samad TA, Jin SX, Schmoll R, Woolf CJ (2002) p38 MAPK activation by NGF in primary sensory neurons after inflammation increases TRPV1 levels and maintains heat hyperalgesia. *Neuron* 36: 57–68
101 Hu HJ, Bhave G, Gereau RW (2002) Prostaglandin and protein kinase A-dependent modulation of vanilloid receptor function by metabotropic glutamate receptor 5: potential mechanism for thermal hyperalgesia. *J Neurosci* 22: 7444–7452
102 Yiangou Y, Facer P, Dyer NH, Chan CL, Knowles C, Williams NS, Anand P (2001) Vanilloid receptor 1 immunoreactivity in inflamed human bowel. *Lancet* 357: 1338–1339
103 Groneberg DA, Niimi A, Dinh QT, Cosio B, Hew M, Fischer A, Chung KF (2004) Increased expression of transient receptor potential vanilloid-1 in airway nerves of chronic cough. *Am J Respir Crit Care Med* 170: 1276–1280
104 Spina D, Page CP (2002) Pharmacology of airway irritability. *Curr Opin Pharmacol* 2: 264–272
105 Sanico AM, Atsuta S, Proud D, Togias A (1997) Dose-dependent effects of capsaicin nasal challenge: *in vivo* evidence of human airway neurogenic inflammation. *J Allergy Clin Immunol* 100: 632–641
106 Canning BJ, Mazzone SB (2003) Reflex mechanisms in gastroesophageal reflux disease and asthma. *Am J Med* 115 (Suppl 3A): 45S–48S

Part V
Therapeutic potential of vanilloid agonists and antagonists

TRPV1 agonist-based therapies: mechanism of action and clinical prospects

Keith R. Bley[1] *and Annika B. Malmberg*[2]

[1]NeurogesX, Inc., 981F Industrial Road, San Carlos, CA 94070, USA; [2]Elan Pharmaceuticals, 800 Gateway Boulevard, South San Francisco, CA 94080, USA

Introduction

Capsaicin and other naturally occurring pungent molecules have been used for centuries as topical analgesics to treat a variety of painful conditions. Moreover, since 1989, periodic instillations of high-concentration capsaicin and resiniferatoxin (RTX) solutions have been found useful for the management of persistent bladder pain and the symptoms of overactive bladder. However, only within the last 8 years has it been appreciated that the selective action of capsaicin and similar compounds on nociceptive sensory nerve fibers is mediated by agonism of a ligand-gated ion channel called the transient receptor potential vanilloid receptor 1 (TRPV1). The selective expression of TRPV1 on nociceptors – those nerve fibers specialized for the detection of stimuli associated with tissue injury – in skin, bladder, joints and other tissues, has resulted in this receptor becoming an important target for rational drug development. Two different, but non-mutually exclusive, strategies are being pursued: optimization of TRPV1 agonist-based therapies which can functionally inactivate nociceptive nerve fibers for extended periods, and identification of receptor antagonists which would prevent nociceptive fibers from being activated by ongoing inflammatory stimuli.

This chapter will focus on recent advances in the understanding of drugs and treatments which attempt to use naturally occurring or synthetic TRPV1 agonists to alter the function of nociceptive sensory nerves and consequently cause pain relief. Consistent with the experimental results reviewed, the terms nociceptor and C-fiber will often be used interchangeably, even though there is not always a complete overlap of the two categories; nociceptors consist of not only C-fibers, but also some Aδ-fibers. Evidence for hypotheses regarding the basis of nociceptor hyperactivity in pain syndromes will be reviewed, and the prospects for efficacy of locally administered therapies against various indications will be evaluated.

Turning up the Heat on Pain: TRPV1 Receptors in Pain and Inflammation, edited by
Annika B. Malmberg and Keith R. Bley

TRPV1 agonist-induced nociceptor desensitization

When activated by a combination of heat, acidosis or endogenous agonists, TRPV1 initiates signal transmission to the spinal cord by depolarizing sensory nerve endings and generating action potentials which may be experienced by the brain as either a warming or a burning sensation (Fig. 1). However, if TRPV1 is activated continuously by on-going exposure to an exogenous agonist (e.g. capsaicin), a local biochemical signal can also be generated in nerve fibers, which produces long-term effects on nociceptive fiber functionality [1]. The TRPV1 channel is highly calcium-permeable, allowing calcium to flow down its steep electrochemical gradient into the cell. Furthermore, as TRPV1 is also expressed on intracellular organelles, external capsaicin application can cause release of calcium from the endoplasmic reticulum and may even induce additional intracellular calcium release from internal stores via calcium-dependent calcium release [2]. If TRPV1 is activated in this continuous fashion, high levels of intracellular calcium and associated enzymatic and osmotic changes can induce processes that impair nociceptor function for extended periods [1].

A persistent lack of responsiveness to stimuli that would normally cause nociceptor activation has been termed 'desensitization'. The use of this term in TRPV1 agonist literature arises from psychophysical studies of human subjects who display reduced reactions to painful stimuli applied to areas pretreated with TRPV1 agonists, and is not narrowly confined to a direct desensitization of TRPV1 or its intracellular signaling mechanisms [1].

Nociceptor hyperactivity

Desensitization of nociceptive nerve fibers may constitute an important therapeutic intervention if they are hyperactive. Following acute injury, the basis for the hyperactivity and/or hypersensitivity of nociceptive nerve endings in affected tissues is well established [3], and the protective behaviors which result from nociceptor hyperactivity are considered fundamental to tissue repair and avoidance of additional damage. Still, TRPV1-agonist-based therapies are being developed for acute traumatic pain syndromes (see below) because it is widely recognized that even acute pain can be non-productive and interfere with healing processes.

In chronically painful conditions, particularly neuropathic pain syndromes, clinical and nonclinical research show collectively that the most peripheral aspects of damaged sensory nerves often display aberrant 'pathophysiological' electrical hyperactivity [4]. However, direct correlations between aberrant activity of nociceptors and patient pain reports have proven difficult to demonstrate, due to the technical complexity of measuring electrical activity in small-diameter nerve fibers. A technique known as microneurography – which measures action potential extracellular-

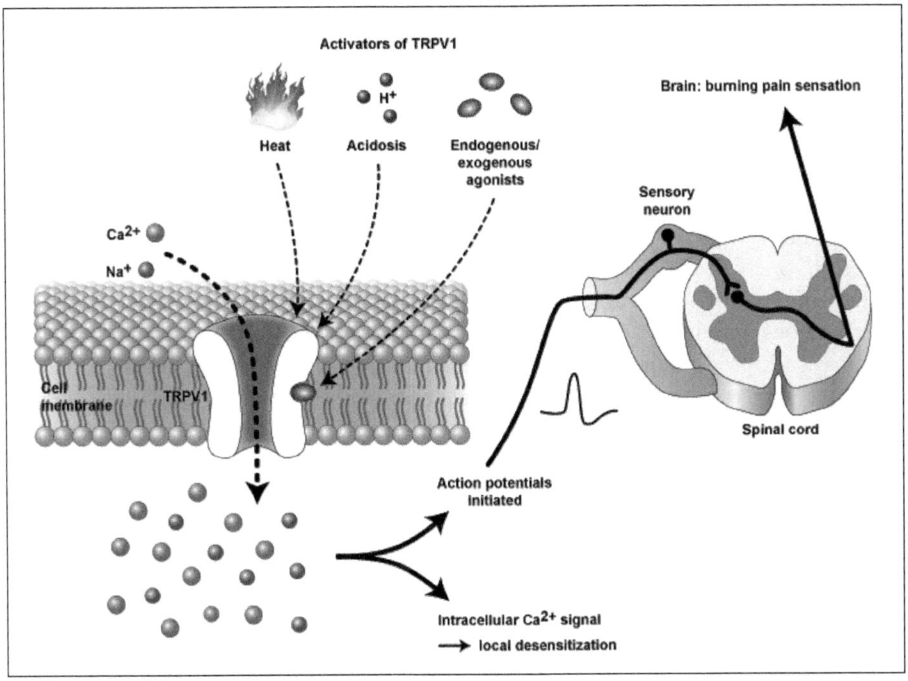

Figure 1
Activation of TRPV1 leads the sensation of heat or burning pain and can also result in localized nociceptor desensitization.

ly – can be used, but this diagnostic procedure is somewhat invasive and may cause discomfort [5]. Moreover, microneurography has the greatest utility for the long nerves of the legs and arms, and thus those chronic pain syndromes with primary presentations in the trunk or face (e.g. lower-back pain or post-herpetic neuralgia) are difficult to analyze. Consequently, there are very few published studies that have correlated successfully nociceptive nerve fiber hyperactivity with patient reports of chronic pain, and they all involve neuropathic pain of the extremities. For instance, a systematic study of hyperactive nociceptors in patients with erythromelalgia (burning pain of the feet) showed altered conduction velocities and spontaneous activity or sensitization in some mechano-insensitive C-fibers [6].

In contrast, spontaneous activity of distal nociceptive fibers following nerve injury has been recorded extensively in nonclinical models, and correlated directly with pain behaviors. For instance, transection of the sciatic nerve in rodents is a long-standing model in which spontaneous or 'ectopic' electrical activity of the resulting neuroma (the injured tip of a nerve fiber) develops, and nerve fibers terminating in the neuroma become extremely sensitive to stimuli [7]. However, although highly instructive regarding basic mechanisms of neural excitability, large nerve neuromas represent only a very small fraction of clinically presenting peripheral neuropathies. Polyneuropathies in which some innervation of the skin or other target organ remains intact – such as due to diabetes – are much more common and clinically important [8].

Accordingly, several recent nonclinical studies have focused on the excitability of *intact* nociceptors following mechanical injuries to surrounding nerve fibers. For instance, one day following ligation and transaction of the L5 spinal nerve in rats, about one-half of the uninjured C-fiber nociceptors in the L4 spinal nerve develop spontaneous activity [9]. Similarly, 7 days following rhizotomy of L5 ventral roots (which leads predominantly to degeneration in myelinated fibers) in rats, a marked decrease in paw withdrawal thresholds occurred concomitantly with increased low-frequency C-fiber spontaneous activity [10]. Furthermore, after partial denervation of the dorsum of the foot was induced by tight ligations of spinal nerve L6 in primates [11], there is a significantly higher incidence of spontaneous activity observed in uninjured single C-fibers in the superficial peroneal nerve recorded using an *in vitro* skin/nerve preparation.

Bases for nociceptor hyperactivity

The projections of nociceptors into target organs can be visualized and quantified by immunostaining of antigens selectively expressed in neurons. Protein gene product 9.5 (PGP 9.5) is the most commonly studied marker, although substance P, calcitonin-gene-related peptide (CGRP) or others have also been used. Antibodies to PGP 9.5 stain all nerve fibers, but because nerve fibers in the epidermis of the skin, uroepithelium of the bladder and mucosa are almost exclusively nociceptors (see below), changes in the density of these fibers may be quantified [12].

In order to understand the mechanisms underlying aberrant activity of nociceptive nerve fibers in chronic pain syndromes, one key factor may be changes in the density of innervation of the target organ. That is, in most pain syndromes, which would be considered neuropathic, sensory neuron axons are lost due either to cell body or nerve fiber damage following viral, metabolic, traumatic or chemical insults. In contrast, in chronically painful conditions, which do not involve direct injury to axons, there may be increased nociceptor innervation of target organs (see below).

Immunohistochemical analyses have indicated that the density of epidermal nerve fibers in skin is decreased in a wide range of neuropathic pain syndromes, including post-herpetic neuralgia [13, 14], diabetic neuropathy [15], painful HIV-associated neuropathy [16], Fabry disease [17] and small-fiber neuropathy [18]. Moreover, data suggest a positive correlation between the extent of epidermal-nerve-fiber loss and the severity of pain in diabetic neuropathy [15] and HIV-associated neuropathy [16]. This observation also extends to nonclinical models: in one study with rats following a chronic constriction injury, there was a very significant reduction of PGP 9.5- and CGRP-immunoreactive fibers in the epidermis of the footpad on the same day as maximum pain behavior [19].

Conversely, there are chronically painful syndromes associated with increased nociceptor density; examples include interstitial cystitis (IC) [20], nostalgia paresthetica [21], vulvodynia [22, 23], rectal hypersensitivity [24] and gastroesophageal reflux disease [25]. As these conditions do no reflect nerve injury *per se*, they could be classified as chronic inflammatory – and not neuropathic – syndromes.

There is evidence that intact nociceptor nerve terminals may develop abnormal electrophysiological properties due to exposure to abnormal concentrations of neurotrophins such as nerve growth factor (NGF) and pro-inflammatory cytokines (Fig. 2). In chronic pain syndromes associated with deinnervation, pain intensity may correlate with reduced nociceptor immunostaining because the fewer the number of intact nociceptive endings, the more likely those endings are to be pathologically active [26]. Although NGF has an important role in controlling the survival and development of small-diameter neurons – both sensory and sympathetic – it has recently become clear that this molecule also serves as an important signal for neuroimmune and inflammatory processes in mature organisms [27]. In the skin, NGF and other neurotrophins are constantly produced by keratinocytes and possibly Langerhans cells [28]. NGF is also produced by human bladder smooth muscle cells [29], and increased levels of NGF have been reported in bladder tissues from both human and rodents when there is outlet obstruction [30]. Human chondrocytes synthesize NGF and production is up-regulated in osteoarthritic chondrocytes [31, 32]. In response to enhanced NGF supply (e.g. in inflamed bladders or osteoarthritis), residual nociceptors may respond by becoming hyperactive and possibly sprouting. Support for this hypothesis can be found in the well-characterized immediate and delayed excitatory effects of NGF. As a direct excitatory stimulant, NGF causes immediate excitation of nociceptors [33], resulting in prolonged hyperalgesia and allodynia [34] and bladder overactivity [35]. In addition to this direct and rapid effect, retrograde transport of NGF to sensory neuron cell bodies may lead to the up-regulation of pro-excitatory proteins such as TRPV1 and voltage-activated sodium channels [36] and down-regulation of anti-excitatory proteins such as voltage-activated potassium channels [37]. Consistently, neutralization of NGF has been reported to reduce pain behaviors in some rodent models of neuropathic pain [38] and bladder overactivity [39].

Pro-inflammatory cytokines can also directly activate and modify gene expression in sensory neurons. There are several sources of these molecules in close proximity to peripheral nerves. Schwann cells, which have often been thought of as having a passive support role for peripheral nerves, are able to secrete pro-inflammatory cytokines [40, 41]. Wallerian degeneration is a post-traumatic process of the

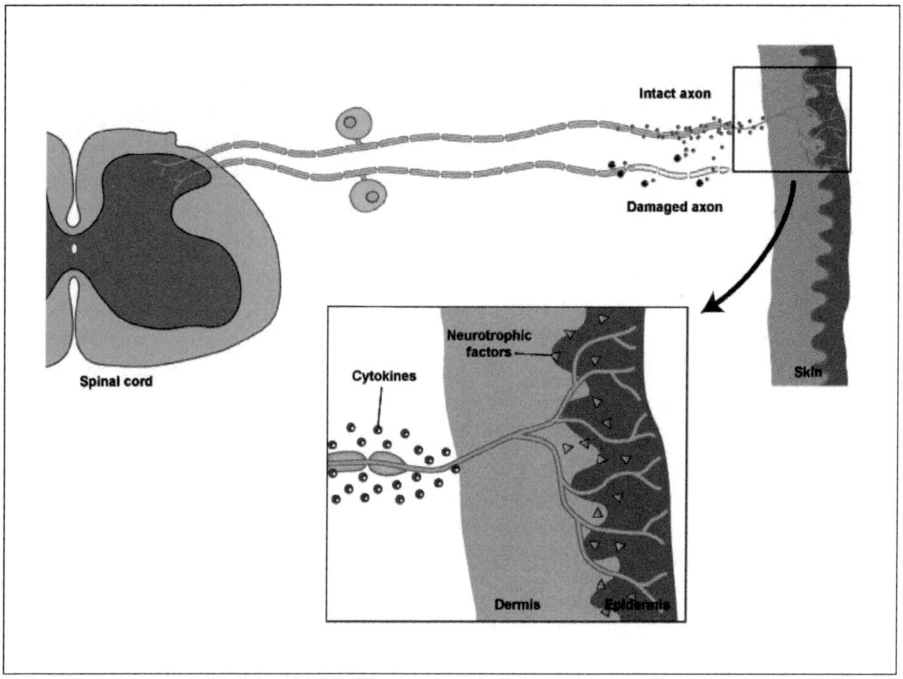

Figure 2
Overexposure to neurotrophins and pro-inflammatory cytokines may underlie the hyperactivity of residual nociceptors.

peripheral nervous system whereby damaged axons and their surrounding myelin sheaths are phagocytosed by infiltrating macrophages or leukocytes. During the process of infiltration of inflamed or damaged peripheral nerves, these immune system cells are known to secrete pro-inflammatory cytokines [42]. Additionally, in the skin, keratinocytes can also release pro-inflammatory cytokines [43].

TRPV1 agonist-induced disruption of nociceptor hyperactivity

The long-term desensitization of nociceptors induced by TRPV1 agonist exposure may not only reduce the ability of nociceptor tips to initiate electrical signals, but it has been long known that nociceptor nerve terminals exposed to capsaicin lose the capacity to take up and retrogradely transport neurotrophic factors such as NGF to the cell body [44, 45]. Without a constant supply of NGF, many nociceptors lose their ability to maintain a hyperexcitable phenotype [27]. Hence, TRPV1 agonist treatments may actually alter the phenotype of nociceptors by depriving them of pro-excitatory influences from their target organs.

The onset of efficacy and duration of pain relief or reduced bladder activity after discontinuation of TRPV1 agonist treatment may be related to the rate at which nociceptive nerve endings lose and then regain functionality [46, 47]. Although nociceptors can remain functionally inactive for weeks following the topical application of capsaicin, they eventually return to pre-treatment levels of sensitivity. Changes in nociceptor immunostaining also occur with the same time course as the effect. As is the case with impaired functionality, immunostaining reductions are reversible, probably through the natural processes of nerve-fiber elongation and protein recycling.

Potential indications for locally administered TRPV1 agonists

Given the ubiquitous distribution of TRPV1-expressing sensory nerve fibers throughout the body, systemic TRPV1 agonist-based therapies are unlikely to be viable, as widespread and persistent nociceptor desensitization may produce an unacceptable side-effect profile. However, directed TRPV1 agonist therapies – in which nociceptor desensitization is restricted to discrete target organs or regions – remain a tenable and potentially attractive means to control localized pain or neurogenic inflammation.

In contrast to a highly localized desensitization of nociceptive terminals in skin or other peripheral sites, another approach to a TRPV1 agonist-based therapy is possible: use of TRPV1 agonists in a neurolytic procedure which targets with high selectivity nociceptive neuron cell bodies. Non-selective neurolytic procedures are still commonly utilized to control intractable chronic pain syndromes, and nonclinical studies have confirmed the efficacy and limited side effects of capsaicin or RTX when these substances are injected into the vicinity of sensory ganglia [48]. To debate the justifications for such a highly invasive procedure is beyond the scope of this chapter, but there can be no doubt that such a therapy would face intense scrutiny from regulatory authorities and, even if approved, would almost certainly be relegated to use as a last resort.

Topical capsaicin for peripheral neuropathies

The skin is innervated with approximately 1 million sensory nerve fibers, with their cell bodies in sensory ganglia (i.e. dorsal root, trigeminal or nodose ganglia). The main nerve branches enter the subdermal fatty tissue, and then divide into smaller bundles that fan out laterally to form a branching network which ascends – often accompanying blood vessels – to form a mesh of interlacing nerves in the superficial dermis, with some fibers extending into the epidermis [12]. The epidermis contains almost exclusively unmyelinated C-fiber nociceptive nerve endings, which specialize in the detection of noxious stimuli [49, 50]. As there are obviously many nociceptive nerve fibers in the dermis too, the term cutaneous nociceptor is used to refer collectively to those nociceptors that may terminate in the epidermis or course through the dermis. As discussed above, pathological activity of cutaneous nociceptors is implicated in chronic pain syndromes.

Capsaicin and closely related vanilloid compounds have been used as topical analgesic agents for centuries. The first formal report of the pain-reducing properties of capsaicin in the West appeared in 1850 as a recommendation to use an alcoholic hot pepper extract on burning or itching extremities [51]. Creams, lotions and patches containing capsaicin generally in the range of 0.025–0.1% by weight are now sold, usually without the requirement of a prescription, for the treatment of neuropathic and musculoskeletal pain. Clinical studies of these low-concentration medications, usually involving three to five daily topical applications for periods of 2–6 weeks, have often suggested beneficial effects for the treatment of many disorders, including post-herpetic neuralgia, diabetic neuropathy, osteoarthritis and even psoriasis [52]. As low-concentration, capsaicin-based products often result in contamination of the patient's environment (clothing, bedding, contact lenses, etc.) and each application is associated with a burning sensation, poor patient compliance with these products is often cited as a likely contributor to limited efficacy [53].

Much as in the case of bladder instillations (to be discussed below), support for the potential utility of single or episodic high-concentration topical exposures was derived from compassionate treatments of 10 patients with intractable pain syndromes [54]. A high-concentration topical capsaicin-containing patch is in clinical development for the management of pain associated with peripheral neuropathies. Phase 1 data suggest that a single 60-min patch application is adequate to induce substantial nociceptor desensitization, as measured by reductions of thermal thresholds and epidermal PGP 9.5 immunostaining [55]. Preliminary efficacy data suggest significant potential for efficacy against post-herpetic neuralgia [56] and painful HIV-associated neuropathy (AN) [57].

Control of bladder pain and overactivity

Bladders are richly innervated with nociceptive sensory nerve fibers, which can detect bladder distension or the presence of irritant chemicals; activation of these fibers triggers reflex bladder contraction and emptying [58]. However, under normal conditions, C-fiber-initiated reflexes are not the primary control mechanism for the bladder: a long pathway passing through the pontine micturition center and initiated by activation of capsaicin-resistant Aδ-fibers is thought to control most normal bladder contractions. There is a short neuronal pathway contained entirely within the sacral spinal cord and initiated by activation of capsaicin-sensitive bladder C-fibers, but this pathway is usually inhibited in adult mammals unless there are pathologies such as inflammation or spinal transection [59]. Thus, under pathophysiological conditions, bladder contractions triggered by capsaicin-sensitive C-fibers and mediated by the sacral reflex are involuntary and can be triggered by small volumes of urine, characteristics that generate an urge to urinate, urinary incontinence and a high urinary frequency. Moreover, these C-fiber-initiated contractions lack coordination with urethral sphincter muscle relaxation and can lead to increased intravesicular pressure and potential harm to the upper urinary tract [58]. Interestingly, TRPV1 expression occurs not only in nociceptive fibers that form close contacts with bladder epithelial (uroepithelial) cells but also in uroepithelial cells themselves [60]; this suggests that these cells may work in concert with underlying afferent nerves to detect the presence of irritating stimuli.

In conditions characterized as IC, overactive or neurogenic bladder and detrusor instability (with the exception of those due to spinal cord injury) an increased density of afferent/nociceptive nerve fibers has been reported, based upon increased levels of immunostaining for PGP 9.5 and other markers [20, 61]. As TRPV1-immunostaining is clearly co-localized with nociceptive fibers in the bladder [62], it is possible that proliferating nociceptors may express enhanced levels of TRPV1 as well.

Interest in TRPV1 agonists for the treatment of overactive bladder and chronic bladder pain began with the need to treat patients with neurogenic detrusor overactivity, frequently due to spinal-cord injury. In order to provide compassionate treatments for patients without alternatives, capsaicin was instilled intravesicularly for the first time in 1989 [63]. These and subsequent unblinded investigations have used single administrations of high-concentration (usually 1–2 mM) solution for about 30 min. Efficacy from single capsaicin instillations has been reported to last from 1 to 3 months, with some patients showing clinical benefit for up to 1 year [64]. In a meta-analysis of unblinded treatments for 131 patients, mean bladder capacity increased by ~50% at various times after treatment and mean symptomatic improvement was ~70%. The primary acute adverse effects reported were suprapubic pain, increased incontinence or macroscopic hematuria; all of these usually resolved within 1 week of therapy. No long-term adverse events or safety issues have been reported [64].

RTX is naturally occurring tricyclic diterpene from *Euphorbia* cactus which is an ultra-potent TRPV1 agonist [65] and is often referred to as a non-pungent TRPV1 agonist, even though clinical experience still shows some pungency [64]. RTX was licensed from the US National Institutes of Health in 1995, with the goal being to develop well-tolerated treatment for neurogenic/overactive bladder or IC. Initial clinical trials suggested that some patients experienced significant relief of incontinence or pain-associated IC with few side effects [66]. Additional studies examined the utility in patients with refractory and stable detrusor hyperreflexia. Initial results showed improvements in bladder capacity, the number of incontinence episodes and patients' subjective ratings [67]. Consistent with previous reports, RTX appeared to be much better tolerated than capsaicin with respect to acute discomfort. Subsequently, a Phase 2 trial for IC has been completed, but RTX treatment failed to meet the primary endpoints [68]. However, academic research studies for patients with refractory bladder disorders continue, and RTX still shows promise for that subset of patients [69].

Quite similar to effects observed in skin, single instillations of high-concentration capsaicin also reduce immunohistochemical markers for suburothelial nerve fibers [70]. Interestingly, there appears to be reduction of nerve-fiber immunostaining only in patients who respond to the treatment. Similarly, in patients with idiopathic detrusor overactivity, instillation of RTX increased the current perception threshold values of C-fibers in all patients who showed symptomatic improvements [71].

Vulvodynia and mucosal disorders

Vulvodynia is a pain syndrome characterized by painful burning sensations, allodynia, hyperalgesia and itching, usually localized in the region of the vulvar vestibules [72]. Vulvar tissue arises from the same urogenital progenitors as bladder; hence it might not be surprising to find parallels involving hyperproliferation of nociceptive fibers. In vulvodynia patients, the hypersensitivity of vulvar C-fibers is well documented [73, 74], and immunohistological evaluation of small-diameter nociceptive nerve fibers shows increased densities relative to normal subjects [22, 26]. Moreover, TRPV1 expression appears to be significantly increased in these proliferated nociceptors [75]. Low-concentration topical capsaicin has been evaluated as a treatment for vulvar vestibulitis, the most common form of vulvodynia. Following 12 weeks of daily applications, significant improvements in pain, irritation and ability to engage in sexual intercourse were reported [76].

Similar patterns of enhanced TRPV1 expression occur in rectal hypersensitivity syndrome, which includes fecal urgency and incontinence as symptoms [23]. Increases in TRPV1 expression appear to correlate with decreases in heat and distension sensory thresholds. In a similar application area, topical capsaicin has been used with success to treat intractable anal pruritis [77].

Traumatic and musculoskeletal pain

Until recently, TRPV1 agonist-induced desensitization for surgical or traumatic injuries has not been widely considered as a therapeutic alternative, as the pungency of these molecules would be perceived as increasing patients' pain burdens. But now a gel formulation containing a high concentration of capsaicin is undergoing clinical evaluation for use in a variety of surgical procedures, including bunion-removal surgery, total knee replacement and abdominal surgeries (such as hernia repair or hysterectomy) [78]. During a surgical procedure, this gel formulation is delivered directly onto the cut surfaces of skin, muscle and bone, with the expectation that desensitization will occur before local and general anesthetics have worn off.

Pain associated with degenerative bone and joint diseases is usually attributed to sensitized nociceptors in inflamed peri-articular soft tissues [79]. Indeed, injection of irritants or activating substances into peri-articular tissues is a standard model for studying the mechanisms underlying nociceptor hypersensitization during inflammation. However, recent investigations have drawn attention to the likely contributions of intrinsic bone innervation in chronic pain syndromes. In particular, during the course of developing rodent bone-cancer pain models, it has been shown that the femur and other bones are richly innervated by nociceptive nerve fibers [80]. While the periosteum is the most densely innervated tissue, when the total volume of each tissue is considered, the bone marrow receives the greatest total number of sensory fibers, followed by mineralized bone and then the periosteum. Thus the innervation of bones may be primarily nociceptive in function and hyperactivity of these nerve fibers may have an important role in both post-orthopedic surgical pain and chronically painful joint and bone diseases such as osteoarthritis (OA), rheumatoid arthritis and metastatic cancer. For instance, it has been proposed that the erosion of articular cartilage in OA uncovers subchondral bone, allowing nociceptive nerve fibers within the bone to be activated by the mechanical forces of normal weight bearing [81]. At this time, there is no published information regarding TRPV1 expression in joints or changes in expression with disease. However, given the potential contribution of NGF to OA [31, 32], increased TRPV1 expression and/or C-fiber sprouting would not be surprising.

An injectable formulation of capsaicin that provides for direct delivery to joints is under clinical evaluation for the treatment of OA and tendonitis; such a therapy would provide direct desensitization of nociceptors in joint tissues. In contrast, topical low-concentration capsaicin has been evaluated as a treatment for osteoarthritis in four double-blind vehicle-controlled clinical trials. From these data – limited by the potential for inadequate blinding due to the lack of vehicle pungency – it can be inferred that topical capsaicin appears effective, either as a monotherapy or adjunctive therapy [82]. Similarly, topical 0.075% civamide (*cis*-capsaicin; see below) cream has recently been shown to be more effective than an active control in a large double-blind, placebo-controlled Phase 3 trial for OA [83]. Assuming that the highly lipophilic capsaicin isomers are not working via transdermal delivery and selective accumulation in joints, the basis for the efficacy of topical TRPV1 agonists in OA may seem mysterious, unlike conditions with evidence for involvement of nociceptors in skin. One intriguing clue is that cutaneous nociceptors are also affected in OA (and rheumatoid arthritis), as patients report regions of both tactile allodynia and hypoesthesia in the skin superficial to diseased joints, much like those reported with many peripheral neuropathies [84].

Migraine and headache

Civamide is the generic name for *cis*-capsaicin (which is also known as zucapsaicin; see Fig. 3). *trans*-Capsaicin is the naturally occurring form of capsaicin, whereas *cis*-capsaicin must be synthesized [85]. Following systemic administration, civamide has been reported to be active in rodent models of nociceptive and neuropathic pain [86]. Subsequently, civamide has been investigated clinically for a number of indications, including prophylaxis of migraine, episodic cluster headache and OA (see above). Phase 2 data regarding a 0.025% (w/v) intranasal liquid spray dosed daily for migraine [87] and cluster headache [88] have been encouraging. As expected, civamide displays pungency similar to capsaicin, and thus nasal irritation and lacrimation are listed as frequent adverse events. Phase 2 data for topical civamide (0.075%, w/w) formulation to be applied several times a day for the management of OA pain were also positive [83].

Although called ‘an orally active capsaicin analog’ in an early publication [86], there are few data to suggest that the pharmacological actions of *cis*-capsaicin are significantly different from *trans*-capsaicin. Indeed, the structure-activity relationship of capsaicin analogs would suggest that the double bond at this position on the aliphatic chain should make little impact on interactions with TRPV1 [89]. At this time, no peer-reviewed data have appeared which support any significant pharmacological or pharmacokinetic differences between *cis*- and *trans*-capsaicin.

Conclusions

Evidence for the hyperactivity of nociceptive nerve fibers in acute and chronic pain syndromes is robust, and thus it should come as no surprise that TRPV1 receptor agonist-mediated desensitization is a long-standing approach to manage pain. Instead of focusing on damaged or lost nerve fibers, explanations for persistent neuropathic pain should focus on the altered properties of residual or intact nociceptors, which share the same innervation territory as nociceptors absent due to damage. Chronic inflammatory pain syndromes seem to be associated with nociceptor proliferation and up-regulation of TRPV1 and other pro-excitatory proteins. The potential contributions of excessive exposure to neurotrophins and cytokines to either nociceptor hyperactivity or sprouting, although suggested by existing data, await further elucidation.

When TPRV1 agonists are administered in such a way that they gain immediate access to nociceptive fibers, which have been sensitized by acute trauma, the prospects for efficacy via the induction of long-term nociceptor desensitization are high. For chronic pain conditions involving increased TRPV1 expression and nociceptive fiber proliferation, and sprouting in some indications (e.g. vulvodynia and IC), significant efficacy is also likely, as those fiber tips should display enhanced vul-

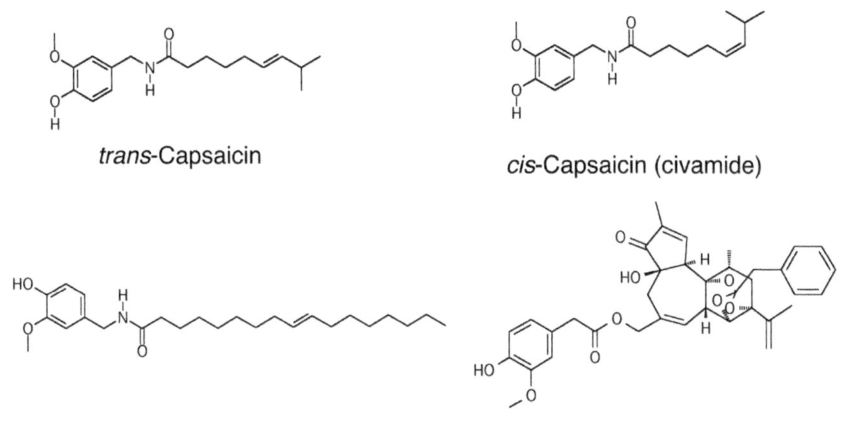

Figure 3
Structures of selected TRPV1 agonists.

nerability to desensitization. Finally, a substantial portion of peripheral neuropathic patients are likely to respond to topical TRPV1 agonist therapies.

All commercially available TRPV1 agonist-based products or treatments are based on low-concentration capsaicin, whereas high-concentration capsaicin and RTX products are in advanced clinical evaluation. The former are more appropriate for patient self-administration, whereas the latter appear to be better suited for physician administration. Directed high-concentration TRPV1 agonist therapies – in which episodic nociceptor desensitization is restricted to discrete regions – may emerge as an attractive means to control localized pain or inflammation. However, given the ubiquitous distribution of TRPV1-expressing sensory nerve fibers throughout the body, systemic TRPV1 agonist-based therapies are unlikely to be viable, as the side-effect profile of widespread desensitization is unlikely to be acceptable.

References

1 Szallasi A, Blumberg PM (1999) Vanilloid (capsaicin) receptors and mechanisms. *Pharmacol Rev* 51: 159–212
2 Liu M, Liu MC, Magoulas C, Priestley JV, Willmott NJ (2003) Versatile regulation of

cytosolic Ca^{2+} by vanilloid receptor I in rat dorsal root ganglion neurons. *J Biol Chem* 278: 5462–5472
3 Julius D, Basbaum AI (2001) Molecular mechanisms of nociception. *Nature* 413: 203–210
4 Michaelis M (2002) Electrophysiological characteristics of injured peripheral nerves. In: AB Malmberg, SR Chaplan (eds): *Mechanisms and Mediators of Neuropathic Pain.* Birkhäuser Verlag, Basel, 3–22
5 Hagbarth KE (2002) Microelectrode recordings from human peripheral nerves (microneurography). *Muscle Nerve* (Suppl 11): S28–S35
6 Orstavik K, Weidner C, Schmidt R, Schmelz M, Hilliges M, Jorum E, Handwerker H, Torebjork E (2003) Pathological C-fibres in patients with a chronic painful condition. *Brain* 126(Pt 3): 567–578
7 Wall PD, Devor M (1983) Sensory afferent impulses originate from dorsal root ganglia as well as from the periphery in normal and nerve injured rats. *Pain* 17: 321–339
8 Bennett GJ (1998) Neuropathic pain: new insights, new interventions. *Hosp Pract (Off Ed)* 33: 95–110
9 Wu G, Ringkamp M, Hartke TV, Murinson BB, Campbell JN, Griffin JW, Meyer RA (2001) Early onset of spontaneous activity in uninjured C-fiber nociceptors after injury to neighboring nerve fibers. *J Neurosci* 21: RC140
10 Wu G, Ringkamp M, Murinson BB, Pogatzki EM, Hartke TV, Weerahandi HM, Campbell JN, Griffin JW, Meyer RA (2002) Degeneration of myelinated efferent fibers induces spontaneous activity in uninjured C-fiber afferents. *J Neurosci* 22: 7746–7753
11 Ali Z, Ringkamp M, Hartke TV, Chien HF, Flavahan NA, Campbell JN, Meyer RA (1999) Uninjured C-fiber nociceptors develop spontaneous activity and alpha-adrenergic sensitivity following L6 spinal nerve ligation in monkey. *J Neurophysiol* 81: 455–466
12 Kennedy WR (2004) Opportunities afforded by the study of unmyelinated nerves in skin and other organs. *Muscle Nerve* 29: 756–767
13 Rowbotham MC, Yosipovitch G, Connolly MK, Finlay D, Forde G, Fields HL (1996) Cutaneous innervation density in the allodynic form of postherpetic neuralgia. *Neurobiol Dis* 3: 205–214
14 Oaklander AL (2001) The density of remaining nerve endings in human skin with and without postherpetic neuralgia after shingles. *Pain* 92: 139–145
15 Kennedy WR, Wendelschafer-Crabb G, Johnson T (1996) Quantitation of epidermal nerves in diabetic neuropathy. *Neurology* 47: 1042–1048
16 Polydefkis M, Yiannoutsos CT, Cohen BA, Hollander H, Schifitto G, Clifford DB, Simpson DM, Katzenstein D, Shriver S, Hauer P et al (2002) Reduced intraepidermal nerve fiber density in HIV-associated sensory neuropathy. *Neurology* 58: 115–119
17 Scott LJ, Griffin JW, Luciano C, Barton NW, Banerjee T, Crawford T, McArthur JC, Tournay A, Schiffmann R (1999) Quantitative analysis of epidermal innervation in Fabry disease. *Neurology* 52: 1249–1254
18 Pittenger GL, Ray M, Burcus NI, McNulty P, Basta B, Vinik AI (2004) Intraepidermal

nerve fibers are indicators of small-fiber neuropathy in both diabetic and nondiabetic patients. *Diabetes Care* 27: 1974–1979
19 Lindenlaub T, Sommer C (2002) Epidermal innervation density after partial sciatic nerve lesion and pain-related behavior in the rat. *Acta Neuropathol (Berl)* 104: 137–143
20 Christmas TJ, Rode J, Chapple CR, Milroy EJ, Turner-Warwick RT (1990) Nerve fibre proliferation in interstitial cystitis. *Virchows Arch A Pathol Anat Histopathol* 416: 447–451
21 Springall DR, Karanth SS, Kirkham N, Darley CR, Polak JM (1991) Symptoms of notalgia paresthetica may be explained by increased dermal innervation. *J Invest Dermatol* 97: 555–561
22 Bohm-Starke N, Hilliges M, Falconer C, Rylander E (1998) Increased intraepithelial innervation in women with vulvar vestibulitis syndrome. *Gynecol Obstet Invest* 46: 256–260
23 Tympanidis P, Terenghi G, Dowd P (2003) Increased innervation of the vulval vestibule in patients with vulvodynia. *Br J Dermatol* 148: 1021–1027
24 Chan CL, Facer P, Davis JB, Smith GD, Egerton J, Bountra C, Williams NS, Anand P (2003) Sensory fibres expressing capsaicin receptor TRPV1 in patients with rectal hypersensitivity and faecal urgency. *Lancet* 361: 385–391
25 Matthews PJ, Aziz Q, Facer P, Davis JB, Thompson DG, Anand P (2004) Increased capsaicin receptor TRPV1 nerve fibres in the inflamed human oesophagus. *Eur J Gastroenterol Hepatol* 16: 897–902
26 Campbell JN (2001) Nerve lesions and the generation of pain. *Muscle Nerve* 24: 1261–1273
27 Petruska JC, Mendell LM (2004) The many functions of nerve growth factor: multiple actions on nociceptors. *Neurosci Lett* 361: 168–171
28 Pincelli C (2000) Nerve growth factor and keratinocytes: a role in psoriasis. *Eur J Dermatol* 10: 85–90
29 Tanner R, Chambers P, Khadra MH, Gillespie JI (2000) The production of nerve growth factor by human bladder smooth muscle cells *in vivo* and *in vitro*. *BJU Int* 85: 1115–1119
30 Gosling JA, Gilpin SA, Dixon JS, Gilpin CJ (1986) Decrease in the autonomic innervation of human detrusor muscle in outflow obstruction. *J Urol* 136: 501–504
31 Iannone F, De Bari C, Dell'Accio F, Covelli M, Patella V, Lo Bianco G, Lapadula G (2002) Increased expression of nerve growth factor (NGF) and high affinity NGF receptor (p140 TrkA) in human osteoarthritic chondrocytes. *Rheumatology (Oxford)* 41: 1413–1418
32 Manni L, Lundeberg T, Fiorito S, Bonini S, Vigneti E, Aloe L (2003) Nerve growth factor release by human synovial fibroblasts prior to and following exposure to tumor necrosis factor-alpha, interleukin-1 beta and cholecystokinin-8: the possible role of NGF in the inflammatory response. *Clin Exp Rheumatol* 21: 617–624
33 Bonnington JK, McNaughton PA (2003) Signalling pathways involved in the sensitisa-

tion of mouse nociceptive neurones by nerve growth factor. *J Physiol* 551(Pt 2): 433–446
34 Dmitrieva N, McMahon SB (1996) Sensitisation of visceral afferents by nerve growth factor in the adult rat. *Pain* 66: 87–97
35 Lamb K, Gebhart GF, Bielefeldt K (2004) Increased nerve growth factor expression triggers bladder overactivity. *J Pain* 5: 150–156
36 Ji RR, Samad TA, Jin SX, Schmoll R, Woolf CJ (2002) p38 MAPK activation by NGF in primary sensory neurons after inflammation increases TRPV1 levels and maintains heat hyperalgesia. *Neuron* 36: 57–68
37 Zhang WY, Li Wan Po A (1994) The effectiveness of topically applied capsaicin. A meta-analysis. *Eur J Clin Pharmacol* 46: 517–522
38 Ramer MS, Bisby MA (1999) Adrenergic innervation of rat sensory ganglia following proximal or distal painful sciatic neuropathy: distinct mechanisms revealed by anti-NGF treatment. *Eur J Neurosci* 11: 837–846
39 Jaggar SI, Scott HC, Rice AS (1999) Inflammation of the rat urinary bladder is associated with a referred thermal hyperalgesia which is nerve growth factor dependent. *Br J Anaesth* 83: 442–448
40 Rutkowski JL, Tuite GF, Lincoln PM, Boyer PJ, Tennekoon GI, Kunkel SL (1999) Signals for proinflammatory cytokine secretion by human Schwann cells. *J Neuroimmunol* 101: 47–60
41 Watkins LR, Maier SF (2002) Beyond neurons: evidence that immune and glial cells contribute to pathological pain states. *Physiol Rev* 82: 981–1011
42 Shamash S, Reichert F, Rotshenker S (2002) The cytokine network of Wallerian degeneration: tumor necrosis factor-alpha, interleukin-1alpha, and interleukin-1beta. *J Neurosci* 22: 3052–3060
43 Southall MD, Li T, Gharibova LS, Pei Y, Nicol GD, Travers JB (2003) Activation of epidermal vanilloid receptor-1 induces release of proinflammatory mediators in human keratinocytes. *J Pharmacol Exp Ther* 304: 217–222
44 Miller MS, Buck SH, Sipes IG, Yamamura HI, Burks TF (1982) Regulation of substance P by nerve growth factor: disruption by capsaicin. *Brain Res* 250: 193–196
45 Taylor DC, Pierau FK, Szolcsanyi J (1984) Long lasting inhibition of horseradish peroxidase (HRP) transport in sensory nerves induced by capsaicin pretreatment of the receptive field. *Brain Res* 298: 45–49
46 Bjerring P, Arendt-Nielsen L, Soderberg U (1990) Argon laser induced cutaneous sensory and pain thresholds in post-herpetic neuralgia. Quantitative modulation by topical capsaicin. *Acta Derm Venereol* 70: 121–125
47 Nolano M, Simone DA, Wendelschafer-Crabb G, Johnson T, Hazen E, Kennedy WR (1999) Topical capsaicin in humans: parallel loss of epidermal nerve fibers and pain sensation. *Pain* 81: 135–145
48 Karai L, Brown DC, Mannes AJ, Connelly ST, Brown J, Gandal M, Wellisch OM, Neubert JK, Olah Z, Iadarola MJ (2004) Deletion of vanilloid receptor 1-expressing primary afferent neurons for pain control. *J Clin Invest* 113: 1344–1352

49 Simone DA, Nolano M, Johnson T, Wendelschafer-Crabb G, Kennedy WR (1998) Intradermal injection of capsaicin in humans produces degeneration and subsequent reinnervation of epidermal nerve fibers: correlation with sensory function. *J Neurosci* 18: 8947–8959

50 Dux M, Sann H, Schemann M, Jancso G (1999) Changes in fibre populations of the rat hairy skin following selective chemodenervation by capsaicin. *Cell Tissue Res* 296: 471–477

51 Turnbull A (1850) Tincture of capsaicin as a remedy for chilblains and toothache. *Dublin Free Press* 1: 95–96

52 Mason L, Moore RA, Derry S, Edwards JE, McQuay HJ (2004) Systematic review of topical capsaicin for the treatment of chronic pain. *BMJ* 328(7446): 991

53 Paice JA, Ferrans CE, Lashley FR, Shott S, Vizgirda V, Pitrak D (2000) Topical capsaicin in the management of HIV-associated peripheral neuropathy. *J Pain Symptom Manage* 19: 45–52

54 Robbins WR, Staats PS, Levine J, Fields HL, Allen RW, Campbell JN, Pappagallo M (1998) Treatment of intractable pain with topical large-dose capsaicin: preliminary report. *Anesth Analg* 86: 579–583

55 Malmberg AB, Mizisin AP, Calcutt NA, Von Stein T, Robbins WR, Bley KR (2004) Reduced heat sensitivity and epidermal nerve fiber immunostaining following single applications of a high-concentration capsaicin patch. *Pain* 111: 360–367

56 Malan TP, Tuchman M, Mollen M, Brady S, Backonja MM, Michna E, Robbins W, Flood M, von Stein T (2004) A single one-hour application of high-concentration trans-capsaicin patches leads to four weeks of pain relief in postherpetic neuralgia (PHN) patients. Poster presented at World Congress of Anesthesiology, Paris, France

57 Simpson D, Brown S, Sampson J, Estanislao L, Vilahu C, Ramanathan S, Jermano J, von Stein T (2004) Novel high-concentration capsaicin patch for the treatment of painful HIV-associated distal symmentrical polyneuropathy: results of an open-label trial: http://www.retroconference.org/2004/cd/PDFs/490.pdf (Accessed February 2005)

58 Cruz F (2004) Mechanisms involved in new therapies for overactive bladder. *Urology* 63 (3 Suppl 1): 65–73

59 de Groat WC (1997) A neurologic basis for the overactive bladder. *Urology* 50 (6A Suppl): 36–52

60 Birder LA, Kanai AJ, de Groat WC, Kiss S, Nealen ML, Burke NE, Dineley KE, Watkins S, Reynolds IJ, Caterina MJ (2001) Vanilloid receptor expression suggests a sensory role for urinary bladder epithelial cells. *Proc Natl Acad Sci USA* 98: 13396–13401

61 Brady CM, Apostolidis A, Yiangou Y, Baecker PA, Ford AP, Freeman A, Jacques TS, Fowler CJ, Anand P (2004) P2X3-immunoreactive nerve fibres in neurogenic detrusor overactivity and the effect of intravesical resiniferatoxin. *Eur Urol* 46: 247–253

62 Yiangou Y, Facer P, Ford A, Brady C, Wiseman O, Fowler CJ, Anand P (2001) Capsaicin receptor VR1 and ATP-gated ion channel P2X3 in human urinary bladder. *BJU Int* 87: 774–779

63 Maggi CA, Barbanti G, Santicioli P, Beneforti P, Misuri D, Meli A, Turini D (1989) Cys-

tometric evidence that capsaicin-sensitive nerves modulate the afferent branch of micturition reflex in humans. *J Urol* 142: 150–154
64 Chancellor MB, de Groat WC (1999) Intravesical capsaicin and resiniferatoxin therapy: spicing up the ways to treat the overactive bladder. *J Urol* 162: 3–11
65 Szallasi A, Blumberg PM (1989) Resiniferatoxin, a phorbol-related diterpene, acts as an ultrapotent analog of capsaicin, the irritant constituent in red pepper. *Neuroscience* 30: 515–520
66 Lazzeri M, Beneforti P, Turini D (1997) Urodynamic effects of intravesical resiniferatoxin in humans: preliminary results in stable and unstable detrusor. *J Urol* 158: 2093–2096
67 Chancellor MB (2000) Discussion: resiniferatoxin-preliminary data. *Urology* 55 (5A Suppl): 65
68 ICOS Corporation Reports Results for 2003 (2003) http://www.corporate-ir.net/ireye/ir_site.zhtml?ticker=ICOS&script=410&layout=0&item_id=491321&sstring=resiniferatoxin (Accessed February 2005)
69 Kim JH, Rivas DA, Shenot PJ, Green B, Kennelly M, Erickson JR, O'Leary M, Yoshimura N, Chancellor MB (2003) Intravesical resiniferatoxin for refractory detrusor hyperreflexia: a multicenter, blinded, randomized, placebo-controlled trial. *J Spinal Cord Med* 26: 358–363
70 Dasgupta P, Chandiramani VA, Beckett A, Scaravilli F, Fowler CJ (2000) The effect of intravesical capsaicin on the suburothelial innervation in patients with detrusor hyperreflexia. *BJU Int* 85: 238–245
71 Yokoyama T, Nozaki K, Fujita O, Nose H, Inoue M, Kumon H (2004) Role of C afferent fibers and monitoring of intravesical resiniferatoxin therapy for patients with idiopathic detrusor overactivity. *J Urol* 172: 596–600
72 Masheb RM, Nash JM, Brondolo E, Kerns RD (2000) Vulvodynia: an introduction and critical review of a chronic pain condition. *Pain* 86: 3–10
73 Sonni L, Cattaneo A, De Marco A, De Magnis A, Carli P, Marabini S (1995) Idiopathic vulvodynia. Clinical evaluation of the pain threshold with acetic acid solutions. *J Reprod Med* 40: 337–341
74 Bohm-Starke N, Hilliges M, Brodda-Jansen G, Rylander E, Torebjork E (2001) Psychophysical evidence of nociceptor sensitization in vulvar vestibulitis syndrome. *Pain* 94: 177–183
75 Tympanidis P, Casula MA, Yiangou Y, Terenghi G, Dowd P, Anand P (2004) Increased vanilloid receptor VR1 innervation in vulvodynia. *Eur J Pain* 8: 129–133
76 Steinberg AC, Oyama IA, Rejba A, Kellog-Spadt S, Whitmore KE (2005) Capsaicin for the treatment of vulvar vestibulitis. *Am J Obstet Gynecol* 192: 1549–1553
77 Lysy J, Sistiery-Ittah M, Israelit Y, Shmueli A, Strauss-Liviatan N, Mindrul V, Keret D, Goldin E (2003) Topical capsaicin – a novel and effective treatment for idiopathic intractable pruritus ani: a randomised, placebo controlled, crossover study. *Gut* 52: 1323–1326
78 AlgoRx Pharmaceuticals, Inc. (2004) ALGRX 4975 – Capsaicin for Injection and Cap-

saicin Gel for Intraoperative Use. http://algorx.com/products/4975.htm (Accessed February 2005)
79 Schmidt RF (1996) The articular polymodal nociceptor in health and disease. *Prog Brain Res* 113: 53–81
80 Mach DB, Rogers SD, Sabino MC, Luger NM, Schwei MJ, Pomonis JD, Keyser CP, Clohisy DR, Adams DJ, O'Leary P, Mantyh PW (2002) Origins of skeletal pain: sensory and sympathetic innervation of the mouse femur. *Neuroscience* 113: 155–166
81 Niv D, Gofeld M, Devor M (2003) Causes of pain in degenerative bone and joint disease: a lesson from vertebroplasty. *Pain* 105: 387–392
82 Rains C, Bryson HM (1995) Topical capsaicin. A review of its pharmacological properties and therapeutic potential in post-herpetic neuralgia, diabetic neuropathy and osteoarthritis. *Drugs Aging* 7: 317–328
83 Winston Laboratories, Inc. (2005) Press release, January 25, 2005: http://www.pharmalexicon.com/medicalnews.php?newsid=19244
84 Hendiani JA, Westlund KN, Lawand N, Goel N, Lisse J, McNearney T (2003) Mechanical sensation and pain thresholds in patients with chronic arthropathies. *J Pain* 4: 203–211
85 Cordell GA, Araujo OE (1993) Capsaicin: identification, nomenclature, and pharmacotherapy. *Ann Pharmacother* 27: 330–336
86 Hua XY, Chen P, Hwang J, Yaksh TL (1997) Antinociception induced by civamide, an orally active capsaicin analogue. *Pain* 71: 313–322
87 Diamond S, Freitag F, Phillips SB, Bernstein JE, Saper JR (2000) Intranasal civamide for the acute treatment of migraine headache. *Cephalalgia* 20: 597–602
88 Saper JR, Klapper J, Mathew NT, Rapoport A, Phillips SB, Bernstein JE (2002) Intranasal civamide for the treatment of episodic cluster headaches. *Arch Neurol* 59: 990–994
89 Janusz JM, Buckwalter BL, Young PA, LaHann TR, Farmer RW, Kasting GB, Loomans ME, Kerckaert GA, Maddin CS, Berman EF et al (1993) Vanilloids. 1. Analogs of capsaicin with antinociceptive and antiinflammatory activity. *J Med Chem* 36: 2595–2604

TRPV1 agonist therapies in bladder diseases

Francisco Cruz, Carlos Silva and Paulo Dinis

Department or Urology, Hospital S. João and Faculty of Medicine of Porto, 4200-076 Porto, Portugal

Rationale for intravesical application of TRPV1 agonists

Clinical interest on TRPV1 agonists for the treatment of bladder diseases started with the observation that reflex bladder contractions triggered by bladder filling in intact or spinally transected cats had distinct sensibilities to systemic capsaicin. In spinally transected cats, but not in intact cats, reflex bladder contractions were completely suppressed by the neurotoxin [1]. The identification of two neuronal pathways involved in micturition control offered a solid explanation to this finding. A long pathway passing through the pontine micturition center and initiated in Aδ-, capsaicin-resistant, bladder sensory fibers controls reflex bladder contractions in mammals with intact spinal cord. A short neuronal pathway entirely lodged in the sacral spinal cord and initiated in type C, capsaicin-sensitive, bladder sensory fibers, usually inactive in adult mammals, is enhanced after spinalization and replaces the supra-spinal reflex [1]. Bladder contractions triggered by the sacral reflex usually call for treatment since they are distinct from those controlled by the supra-spinal reflex. First, they are totally involuntary and triggered by small volumes of bladder filling, characteristics that generate urge to urinate, urinary incontinence and high urinary frequency. Second, contractions driven by the sacral reflex lack coordination with urethral sphincter relaxation leading to bladder–sphincter dyssinergia, a situation that may increase intravesical pressure to potentially harmful levels for the upper urinary tract [1].

Additional experimental and clinical studies revealed that the capsaicin-sensitive sacral micturition reflex is enhanced in pathological conditions other than spinal-cord transection, including obstructive and idiopathic forms of overactive bladder. Local anaesthesia of the prostatic urethra increases bladder volume at which first sensation to urinate and maximal bladder capacity occur in patients with bladder-outlet obstruction caused by benign prostatic enlargement [2]. As lidocaine is more effective to anesthetize C- than Aδ-fibers, the contribution of C-fiber input to abnormal bladder reflex activity observed in some prostatic patients was strongly suggested [2]. This hypothesis was further reinforced by the report of a positive ice-

Turning up the Heat on Pain: TRPV1 Receptors in Pain and Inflammation, edited by Annika B. Malmberg and Keith R. Bley

water test in a high percentage of these patients. In fact, cold-induced bladder contractions depend upon a C-fiber-driven sacral reflex usually inactive in normal adults [3]. Also in accordance with this, a high density of nerve fibers staining for substance P and calcitonin-gene-related peptide was reported in the bladder mucosa of patients with idiopathic overactive bladder [4]. Observations carried out in rat models of chronic bladder-outlet obstruction similarly conclude for the facilitation of the capsaicin-sensitive micturition reflex. In normal rats, pelvic nerve stimulation evoked a long-latency parasympathetic output through the activation of the supraspinal micturition reflex. However, in rats with chronic bladder-outlet obstruction, stimulation of the pelvic nerve induced a short-latency parasympathetic outflow preceding the long-latency response, suggesting enhancement of the sacral micturition reflex [5].

TRPV1 agonists, by desensitizing capsaicin-sensitive bladder afferents, decrease their responsiveness to bladder distention [6, 7]. Since topical bladder application of capsaicin and resiniferatoxin (RTX) avoid the severe neurotoxicity observed during their systemic application [8], these two TRPV1 agonists were assayed in patients with neurogenic and non-neurogenic forms of overactive bladder. The intention of this chapter is to review the results of the clinical studies carried on with intravesical TRPV1 agonists and to discuss possible mechanisms of action.

Clinical experience with intravesical TRPV1 agonists in neurogenic forms of bladder overactivity

Spinal-cord injury patients presenting high urinary frequency, urgency and incontinence frequently require vigorous treatment to improve their social integration and quality of life. Anticholinergic or smooth muscle-relaxant drugs are among the first-line therapeutic options. However, as these compounds do not completely avoid frequent voidings or non-voluntary urine loss without giving rise to unacceptable side effects [9], clinical trials with intravesical TRPV1 agonists were pushed forward in several medical centers devoted to spinal patients.

More than 100 patients with different types of spinal-cord lesions received intravesical capsaicin in seven non-controlled [10–16] and one controlled [17] clinical trial. Treatments followed a similar methodology. Capsaicin was dissolved in 30% ethanol and 100–125 ml (or half of the bladder capacity if lower than that volume) of 1–2 mM solutions were instilled into the bladder and left in contact with the mucosa for 30 min.

Best clinical results were found among patients who maintained some degree of bladder sensation and emptied the bladder by micturition. Most of these patients had incomplete spinal-cord lesions caused by multiple sclerosis, trauma or infectious diseases. After an initial period of 2 weeks during which urinary symptoms worsened, complete continence or satisfactory improvement was reported by 70–90% of the patients [11, 15, 16]. In addition, capsaicin also decreased their urinary frequency and urge to urinate [11, 15, 16]. The duration of capsaicin's effect on micturition symptoms was long-lasting, exceeding 6 or even 9 months in some cases [11, 15, 16]. Upon reinstillation, capsaicin maintained the efficacy found at the first administration [11, 15, 16]. In patients who lacked bladder sensation or who emptied the bladder without warning due to totally reflex bladder contractions the success rate was, on the contrary, very poor. Among patients with complete cervical cord lesions full urinary continence was achieved in 20% and another 20% found partial relief of this problem [12].

Cystometric studies showed that intravesical capsaicin increased bladder capacity in 70–90% of the patients, the increase above pretreatment capacity ranging between 47 and 156%. In three studies maximal pressure during bladder contractions was also significantly decreased [11, 16, 17], an important aspect in the management of spinal patients in order to prevent damage to the upper urinary tract.

A randomized, controlled study comparing capsaicin with 30% ethanol confirmed the superiority of the TRPV1 agonist against the vehicle solution [17]. Twenty cases were randomized to receive either capsaicin or 30% ethanol. All patients treated with capsaicin reported improvement of urge sensation and urinary incontinence whereas only one ethanol-treated patient noticed some improvement. On cystometry, only capsaicin-treated patients had an increase in the bladder volume to micturition [17].

The first non-controlled trials with RTX included 54 patients, most of them with incomplete spinal cord lesions [18–22] in which different RTX concentrations were tested. Vehicle solutions included 10% ethanol [18, 21, 22] or saline [19, 20]. RTX instillation in patients with preserved bladder sensation did not cause severe suprapubic burning pain. In addition, also in contrast with capsaicin did RTX not exacerbate the urinary symptoms transiently. In most studies [18–21], 50–100 nM RTX brought an immediate improvement or disappearance of urinary incontinence in a large number of patients (67–85%), and a 30% decrease in daily urinary frequency. The effect was long-lasting, being observed up to 12 months after the instillation [18, 21]. The only exception came from a recent report [22] where improvement was only found in 32% of the patients. However, in this study the concentration of RTX effectively administered was imprecise due to long storage of the solution in plastic containers that absorbed the compound [22]. With 10 nM doses a very short-lasting improvement of urinary symptoms was obtained [19]. In patients treated with doses so high as 10 μM, a transient urinary retention due to detrusor hipoactivity occurred [20]. Thus, at the moment 50 or 100 nM solutions seem ideal to start intravesical RTX in neurogenic patients. Urodynamic improvement was observed in most of the RTX-treated patients. The volume to first detrusor contraction increased 40% after 50–100 nM RTX. Maximal bladder capacity increased 25% after 10 nM [19], 80% after 50–100 nM [18, 21] and 120% after 10 μM solutions [20]. Maximal detrusor pressure was decreased only after 10 μM solutions [20].

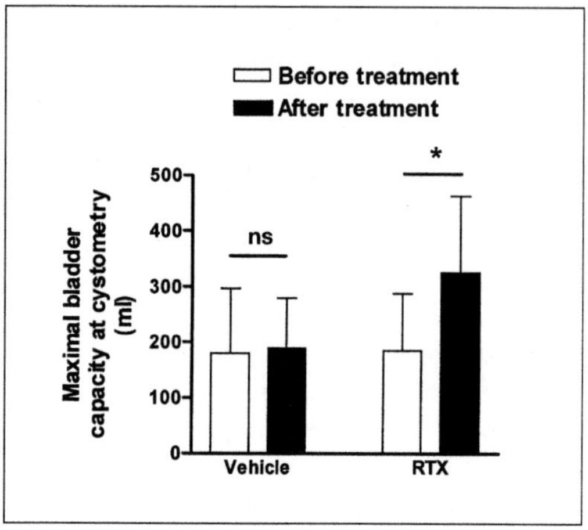

Figure 1
Maximal bladder capacity in a randomized clinical trial in which patients received intravesical 50 nM RTX or its vehicle solution (10% ethanol). At baseline maximal bladder capacity was identical in both groups (open bars). Maximal bladder capacity increased 3 months later (black bars) only in patients who received the TRPV1 agonist. *$P<0.05$; *ns, non-significant.*

RTX was compared against the vehicle solution (10% ethanol in saline) in a randomized study [23]. The discomfort reported by patients was similarly low in both arms. A significant improvement of urinary frequency and incontinence was found only in the RTX arm. In addition, only patients receiving RTX had a significant increase of the volume to first detrusor contraction and of maximal bladder capacity; 35 and 80% above the pre-treatment values, respectively (Fig. 1). Three studies compared 50–100 nM RTX with 1–2 mM capsaicin [24–26]. In general, RTX solutions were better tolerated and their effects on continence and urodynamic parameters were equivalent [26] or superior [25] to those induced by capsaicin solutions.

Clinical experience with intravesical TRPV1 agonists in obstructive forms of overactive bladder

Symptoms of urgency, frequency, nocturia and urge incontinence are highly prevalent and bothersome among patients with prostatic enlargement due to benign prostatic hyperplasia (BPH) [27]. In addition, they are more refractory to α-blockers and 5α-reductase inhibitors, two commonly prescribed medications for BPH patients, than poor urinary stream or the sensation of incomplete bladder emptying.

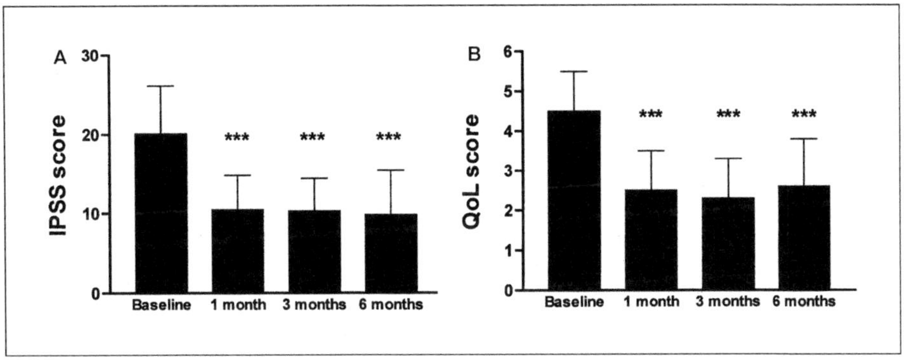

Figure 2
*Mean International Prostate Score Symptom (IPSS; A) and quality of life (QoL; B) scores in 12 patients with benign prostatic hyperplasia at baseline and at 1, 3 and 6 months after intravesical 50 nM RTX. ***P<0.001 against baseline values. Reproduced from [28], with permission.*

As some studies have shown the enhancement of the spinal micturition reflex by chronic bladder-outlet obstruction (see above), we decided to assay the effect of C-fiber desensitization on the intensity of urinary symptoms in BPH patients [28].

Twelve males with a mean age of 60 years and a mean prostate volume of 45 ml received a single bladder instillation of a 50 nM RTX solution in 10% ethanol over 30 min [28]. All of them complained of urgency, frequency and nocturia, and half of them had episodes of urge incontinence. The intensity and bother of urinary symptoms were quantified by the International Prostate Score Symptom (IPSS) questionnaire and associated quality of life (QoL) questions. On average, RTX decreases IPSS and QoL scores by half (Fig. 2), with 91% of the patients having reported at least a 25% improvement [28]. Mean urinary frequency decreased from 15 to about 10 micturitions per day. Urge incontinence, which was present in six cases, disappeared in four and decreased to less than half in the other two. In addition, at cystometry, mean volume at which patients reported the first desire to void and the urge sensation to urinate increased 90 and 55%, respectively [28].

Another trial, involving 18 BPH patients in whom urge incontinence due to bladder overactivity persisted in spite of having been submitted to prostatectomy, confirmed the effect of RTX in BPH-associated urge incontinence [29]. The treatment dose was 100 nM in 10% ethanol. In spite of being a subgroup of patients particularly difficult to handle, 11 patients (61%) improved due to a decrease of 50% or more in the number of urge-incontinence episodes [29]. At present a randomized, clinical trial comparing the effect of intravesical 50 nM RTX with its vehicle solution (10% ethanol) is being organized in order to confirm these preliminary data.

Clinical experience with intravesical TRPV1 agonists in idiopathic forms of overactive bladder

Idiopathic overactive bladder is a chronic condition which is probably the leading cause of high urinary frequency, urge sensation to urinate and urge incontinence among the adult population [30]. At urodynamic testing, the bladder of many of these patients will display involuntary contractions during filling [30]. In spite of the considerable efforts to clarify the origin of such abnormal smooth muscle activity, the pathologic mechanisms underlying idiopathic overactive bladder remain unknown.

In a recent study we applied intravesically RTX to a small group of these patients [31]. The objectives were to clarify the role of C-fiber input in the emergence of urinary symptoms and, at the same time, investigate the potential therapeutic role of TRPV1 agonists in idiopathic overactive bladder. Eleven females and two males with a mean age of 50 years and urinary symptoms for more than 1 year, all presenting involuntary bladder smooth muscle contractions during urodynamic testing, were treated with 50 nM RTX in 10% ethanol [31]. RTX delayed or suppressed involuntary detrusor contractions during filling cystometry (Fig. 3A). In consequence, the volume of bladder filling at which first involuntary bladder contraction occurred increased in more than 90% of the patients, the mean post-treatment volume exceeding the pre-treatment one by almost 200% (Fig. 3B) [31]. Maximal bladder filling tolerated also increased [31]. Both changes were sustained up to 6 months (Fig. 3B). Intravesical RTX improved urinary incontinence, 25% achieving full continence and 50% having a decrease in the daily number of incontinence episodes superior to 50%. Mean urinary frequency also decreased [31].

A similar study conducted by Palma and co-workers [32], which included 12 women with an average age of 65 years, reproduced our findings. Episodes of urinary incontinence decreased in average from four to one during the daytime and from three to one during sleep. This was accompanied by a decrease in urinary frequency and by a 125% increase in the volume at which first involuntary bladder smooth muscle contraction occurred.

Figure 3
(A) Filling cystometry of a patient with idiopathic overactive bladder at baseline (left) and after instillation of 50 nM RTX (right). The top line (P_{det}), which indicates pressure dependent upon bladder smooth muscle contraction, shows that the volume at which the first involuntary contraction occurred increased from 135 to 483 ml. P_{det}, detrusor pressure; P_{ves}, vesical pressure; P_{abd}, abdominal pressure; $P_{det} = P_{ves} - P_{abd}$; V_{infus}, volume infused. (B) Mean volume to first involuntary bladder contraction (open bars) and maximal bladder capacity (black bars) at baseline (0) and at 30 and 90 days after 50 nM RTX in 13 patients with idiopathic overactive bladder. Reproduced from [31], with permission.

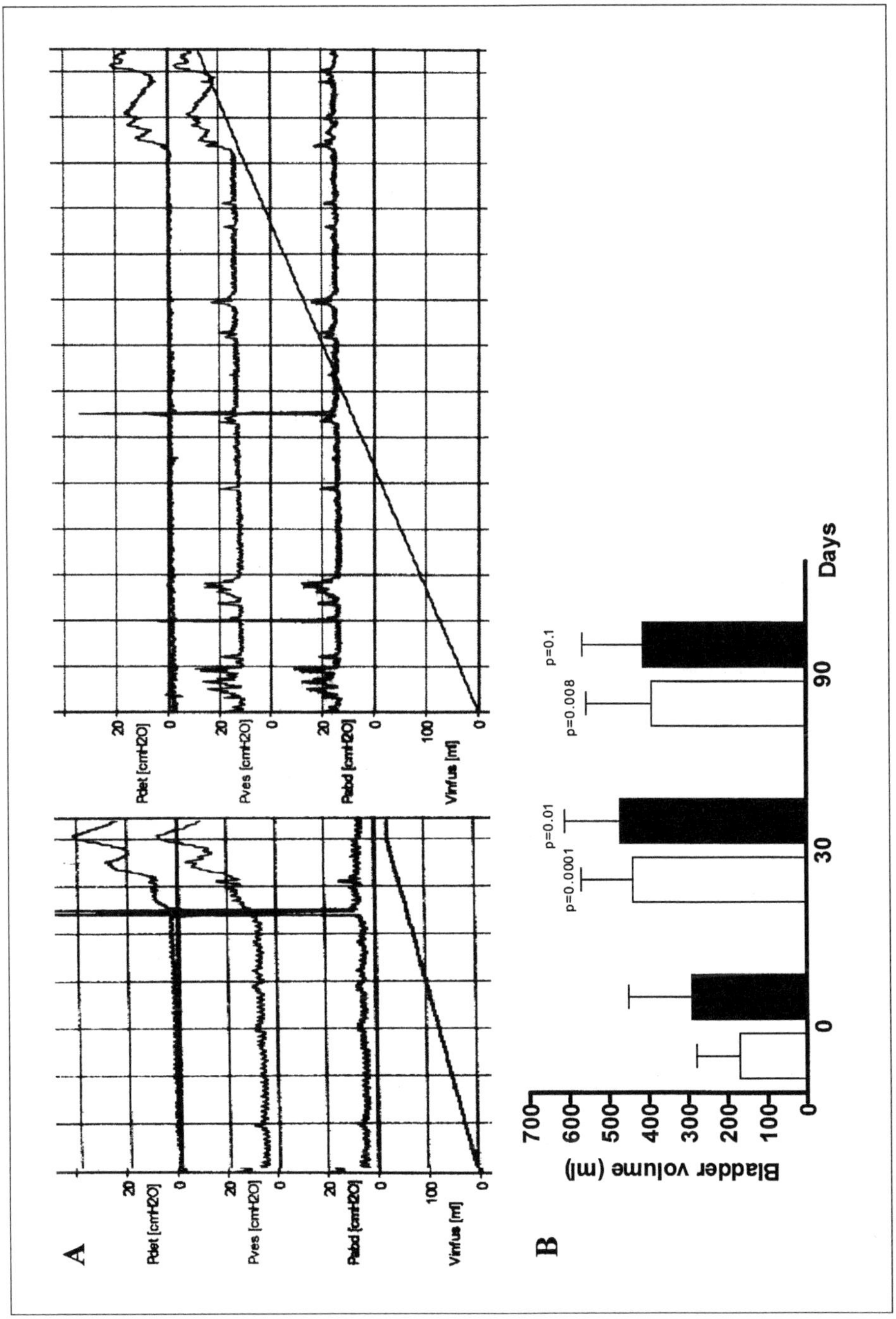
A
Pdet [cmH2O]
Pves [cmH2O]
Pabd [cmH2O]
Vinfus [ml]
B
Bladder volume (ml)
Days
p=0.0001
p=0.01
p=0.008
p=0.1

Hence, from these two studies it should be concluded that C-fiber input plays an important role in idiopathic overactive bladder and that TRPV1 agonists should be considered in the treatment of this bladder disease.

Mechanisms of action of intravesical vanilloids

Capsaicin and RTX cause a long-lasting suppression of sensory activity in type-C primary afferent fibers by a process that requires binding to TRPV1 receptors and Ca^{2+} inflow into the sensory neuron [33]. In bladder afferents this process is accompanied by a loss of immunoreactivity to TRPV1 [22, 34] and P2X3 receptors [35] and to neuropeptides like substance P and calcitonin-gene-related peptide [8]. By contrast, immunoreactivity to galanin, a peptide usually not expressed by bladder sensory neurons, is increased, possibly related to the up-regulation of the c-*jun* gene [36]. Eventually, loss of peripheral sensory terminal may occur [22, 35], although an ultrastructural inspection of rat bladders 24 h after intravesical RTX was unable to detect acutely degenerated nerve fibers [8].

A few studies suggested that local application of TRPV1 agonists decreases nerve growth factor (NGF) in sensory fibers [36, 37]. Possible mechanisms include a reduced retrograde transport of NGF from the periphery into dorsal root ganglions or a reduced uptake of the neurotrophic factor by the peripheral sensory endings [38]. Whatever the mechanism, this is an interesting finding taking into consideration that high levels of NGF have been identified in bladders after spinal-cord transection [39, 40] or after prolonged bladder-outlet obstruction [41]. As NGF decreases bladder C-fiber threshold and turns normal bladders hyperactive [42], reducing the bladder levels of this neurotrophic factor has emerged as an interesting concept to prevent the enhancement of the C-fiber-driven spinal micturition reflex [43]. First attempts showed that NGF inactivation was able to reduce bladder overactivity in animal models of spinalization or bladder-outlet obstruction [40, 43]. If future studies will confirm the possibility of reducing the deleterious effects of an excess of bladder NGF by intravesical TRPV1 agonits, then the therapeutic indications of these compounds will be substantially expanded.

Safety of intravesical TRPV1 agonists

The most frequent problem seen with capsaicin instillation is supra-pubic burning pain felt by patients with preserved bladder sensation. It starts immediately after the beginning of the treatment and may require vigorous analgesic medication and prompt capsaicin evacuation. Preliminary bladder anesthesia with lidocaine may provide partial relief [11, 15]. In addition, a transient worsening of the urinary symptoms may occur during the first 1–2 weeks after instillation, which should not be interpreted as a treatment failure [11, 15]. Another side effect of capsaicin needs to be stressed. In patients with high, complete spinal-cord lesions, capsaicin can trigger severe, life-threatening episodes of autonomic dysreflexia [12, 15].

In contrast to capsaicin, RTX does not evoke significant bladder pain. In spinal patients that had already received intravesical capsaicin, discomfort caused by 50 or 100 nM RTX instillation was one-fifth of that induced by 1 mM capsaicin [21] and similar to that caused by instillation of the vehicle solution, 10% ethanol [23]. In addition, no cases of autonomic dysreflexia were reported with these doses [18–23]. In patients with non-neurogenic forms of bladder overactivity, who in contrast to spinal patients conserve normal bladder-pain sensation, instillation of 50 nM RTX also evoked a minimal discomfort which did not require analgesic treatment or bladder evacuation of the solution [28, 31, 32].

Human contact with capsaicin and RTX was not initiated with intravesical application of these substances. Capsaicin spices the diet of millions of people and historical reports state that unguents prepared with the latex of *Euphorbia resinifera*, the cactus plant from which RTX is extracted, have analgesic properties [33]. Nevertheless, the consequence for the bladder mucosa of the instillation of TRPV1 agonists needs careful evaluation. Until the moment, bladder biopsies obtained from patients repeatedly instilled with capsaicin [44] or RTX [45] and examined under the conventional and electronic microscope did not detect any significant modification of the bladder wall brought about by intravesical TRPV1 agonists.

Other possible applications of intravesical TRPV1 agonists

Intravesical administration of capsaicin [6] or RTX [7] can prevent pain induced by acute bladder inflammation in the rat, as shown by the suppression of c-*fos* gene expression in the spinal cord. An eventual analgesic effect of intravesical vanilloids in pain generated by chronic inflammatory bladder conditions would, therefore, be extremely relevant in a clinical setting. Chronic inflammatory pain is frequently disproportional to the intensity of the stimulus and is more resistant to common analgesics than acute pain. To investigate the effect of intravesical vanilloids in chronic inflammatory pain at the experimental level, a rat model of chronic cystitis induced by cyclophosphamide was used [46]. Instillation of a 100 nM RTX solution in cyclophosphamide-inflamed bladders prevented spinal c-*fos* overexpression induced by physiologic bladder contractions (Fig. 4) [46]. In addition, increased micturition frequency accompanying bladder inflammation was analogously suppressed. The potential therapeutic application of these observations remains unknown. Maggi and co-workers [47], however, recognized that intravesical capsaicin could reduce pain in patients with bladder hypersensitivity disorders. Further studies with capsaicin [15, 48, 49] or RTX [50, 51] in these patients, including two placebo-con-

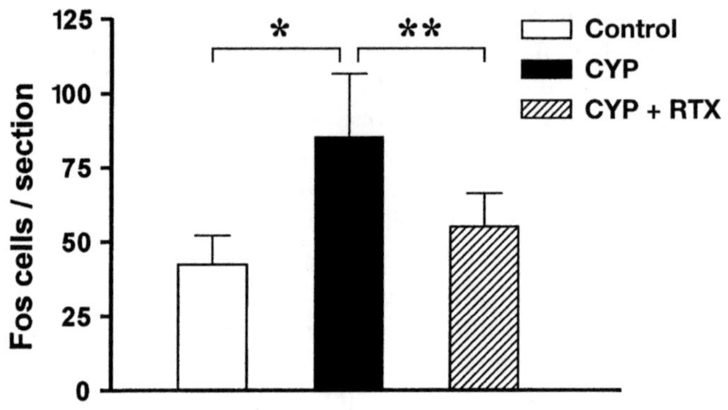

Figure 4
Number of Fos-immunoreactive cells per L6 spinal-cord section in sham-inflamed (open bar), cyclophosphamide-inflamed (CYP; black bar) and cyclophosphamide-inflamed+10 nM RTX-treated (hatched bar) rats after physiological bladder distention. *$P<0.05$, **$P<0.01$.

trolled trials, confirmed the clinical utility forwarded by Maggi's initial observations. Future research will clarify whether these compounds can be useful in the treatment of severe bladder pain in patients with interstitial cystitis.

Conclusions

Findings of several clinical trials in which TRPV1 agonists have been used intravesically show that desensitization of bladder sensory innervation may have a place in the treatment of urge incontinence, high urinary frequency and bladder pain. In addition, the same clinical trials gave an important contribution to the pathophysiology of the overactive bladder by confirming the pivotal role of sensory input conveyed in capsaicin-sensitive fibers to this condition. At the moment RTX seems the ideal TRPV1 agonist for clinical use. It is not harmful to the bladder mucosa and its instillation does not evoke significant bladder pain. However, it is probable that future TRPV1 agonists still to be developed will prove even more effective than RTX.

It should also be kept in mind that C-fiber desensitization induced by intravesical RTX or capsaicin is a crude phenomenon during which C-fibers lose many of their neurochemical and receptor characteristics or even degenerate. Hence, among the receptors expressed by C-fibers, the identification of those encoding for the degree of bladder filling and urge sensation to urinate looks like an indispensable step. Such studies may, additionally, uncover appropriate receptor antagonists for clinical use. At present the most obvious receptors to be studied are TRPV1 and P2X3 receptors, which previous studies have shown to be largely co-expressed in type-C sensory fibers [52]. Knockout mice for these receptor genes have a volume to reflex bladder contraction that is higher than that found in wild-type animals [53, 54]. In addition, in men with neurogenic detrusor overactivity who received intravesical RTX it was shown that bladder capacity is inversely correlated with the density of TRPV1- and P2X3-expressing nerve fibers coursing in the bladder wall [22, 35]. Following this rationale, a recent study [55] showed that capzasepine, the specific TRPV1 antagonist, could decrease the frequency of reflex bladder contractions in a rat model of chronic bladder inflammation. Thus, it is likely that in the near future effective TRPV1 antagonists will also prove effective in the control of micturition symptoms in patients with overactive bladder.

Acknowledgment
This work was funded by the Portuguese government, through the Fundação Ciência Tecnologia, project POCTI/FEDER32466/NSE/00.

References

1 de Groat WC (1997) A neurologic basis for the overactive bladder. *Urology* 50 (Suppl 6A): 36–52
2 Yokoyama O, Nagano K, Kawaguchi K, Ueki O, Mitsuo O (1994) The influence of prostatic urethral anesthesia in overactive detrusor in patients with benign prostatic hyperplasia. *J Urol* 151: 1554–1556
3 Chai TC, Gray ML, Steers W (1998) The incidence of a positive ice water test in bladder outlet obstructed patients: evidence for bladder neural plasticity. *J Urol* 160: 34–38
4 Smet PJ, Moore KH, Jonavicius J (1997) Distribution and colocalization of calcitonin gene-related peptide tachykinins and vasoactive intestinal peptide in normal and idiopathic unstable human urinary bladder. *Lab Invest* 77: 37–49
5 Steers WD, de Groat WC (1988) Effect of bladder outlet obstruction on micturition reflex pathways in the rat. *J Urol* 140: 864–871
6 Cruz F, Avelino A, Coimbra A (1996) Desensitization follows excitation of bladder primary afferents by intravesical capsaicin, as shown by c-fos activation in the rat spinal cord. *Pain* 64: 553–557
7 Avelino A, Cruz F, Coimbra A (1999) Intravesical resiniferatoxin desensitizes rat bladder sensory fibres without causing intense noxious excitation. A c-fos study. *Eur J Pharmacol* 378: 17–22
8 Avelino A, Cruz F (2000) Peptide immunoreactivity and ultrastructure of rat urinary

bladder nerve fibers after topical desensitization by capsaicin or resiniferatoxin. *Auton Neurosci* 86: 37–46

9 Chapple CR (2000) Muscarinic receptor antagonists in the treatment of overactive bladder. *Urology* 55 (Suppl 5A): 33–46

10 Fowler CJ, Jewkes D, McDonald WI, Lynn B, de Groat WC (1992) Intravesical capsaicin for neurogenic bladder dysfunction. *Lancet* 339: 1239

11 Fowler CJ, Beck RO, Gerrard S, Betts CD, Fowler CG (1994) Intravesical capsaicin for the treatment of detrusor hyperreflexia. *J Neurol Neurosurg Psychiatry* 57: 169–173

12 Geirsson G, Fall M, Sullivan L (1995) Clinical and urodynamic effects of intravesical capsaicin treatment in patients with chronic traumatic spinal detrusor hyperreflexia. *J Urol* 154: 1825–1829

13 Das A, Chancellor MB, Watanabe T, Sedor J, Rivas DA (1996) Intravesical capsaicin in neurogenic impaired patients with detrusor hyperreflexia. *J Spinal Cord Med* 19: 190–193

14 Igawa Y, Komiyama I, Nishizawa O, Ogawa A (1996) Intravesical capsaicin inhibits autonomic dysreflexia in patients with spinal cord injury. *Neurourol Urodyn* 15: 374–375

15 Cruz F, Guimarães M, Silva C, Rio ME, Coimbra A, Reis M (1997) Desensitization of bladder sensory fibers by intravesical capsaicin has long lasting clinical and urodynamic effects in patients with hyperactive or hypersensitive bladder dysfunction. *J Urol* 157: 585–589

16 De Ridder D, Chandiramani V, Dasgupta P, Van Poppel H, Baert L, Fowler CJ (1997) Intravesical capsaicin as a treatment for refractory detrusor hyperreflexia: a dual center study with long-term followup. *J Urol* 158: 2087–2092

17 de Séze M, Wiart L, Joseph PA, Dosque JP, Mazaux JM, Barat M (1998) Capsaicin and neurogenic detrusor hyperreflexia. A double blind placebo controlled study in 20 patients with spinal cord lesions. *Neurourol Urodyn* 17: 513–523

18 Cruz F, Guimarães M, Silva C, Reis M (1997) Supression of bladder hyperreflexia by intravesical resiniferatoxin. *Lancet* 350: 640–641

19 Lazzeri M, Beneforti P, Turini D (1997) Urodynamic effects of intravesical resiniferatoxin in humans: preliminary results in stable and unstable detrusor. *J Urol* 158: 2093–2096

20 Lazzeri M, Spinelli M, Beneforti P, Zanollo A, Turini D (1998) Intravesical resiniferatoxin for the treatment of detrusor hyperreflexia refractory to capsaicin in patients with chronic spinal cord diseases. *Scand J Urol Nephrol* 32: 331–334

21 Silva C, Rio ME, Cruz F (2000) Desensitization of bladder sensory fibers by intravesical resiniferatoxin, a capsaicin analog: long-term results for the treatment of detrusor hyperreflexia. *Eur Urol* 38: 444–452

22 Brady CM, Apostolidis AN, Harper M, Yiangou Y, Beckett A, Jacques TS, Freeman A, Scaravilli F, Fowler CJ, Anand P (2004) Parallel changes in bladder suburothelial vanilloid receptor TRPV1 and pan-neuronal marker PGP9.5 immunoreactivity in patients

with neurogenic detrusor overactivity after intravesical resiniferatoxin treatment. *BJU Int* 93: 770–776
23 Cruz F, Silva C, Ribeiro M, Avelino A (2002) The effect of intravesical resiniferatoxin in neurogenic forms of bladder overactivity. Preliminary results of a randomised placebo controlled clinical trial. *Neurourol Urodyn* 21: 426–427
24 Park WH, Kim HG, Park BJ, Kim DY, Oh DJ, Lim YS, Shim HB (1999) Comparison of the effects of intravesical capsaicin and resiniferatoxin for treatment of detrusor hyperreflexia in patients with spinal cord injury. *Neurourol Urodyn* 18: 402
25 Giannantoni A, Di Stasi SM, Stephen RL, Navarra P, Scivoletto G, Mearini E, Porena M (2002) Intravesical capsaicin versus resiniferatoxin in patients with detrusor hyperreflexia: a prospective randomized study. *J Urol* 167: 1710–1714
26 de Seze M, Wiart L, de Seze MP, Soyeur L, Dosque JP, Blajezewski S, Moore N, Brochet B, Mazaux JM, Barat M, Joseph PA (2004) Intravesical capsaicin *versus* resiniferatoxin for the treatment of detrusor hyperreflexia in spinal cord injured patients: a double-blind, randomized, controlled study. *J Urol* 171: 251–255
27 Peters TJ, Donovan JL, Kay HE, Abrams P, de la Rosette JJMCH, Porru D, Thuroff JW (1997) The International Continence Society "Benign Prostatic Hyperplasia" study: the bothersomeness of urinary symptoms. *J Urol* 157: 885–889
28 Dinis P, Silva J, Ribeiro MJ, Avelino A, Reis M, Cruz F (2004) Bladder C-fiber desensitization induces a long-lasting improvement of BPH-associated storage LUTS. A pilot study. *Eur Urol* 46: 88–93; discussion 93–94
29 Kuo H (2003) Effectiveness of intravesical resiniferatoxin for anticholinergic treatment refractory detrusor overactivity due to nonspinal cord lesions. *J Urol* 170: 835–839
30 Debruyne FJM, Heesakkers JPFA, (2004) Clinical and socioeconomic relevance of overactive bladder. *Urology* 63 (3A): 42–44
31 Silva C, Ribeiro MJ, Cruz F (2002) The effect of intravesical resiniferatoxin in patients with idiopathic detrusor instability suggests that involuntary detrusor contractions are triggered by C-fiber input. *J Urol* 168: 575–579
32 Palma P, Dambros M, Thiel M, Fraga R, Riccetto CLZ, Netto NR (2003) Resultados preliminares da resiniferatoxina no tratamento de pacientes com instabilidade detrusora. *Urodinamica Aplicada* 16: 261–267
33 Szallasi A, Blumberg PM (1999) Vanilloid (capsaicin) receptors and mechanisms. *Pharmacol Rev* 51: 159–211
34 Avelino A, Cruz C, Nagy I, Cruz F, (2002) Vanilloid receptor 1 expression in the rat urinary tract. *Neuroscience* 109: 787–798
35 Brady C, Apostolidis AN, Yiangou Y, Baecker PA, Ford AP, Fowler CJ, Anand P (2003) P2X3-immunoreactive nerve fibers in neurogenic detrusor overactivity and the effect of intravesical resiniferatoxin (RTX). *Eur Urol* 2 (Suppl): 93 (abstract)
36 Avelino A, Cruz C, Cruz F (2002) Nerve growth factor regulates galanin and c-jun overexpression occurring in dorsal root ganglion cells after intravesical resiniferatoxin application. *Brain Res* 951: 264–269

37 Miller MS, Buck SH, Sipes IG, Yamamura HI, Burks TF (1982) Regulation of substance P by nerve growth factor: disruption by capsaicin. *Brain Res* 250: 193–196

38 Taylor DCM, Pierau FK, Szolcsányi J (1984) Long lasting inhibition of horseradish peroxidase (HRP) transport in sensory nerves induced by capsaicin treatment of the receptive field. *Brain Res* 298: 45–49

39 Vizzard MA (2000) Changes in urinary bladder neurotrophic factor mRNA and NGF protein following urinary bladder dysfunction. *Exp Neurol* 161: 273–284

40 Seki S, Sasaki K, Fraser MO, Igawa Y, Nishizawa O, Chancellor MB, de Groat WC, Yoshimura N (2002) Immunoneutralization of nerve growth factor in lumbosacral spinal cord reduces bladder hyperreflexia in spinal cord injured rats. *J Urol* 168: 2269–2274

41 Steers WD, Kolbeck S, Creedon D, Tuttle JB (1991) Nerve growth factor in the urinary bladder of the adult regulates neuronal form and function. J Clin Invest 88: 1709–1715

42 Chuang YC, Fraser MO, Yu Y, Chancellor MB, de Groat WC, Yoshimura N (2001) The role of bladder afferent pathways in bladder hyperactivity induced by the intravesical administration of nerve growth factor. *J Urol* 165: 975–979

43 Steers WD, Creedon DJ, Tuttle JB (1996) Immunity to nerve growth factor prevents afferent plasticity following urinary bladder hypertrophy. *J Urol* 155: 379–385

44 Dasgupta P, Chandiramani V , Parkinson MC, Beckett A, Fowler CJ (1998) Treating the human bladder with capsaicin: is it safe? *Eur Urol* 33: 28–31

45 Silva C, Avelino A, Souto-Moura C, Cruz F (2001) A light and electron microscope histopathological study of the human bladder mucosa after intravesical resiniferatoxin application. *BJU Int* 88: 355–360

46 Dinis P, Charrua A, Avelino A, Cruz F (2004) Intravesical resiniferatoxin decreases spinal c-fos expression and increases bladder volume to reflex micturition in rats with chronic inflamed urinary bladders. *BJU Int* 94: 153–157

47 Maggi CA, Barbanti G, Santicioli P, Beneforti P, Misuri D, Meli A, Turini D (1989) Cystometric evidence that capsaicin-sensitive nerves modulate the afferent branch of micturition reflex in humans. *J Urol* 142: 150–154

48 Barbanti G, Maggi CA, Beneforti P, Baroldi P, Turini D (1993) Relief of pain following intravesical capsaicin in patients with hypersensitive disorders of the lower urinary tract. *BJU* 71: 686–691

49 Lazzeri M, Beneforti P, Benaim G, Maggi CA, Lecci A, Turini D (1996) Intravesical capsaicin for treatment of severe bladder pain: a randomized placebo controlled study. *J Urol* 156: 947–952

50 Lazzeri M, Beneforti P, Spinelli M, Zanollo A, Barbagli G, Turini D (2000) Intravesical resiniferatoxin for the treatment of hypersensitive disorder: a randomized placebo controlled study. *J Urol* 164: 676–679

51 Lazzeri M, Spinelli M, Beneforti P, Malaguti S, Giardiello G, Turini D (2004) Intravesical infusion of resiniferatoxin by a temporary in situ drug delivery system to treat interstitial cystitis: a pilot study. *Eur Urol* 45: 98–102

52 Guo A, Vulchanova L, Wang J, Li X, Elde R (1999) Immunocytochemical localization

of the vanilloid receptor 1 (VR1): relationship to neuropeptides, the P2X3 purinoreceptor and IB4 binding sites. *Eur J Neurosci* 11: 946–958
53 Cockayne DA, Hamilton SG, Zhu QM, Dunn PM, Zhong Y, Novakovic S, Malmberg AB, Cain G, Berson A, Kassotakis L et al (2000) Urinary bladder hyporeflexia and reduced pain-related behaviour in P2X3-deficient mice. *Nature* 407: 1011–1015
54 Birder LA, Nakamura Y, Kiss S, Nealen ML, Barrick S, Kanai AJ, Wang E, Ruiz G, De Groat WC, Apodaca G et al (2002) Altered urinary bladder function in mice lacking the vanilloid receptor TRPV1. *Nat Neurosci* 5: 856–860
55 Avelino A, Dinis P, Charrua A., Nagy I, Yacoob M, Cruz F (2003) The endogenous TRPV1 ligand anandamide increases in the rat inflamed urinary bladder and may contribute to inflammatory pain. Program 608.3. Abstract Viewer/Itinerary Planner. Society for Neuroscience, Washington DC

TRPV1 antagonists and chronic pain

Kenneth J. Valenzano[1], *James D. Pomonis*[2] *and Katharine Walker*[3]

[1]Amicus Therapeutics, 6 Cedarbrook Drive, Cranbury, NJ 08512, USA; [2]Algos Therapeutics, 1246 University Ave W, Suite 205, St. Paul, MN 55104, USA; [3]Beatson Institute for Cancer Research, Garscube Estate, Switchback Road, Glasgow G61 1BD, Scotland, UK

Introduction

The receptor mediating the pungent effects of capsaicin, identified as TRPV1 in 1997, has long been of interest as a therapeutic target for drug discovery. Capsaicin is well known for its ability to induce nocifensive and hyperalgesic behaviors in many different species and intense burning pain in humans when injected subcutaneously or applied topically [1–5]. As discussed in the preceding chapters of this book, TRPV1 is not only activated by capsaicin and the ultra-potent resiniferatoxin (RTX), but also by a variety of noxious physical and chemical stimuli, particularly those associated with inflammatory processes such as low extracellular pH, products of the lipoxygenase pathway and heat. These stimuli may act independently or in concert to alter the activity of TRPV1 and, in turn, the sensitivity of TRPV1-expressing sensory neurons. These characteristics suggest a role for TRPV1 as an integrator of multiple pain-producing stimuli, and have stimulated significant efforts to identify small-molecule inhibitors of TRPV1 function as potential therapeutics to treat pain.

Several molecularly diverse TRPV1 antagonists have recently been reported. In some cases these compounds have been examined in pre-clinical models of acute, inflammatory or neuropathic pain. This chapter will focus primarily on the discovery of new TRPV1 modulators and their effects on pain modulation.

From agonists to antagonists

To date, only the TRPV1 agonist capsaicin has been readily available for the treatment of clinical pain syndromes. Topical capsaicin is available in a variety of preparations without prescription and has been shown to provide mild pain relief to osteoarthritis patients as well as to reduce and prevent itching associated with chronic pruritus [6, 7]. Prolonged topical capsaicin treatment has also shown efficacy in a large number of painful syndromes, including post-herpetic neuralgia,

Turning up the Heat on Pain: TRPV1 Receptors in Pain and Inflammation, edited by
Annika B. Malmberg and Keith R. Bley

post-mastectomy pain, phantom limb pain, reflex sympathetic dystrophy, trigeminal neuralgia, fibromyalgia and diabetic neuropathy [6, 8]. Although this list of indications is impressive, the analgesia produced by topical capsaicin appears to be highly variable and dependent on the limited amount of capsaicin that penetrates the skin due to its poor absorption characteristics. Furthermore, the irritation and burning caused by the initial excitation produced by capsaicin has led to poor compliance for this therapy. Nevertheless, even with these drawbacks, the limited efficacy of topical capsaicin in a wide variety of chronic pain syndromes is intriguing. Importantly, from a therapeutic perspective, analgesia following topical capsaicin treatment appears not to be limited to a reduction in thermal hyperalgesia, but also includes a reduction in pain associated with mechanical stimulation.

To date, a variety of naturally occurring and synthetic TRPV1 agonists, including olvanil [9], glyceryl nonamide [9], phorbol 12-phenylacetate 13 acetate 20-homovanillate [10], scutigeral [11] and capsiate [12], have been described that show reduced pungency compared to capsaicin and could thus offer improved patient compliance. Multiple hypotheses have been put forth to explain the diminished pungency, such as the existence of TRPV1 subtypes, slow activation kinetics or pharmacokinetic limitations [9–11]. Regardless of the mechanism, it is clear that low pungency TRPV1 agonists can retain capsaicin-like analgesic and antihyperalgesic properties in rodent models. To this end, the low pungency TRPV1 agonist SDZ 249-665 showed efficacy in the mouse tail-flick and writhing assays of acute nociception as well as in rat and guinea pig models of inflammatory mechanical hyperalgesia [13].

The significance of these TRPV1 agonist studies is two-fold. First, they highlight TRPV1 as a target for a variety of structurally diverse compounds, well beyond its initial narrow classification as a vanilloid receptor. Second, these studies suggest that desensitization or down-regulation of TRPV1 expressed on nociceptive neurons, with or without a strong initial excitatory component, can significantly modify nociception in both animals and humans. Thus, antagonism of TRPV1 may provide a similar therapeutic utility. This hypothesis is further supported by experiments in mice that have had TRPV1 genetically deleted and which show significantly diminished thermal hyperalgesia in models of inflammatory pain [14, 15]. Although initial studies utilizing the classic TRPV1 antagonist capsazepine in rat models of inflammatory pain did not demonstrate antihyperalgesia, with the recent cloning of TRPV1 from multiple species and the advent of new pharmacological tools, explanations for the lack of efficacy in these early studies are now available.

TRPV1 antagonists

It is becoming increasingly apparent that broad *in vitro* antagonist activity against multiple modes of TRPV1 activation (i.e. capsaicn-site agonists, low pH and heat) may be an important predictor of *in vivo* efficacy in animal models of pain (see below). While some of these molecules have been tested against these various mechanisms of TRPV1 activation, many still require further pharmacological characterization. A clear understanding at the molecular level, in conjunction with pharmacokinetic and pharmacodynamic considerations, should ultimately allow a more accurate prediction of TRPV1 antagonist activity *in vivo*. Those TRPV1 antagonists that have been profiled in animal models of pain are presented in Table 1.

Classic TRPV1 antagonists: ruthenium red (RR) and capsazepine

RR is a polycationic dye that was the first reported antagonist of capsaicin's excitatory action on neurons (Tab. 1). RR is an antagonist of capsaicin-stimulated Ca^{2+} influx [16, 17] and transmitter release in cultured dorsal root ganglion neurons [18, 19] and it inhibits capsaicin-induced excitation of primary afferent neurons [20]. Using a [^{3}H]RTX binding assay, RR did not interact with the capsaicin-binding site, suggesting a non-competitive mechanism of channel blockade [21]. Characterization of heterologously expressed rat and human TRPV1 subsequently showed that RR could block channel activation by capsaicin and low pH [22, 23].

A limited number of reports have demonstrated efficacy of RR in models of pain including the formalin test and the capsaicin model of hyperalgesia [24, 25]. The utility of RR as a tool to study TRPV1 function *in vivo* has been severely hampered by off-target side effects in the dose range necessary to antagonize capsaicin's actions [26]. Indeed, recent studies demonstrate that RR shows antagonist activity at multiple TRPV channels as well as many other channels and receptors [27, 28].

Capsazepine was the first reported competitive small-molecule antagonist of capsaicin-evoked activation of sensory neurons [29, 30] (Tab. 1). *In vitro*, capsazepine has been shown to competitively inhibit capsaicin-mediated responses in isolated dorsal root ganglion neurons [29] and tissues from rat [31–34], mouse [35] and guinea pig [2, 3, 36–38]. However, while capsazepine is able to block acid-induced activation of recombinant guinea pig and human TRPV1 *in vitro* [23, 37, 39–41], a similar blockade is not seen against native rat TRPV1 *ex vivo* [31, 42–44] or recombinant rat TRPV1 *in vitro* [23, 44]. The use of capsazepine to study the role of TRPV1 in chronic pain has thus been complicated by its species selectivity with respect to blockade of low-pH-induced TRPV1 activation. Similar discrepancies have also been reported for blockade of heat-induced activation of TRPV1 [23]. In addition, at higher concentrations, capsazepine shows some antagonist activity towards acetylcholine receptors [45], voltage-gated Ca^{2+} channels [46] and hyperpolarization-activated cyclic nucleotide-gated channels [47].

Subcutaneous administration of capsazepine has not been shown to affect normal withdrawal thresholds to noxious thermal or mechanical stimuli in mice, rats or guinea pigs [48]. In contrast, TRPV1-null mice show reduced responses to noxious

Table 1 - In vitro *and* in vivo *pharmacological activities of TRPV1 antagonists*

Antagonist	***In vitro***	**References**	***In vivo***	**References**
Capsazepine	↓ CAP response in rat DRG neurons/ tissues	[29, 31–34]	↓ CAP-induced hyperalgesia in mice, rats and guinea pigs	[48]
	↓ CAP responses in mouse/guinea pig TRPV1	[35–38]	↓ FCA-induced hyperalgesia in guinea pigs (not rats or mice)	
	↓ Low-pH responses in human or guinea pig TRPV1 (not rat)	[31, 39, 40, 42–44]	↓ Carrageenan-induced hyperalgesia in guinea pigs	
	↓ Heat response in human/guinea pig TRPV1	[23]	↓ Hyperalgesia following partial sciatic nerve injury in guinea pigs	
Ruthenium red	↓ CAP-mediated excitation of primary afferents	[20]	↓ Formalin-induced licking ↓ CAP-induced licking	[24]
	Non-competitive block of TRPV1	[21]		
	Reversible inhibition of CAP-induced… · Ca^{2+} ↑ in DRG neurons · currents in DRG neurons · nociceptor activity in rat spinal-cord/tail preparation	[16]	↓ Early- and late-phase formalin- and CAP-induced licking in mice ↓ CAP-induced licking in mice	[25]
	↓ CAP and low-pH responses in rat and human TRPV1	[22, 23]		
Basic hexapeptide RRRRWW	↓ CAP currents in oocytes	[54]	↓ CAP-induced eye irritation	[54]
	↓ CAP- and RTX-mediated Ca^{2+} influx in rat DRG neurons	[55]		
Peptoids DD161515 and DD191515	↓ CAP currents in oocytes ↓ CAP-evoked neural activity in rat knee-joint nociceptive fibers	[56]	↓ Thermal nociception in mouse hot-plate and tail-immersion tests ↓ Licking/flinching behavior after mouse hind-paw CAP injection ↓ CAP-mediated neurogenic inflammation (paw volume assay) Partial ↓ of mustard oil-induced thermal hyperalgesia in mouse hind paw	[56]

Antagonist	*In vitro*	References	*In vivo*	References
5-IodoRTX	↓ [^{125}I]RTX binding to recombinant rat and spinal TRPV1 ↓ CAP currents in oocytes	[57]	↓ CAP-mediated paw licking	[57]
	↓ [^{125}I]RTX binding to rat and human TRPV1 ↓ CAP-mediated action potentials in rat skin/nerve preparation	[58]	↑ Paw flinching in rats, similar to CAP; effect also in TRPV1-knockout mice	[58]
	↓ CAP-mediated... · Ca^{2+} influx in TG neurons · CGRP release from rat spinal cords · contraction of rat and guinea pig bronchus	[59]	↓ CAP-induced plasma extravasation in mouse urinary bladder ↓ Acetic acid-induced writhing in mice	[59]
SC0030	↓ [^{125}I]RTX binding to rat TRPV1 ↓ CAP- and RTX-mediated Ca^{2+} uptake in rat TRPV1-CHO cells and DRG neurons ↓ TRPV1 activation by pH 6.0 and heat (44 °C)	[61]	↓ PBQ-induced writhing in mice ↓ CAP-mediated nociception	[63]
Compounds 40 and 61	↓ [^{125}I]RTX binding to rat TRPV1 ↓ CAP-mediated Ca^{2+} uptake in rat TRPV1-CHO cells	[64]	↓ Acetic acid-induced writhing in mice	[64]
BCTC	↓ CAP and low-pH activation of rat TRPV1 ↓ CAP- and low-pH-mediated action potentials in rat skin/nerve preparation	[44]	↓ Mechanical/thermal hyperalgesia induced by... · intraplantar CAP in rat hind paw · intraplantar FCA in rat hind paw ↓ Mechanical hyperalgesia/tactile allodynia in... · rat Seltzer model · rat model of post-surgical pain	[68]

CAP, capsaicin; CGRP, calcitonin-gene-related peptide; CHO, Chinese hamster ovary; DRG, dorsal root ganglion; FCA, Freund's complete adjuvant; PBQ, phenylbenzoquinone; RTX, resiniferatoxin; TG, trigeminal ganglia.

thermal stimuli [14] and a complete absence of carrageenan-induced thermal hyperalgesia [15]. However, the hyperalgesic effects of heat cannot be explained by the action of TRPV1 alone, since mice lacking TRPV1 retain the ability to escape from noxious heat [14, 15], suggesting the involvement of additional thermal receptors. In keeping with this hypothesis, two TRPV1 homologs, VRL-1 (TRPV2) and TRPV3 have been reported to be insensitive to capsaicin or protons but respond to either high- or low-threshold heat stimulation; neither is blocked by capsazepine [49, 50].

Consistent with its *in vitro* pharmacological profile, capsazepine inhibits the development of mechanical hyperalgesia induced by intraplantar injection of capsaicin, with a similar potency in mice, rats and guinea pigs [25, 48, 51–53]. Most interestingly, although capsazepine does not affect mechanical hyperalgesia in rat or mouse models of inflammatory pain, it does significantly reverse inflammatory hyperalgesia in the guinea pig. Similarly, in a model of neuropathic pain, capsazepine was surprisingly effective in the guinea pig, producing up to 80% reversal of mechanical hyperalgesia, but had no effect in the rat or mouse [48]. These data parallel the observations that capsazepine inhibits capsaicin-induced activation of TRPV1 in both rats and guinea pigs, but only inhibits low-pH-induced activation of guinea pig, but not rat, TRPV1 [23]. As the *in vitro* pharmacological profile for human TRPV1 is most similar to that for the guinea pig TRPV1 [41], these data raise the possibility that blockade of TRPV1 responses activated by low pH, or dual blockade of capsaicin-site- and pH-site-mediated responses, may be important for antihyperalgesic efficacy *in vivo*.

Peptide-based antagonists

Ferrer-Montiel and colleagues [54] hypothesized that, similar to the *N*-methyl-D-aspartate receptor, the TRPV1 high-affinity Ca^{2+}-binding site within the receptor's channel pore could interact with arginine-rich hexapeptides [54]. Indeed, using an oocyte expression system, capsaicin-induced TRPV1 currents were potently and reversibly blocked by the peptide R4W2 in a non-competitive and non-stereoselective manner (Tab. 1). Similarly, it was shown that the naturally occurring arginine-rich peptide, dynorphin A, could also block capsaicin-mediated TRPV1 currents in oocytes. A more recent study however, supports a competitive interaction for these peptides at the capsaicin-binding site [55]. Additional studies will be necessary to understand these discrepancies, investigate their activity against low-pH- and heat-mediated TRPV1 activation, and test their utility in animal models of chronic pain. Along these lines, it has been shown that the irritation associated with topical application of capsaicin to the mouse eye is significantly reduced by co-application of either R_4W_2 or dynorphin A.

Continued work within the Ferrer-Montiel laboratory resulted in the identification of trimeric *N*-alkylglycine peptoid TRPV1 channel blockers [56]. Two molecules were identified (Table 1) that demonstrated non-competitive, low-micromolar ability to block capsaicin activation of rat TRPV1 expressed in oocytes (DD161515 and DD191515, with IC_{50} values of 0.7 and 2.6 μM, respectively). Importantly, DD161515 completely reversed capsaicin-evoked neural activity in rat knee-joint nociceptor fibers and significantly attenuated thermal nociception in the mouse hot-plate and tail-immersion tests (both 52 °C) upon intraperitoneal administration at 0.2 mmol/kg. The characteristic licking and flinching responses elicited by intraplantar injection of capsaicin into the mouse hindpaw were also dose-dependently inhibited by both DD161515 and DD191515. Similarly, capsaicin-induced hindpaw edema and mustard oil-induced thermal hyperalgesia were prevented by treatment with DD161515.

RTX- and capsaicin-based antagonists

The characterization of a 5-iodo-substituted analog of RTX has recently been described [57–59]. Surprisingly, iodine substitution on the 5-position of the vanillyl ring converts the highly potent TRPV1 agonist into a potent antagonist (Tab. 1). In the work by Wahl and colleagues [57], 5-iodoRTX showed K_i values of 6 and 5 nM for recombinant and spinal-cord TRPV1, respectively, as measured by [^{125}I]RTX displacement. Functionally, 5-iodoRTX showed no agonist activity in oocytes expressing recombinant rat TRPV1. However, non-competitive, concentration-dependent inhibition of capsaicin-induced currents was observed with an IC_{50} value of 3.8 nM, compared to 152 nM for capsazepine. *In vivo*, 5-iodoRTX administered intrathecally to mice was more potent at blocking capsaicin-mediated paw-licking behavior (ED_{50} value, 16 ng/mouse) than morphine or nociceptin, while capsazepine was ineffective in this model.

Similarly, in the work by Seabrook and colleagues [58], 5-iodoRTX showed K_i values of 0.39 and 6.7 nM for recombinant rat and human TRPV1, respectively, as measured by [^{3}H]RTX displacement. Functionally, 5-iodoRTX concentration-dependently blocked capsaicin-mediated currents in HEK-293 cells expressing rat or human TRPV1 with IC_{50} values of 0.7 and 5.4 nM, respectively. 5-IodoRTX also completely blocked currents induced by pH 5.5 (IC_{50} value of 550 nM for both rat and human TRPV1), or a heat ramp from 25 to 48 °C. In contrast to the report by Wahl et al. [57], intraplantar administration of 5-iodoRTX increased paw-flinching behavior in rats to a similar magnitude as capsaicin, an effect that was also observed in TRPV1-knockout mice, suggesting a non-TRPV1 mechanism of action for this stimulatory effect.

5-IodoRTX blocked acetic acid-induced writhing in mice with an ED_{50} value of 0.42 μmol/kg compared to 7.9 μmol/kg for capsazepine [59]. In this assay, 5-iodoRTX did cause mild writhing responses in ~33% of the mice injected. The data generated thus far with 5-iodoRTX show potent TRPV1-blocking activity across a number of species and in a number of *in vitro*, *ex vivo* and *in vivo* pain and neurogenic inflammation models, strongly suggesting its potential as a new pharmacological tool to better understand TRPV1 physiology.

Capsaicin-based antagonists have also been identified, with 6-iodo-nordihydrocapsaicin showing competitive blockade of capsaicin-induced Ca^{2+} influx in HEK-293 cells expressing recombinant human TRPV1 (IC_{50} value of 10 nM, four times more potent than capsazepine) [60]. A significantly lower potency was seen using Ca^{2+} influx in cultured rat dorsal root ganglion neurons, with an IC_{50} value of 640 nM, compared to 2.34 μM for capsazepine. This discrepancy may represent a fundamental difference between the two assays, as both the iodo-nordihydrocapsaicin and capsazepine concentration-response curves were shifted approximately 60-fold to the right. Further investigations into the activity against low-pH- and heat-mediated TRPV1 activation, as well as utility in animal models of chronic pain, are necessary with this new series of capsaicin-based antagonists.

A new generation of TRPV1 antagonists

Novel small-molecule TRPV1 antagonists identified by high-throughput screening and combinatorial chemistry efforts at several pharmaceutical companies are now being reported. A series of conformationally more flexible thiourea analogs of capsazepine have recently been described as potent TRPV1 antagonists, with detailed analyses presented for two analogs from this series, MK056 and SC0030 (Table 1) [61, 62]. Both compounds displaced [^{3}H]RTX from rat TRPV1 expressed in Chinese hamster ovary (CHO) cells with K_i values of 63 and 54 nM, respectively, compared to 1300 nM for capsazepine. In addition, competitive blockade of capsaicin-stimulated $^{45}Ca^{2+}$ uptake was demonstrated in the rat TRPV1-CHO cell line with K_i values of 31 and 8 nM, respectively, compared to 430 nM for capsazepine. Similar results were obtained using the Ca^{2+}-sensitive dye fura-2 in both recombinant cells and isolated neurons of the dorsal root ganglion. Importantly, SC0030 completely prevented channel activation in response to elevated temperature (44 °C) and low pH (6.0), whereas MK506 and capsazepine showed only partial blockade against these stimuli. *In vivo*, intraperitoneal administration of SC0030 showed a complete and dose-dependent reversal of phenylbenzoquinone-induced writhing in mice (ED_{50}, ~0.3 mg/kg) [63]. Similarly, SC0030 dose-dependently inhibited capsaicin-induced nociception and pain-related behaviors.

Another series of structurally similar thiourea analogs was discovered as potent TRPV1 antagonists from the same group [64]. From this series, compounds 40 and 61 showed the highest potency (Tab. 1). Both compounds displaced [^{3}H]RTX from rat TRPV1 expressed in CHO cells with K_i values of 29 and 54 nM, respectively. In addition, competitive blockade of capsaicin-stimulated $^{45}Ca^{2+}$ uptake was demonstrated in the rat TRPV1-CHO cell line with K_i values of 67 and 8 nM, respectively. The analgesic activities of both analogs (intraperitoneal administration) were evaluated in the acetic acid-induced writhing model in mice, and were compared to the clinically used non-steroidal anti-inflammatory drug, ketorolac. While compound 40 showed a comparable potency (ED_{50}, 2.6 mg/kg) relative to ketorolac (ED_{50}, 2.8 mg/kg), compound 61 exhibited dramatically enhanced analgesic potency, with an ED_{50} value of 0.007 mg/kg, approximately 350-fold more potent than compound 40.

A number of companies have reported 2-pyridylpiperazine urea analogs as potent TRPV1 antagonists [65–67]. The structure-activity relationship for this series of compounds has been published recently [67]. BCTC [*N*-(4-tertiarybutylphenyl)-4-(3-cholorphyridin-2-yl) tetrahydropryazine-1(2*H*)-carbox-amide] was discovered as one of most potent TRPV1 antagonists among this series (Table 1) and two reports describing the *in vitro* and *in vivo* biological activity of BCTC have published recently [44, 68]. Using a Ca^{2+}-sensing assay, BCTC inhibited capsaicin-induced activation of rat TRPV1 with an IC_{50} value of 35 nM, compared to ~4 μM for capsazepine. The mechanism of BCTC inhibition was competitive with the capsaicin-binding site. Importantly, BCTC also potently inhibited pH 5.5-induced activation of rat TRPV1 with an IC_{50} value of 6.0 nM while capsazepine was inactive in this assay. Similarly, in the rat skin/nerve preparation both BCTC and capsazepine blocked capsaicin-induced action potentials, while the response to acidification was inhibited by BCTC, but not by capsazepine. The mechanism of BCTC inhibition of low-pH-induced channel activation was not investigated; however, similar experiments with capsazepine on human TRPV1 have demonstrated a non-competitive interaction [23]. These *in vitro* pharmacological properties together with oral availability in the rat may make BCTC a better pharmacological tool than capsazepine to investigate the potential role of TRPV1 *in vivo*.

To this end, the effects of BCTC were examined in rat models of inflammatory and neuropathic pain [68]. Like capsazepine, BCTC significantly reduced hyperalgesia induced by intraplantar injection of capsaicin to the hindpaw of rats. However, unlike capsazepine, BCTC significantly reduced thermal and mechanical hyperalgesia in the inflamed rat hind paw. BCTC also reduced the mechanical hyperalgesia and tactile allodynia associated with peripheral nerve injury and incision of the plantar surface of the rat hind paw. These data further support the hypothesis that blockade of low-pH-induced, or both low-pH- and capsaicin-induced, activation of TRPV1 is necessary for reversal of the pain-like behaviors observed in animal models of chronic pain.

SB-366791 is a TRPV1 antagonist discovered and reported by GlaxoSmithKline [69]. Using a Ca^{2+}-sensing assay, the compound showed competitive antagonism of human TRPV1 with an IC_{50} value of 25 nM against capsaicin-mediated channel activation. SB-366791 also blocked low-pH- and heat-mediated activation of human TRPV1. Similar potencies were seen against rat TRPV1 in the capsaicin and low-pH assays, although no data were presented on blockade of heat responses in this species. Based on these pharmacological properties, SB-366791 represents another potentially strong candidate for *in vivo* profiling in animal models of pain and inflammation.

Finally, a recent patent application from Novartis reports a compound that is claimed to have an *in vitro* IC_{50} value of 65 nM against human TRPV1 and 19 nM against rat TRPV1 [70]. A dose-dependent reversal of mechanical hyperalgesia in a rat model of neuropathic pain was also claimed for orally administered compounds from this series, at a single dose from 0.3 to 30 mg/kg.

Mechanisms and perspectives

The advent of systemically available small-molecule TRPV1 antagonists has allowed investigation of the role of TRPV1 in animal models of chronic pain. Surprisingly, these compounds have suggested a role for TRPV1 antagonists as broad-spectrum drugs for the treatment of chronic pain. Administration of TRPV1 antagonists has been shown to reduce not only thermal but mechanical hyperalgesia in rat and guinea pig models of inflammatory pain, and to reduce mechanical hyperalgesia and tactile allodynia in neuropathic pain [48, 68].

The recent findings that TRPV1 antagonists show efficacy in a broad range of chronic pain models and against thermal as well as mechanical endpoints were initially counter-intuitive given the *in vitro* profile of TRPV1 activation and data generated from knockout mice. As discussed above, both noxious heat and low pH are known to activate the channel, but no *in vitro* experiments have demonstrated mechanosensory properties of TRPV1. The lack of these findings does not necessarily exclude the potential for TRPV1-expressing neurons to be at least indirectly involved in the transmission of mechanical sensory information. A significant proportion of small-diameter neurons respond to mechanical as well as heat stimuli (only approximately 30% of C-fiber nociceptors are unresponsive to mechanical stimuli or have very high mechanical thresholds [71–73]). Thus, while there is no evidence to suggest that TRPV1 is directly activated by mechanical stimuli, TRPV1 may be expressed by mechanosensitive primary afferents as more than 30% of small-diameter neurons express TRPV1 [74].

Examination of the effects of capsaicin stimulation may also provide useful information. It has long been known that intradermal or topical application of capsaicin induces not only thermal but also mechanical hyperalgesia in a number of species, including humans [1–3, 5, 75, 76]. There are significant differences in the hyperalgesic areas and duration of hyperalgesia elicited by mechanical versus thermal stimuli. For example, the area of capsaicin-induced mechanical hyperalgesia is greater relative to thermal hyperalgesia [5] and the duration of mechanical hyperalgesia is significantly longer than the duration of thermal hyperalgesia [5]. It has been proposed that capsaicin-induced mechanical hyperalgesia is centrally mediated, as intradermal capsaicin administration to monkeys excites spinothalamic tract neurons in the dorsal horn of the spinal cord, and enhances mechanically evoked responses [77]. Although TRPV1 receptors are present in the dorsal horn of the spinal cord [78], it is unlikely that the effects of capsaicin on dorsal horn neurons are centrally mediated given the localized injection. In addition to small-diameter mechanosensitive primary afferents, there may exist some additional mechanisms for communication between pathways activated by TRPV1 and those involved with mechanosensation.

One possible explanation for these interactions may be changes in the expression pattern of TRPV1 following inflammation and peripheral nerve damage. In naïve animals, TRPV1-immunoreactivity is largely confined to small-diameter neurons. However, following inflammation there is a significant increase in TRPV1-immunoreactivity in medium-diameter neurons that also express neurofilament 200, indicative of expression in Aδ neurons [74]. Furthermore, studies in rat models of neuropathic pain have shown that ligation of the L5 spinal nerve results in increased TRPV1 mRNA and immunoreactivity in A-fiber cell bodies in the uninjured L4 dorsal root ganglia [79, 80], and in undamaged neurons in the partial sciatic nerve ligation model of neuropathy [79]. Furthermore, similar increases in TRPV1-immunoreactivity in myelinated neurons are seen in streptozotocin-induced diabetic rats [81]. Given that inflammation and peripheral nerve injury are associated with decreases in pH and increased local temperature [82, 83], which are both activators of TRPV1, it may be that these potentially mechanosensitive neurons become sensitized to mechanical stimuli due to the altered expression of TRPV1.

In conclusion, the observations that small-molecule TRPV1 antagonists of different structural chemotypes show similar efficacy profiles against mechanical and thermal hyperalgesia in models of inflammatory and chronic pain [48, 68] suggests that these effects are mediated by inhibition of TRPV1. It is also clear that due to important species differences in the pharmacology and function of avian, rodent and human TRPV1, new molecules must be carefully profiled for their activity in the species of interest before *in vivo* results can be interpreted. As additional novel molecules with the appropriate pharmacological profiles become available it will be possible to further test the role of TRPV1 in models of chronic pain. Finally, it is expected that results from the first clinical pain trials for TRPV1 antagonists are drawing closer and that these will provide much anticipated information on the therapeutic efficacy of TRPV1 antagonists in humans.

References

1 Szolcsanyi J (1977) A pharmacological approach to elucidation of the role of different nerve fibres and receptor endings in mediation of pain. *J Physiol (Paris)* 73: 251–259

2 Carpenter SE, Lynn B (1981) Vascular and sensory responses of human skin to mild injury after topical treatment with capsaicin. *Br J Pharmacol* 73: 755–758

3 Simone DA, Ngeow JY, Putterman GJ, Lamotte RH (1987) Hyperalgesia to heat after intradermal injection of capsaicin. *Brain Res* 418: 201–203

4 Simone DA, Baumann TK, Collins JG, Lamotte RH (1989) Sensitization of cat dorsal horn neurons to innocuous mechanical stimulation after intradermal injection of capsaicin. *Brain Res* 486: 185–189

5 Gilchrist HD, Allard BL, Simone DA (1996) Enhanced withdrawal responses to heat and mechanical stimuli following intraplantar injection of capsaicin in rats. *Pain* 67: 179–188

6 Hautkappe M, Roizen MF, Toledano A, Roth S, Jeffries JA, Ostermeier AM (1998) Review of the effectiveness of capsaicin for painful cutaneous disorders and neural dysfunction. *Clin J Pain* 14: 97–106

7 Zhang WY, Li Wan PA (1994) The effectiveness of topically applied capsaicin. A meta-analysis. *Eur J Clin Pharmacol* 46: 517–522

8 Dray A (1999) Vanilloids as analgesics. In: J Sawynok, A Cowan (eds): *Novel aspects of pain management: opioids and beyond.* Wiley-Liss, New York, 117–134

9 Liu L, Lo Y, Chen I, Simon SA (1997) The responses of rat trigeminal ganglion neurons to capsaicin and two nonpungent vanilloid receptor agonists, olvanil and glyceryl nonamide. *J Neurosci* 17: 4101–4111

10 Liu L, Szallasi A, Simon SA (1998) A non-pungent resiniferatoxin analogue, phorbol 12-phenylacetate 13 acetate 20-homovanillate, reveals vanilloid receptor subtypes on rat trigeminal ganglion neurons. *Neuroscience* 84: 569–581

11 Szallasi A, Biro T, Szabo T, Modarres S, Petersen M, Klusch A, Blumberg PM, Krause JE, Sterner O (1999) A non-pungent triprenyl phenol of fungal origin, scutigeral, stimulates rat dorsal root ganglion neurons via interaction at vanilloid receptors. *Br J Pharmacol* 126: 1351–1358

12 Iida T, Moriyama T, Kobata K, Morita A, Murayama N, Hashizume S, Fushiki T, Yazawa S, Watanabe T, Tominaga M (2003) TRPV1 activation and induction of nociceptive response by a non-pungent capsaicin-like compound, capsiate. *Neuropharmacology* 44: 958–967

13 Urban L, Campbell EA, Panesar M, Patel S, Chaudhry N, Kane S, Buchheit K, Sandells B, James IF (2000) *In vivo* pharmacology of SDZ 249-665, a novel, non-pungent capsaicin analogue. *Pain* 89: 65–74

14 Caterina MJ, Leffler A, Malmberg AB, Martin WJ, Trafton J, Petersen-Zeitz KR, Koltzenburg M, Basbaum AI, Julius D (2000) Impaired nociception and pain sensation in mice lacking the capsaicin receptor. *Science* 288: 306–313

15 Davis JB, Gray J, Gunthorpe MJ, Hatcher JP, Davey PT, Overend P, Harries MH, Latcham J, Clapham C, Atkinson K et al. (2000) Vanilloid receptor-1 is essential for inflammatory thermal hyperalgesia. *Nature* 405: 183–187

16 Dray A, Forbes CA, Burgess GM (1990) Ruthenium red blocks the capsaicin-induced increase in intracellular calcium and activation of membrane currents in sensory neu-

rones as well as the activation of peripheral nociceptors *in vitro*. *Neurosci Lett* 110: 52–59

17 Wood JN, Winter J, James IF, Rang HP, Yeats J, Bevan S (1988) Capsaicin-induced ion fluxes in dorsal root ganglion cells in culture. *J Neurosci* 8: 3208–3220

18 Maggi CA, Santicioli P, Geppetti P, Parlani M, Astolfi M, Pradelles P, Patacchini R, Meli A (1988) The antagonism induced by ruthenium red of the actions of capsaicin on the peripheral terminals of sensory neurons: further studies. *Eur J Pharmacol* 154: 1–10

19 Maggi CA, Patacchini R, Santicioli P, Giuliani S, Geppetti P, Meli A (1988) Protective action of ruthenium red toward capsaicin desensitization of sensory fibers. *Neurosci Lett* 88: 201–205

20 Amann R, Donnerer J, Lembeck F (1990) Activation of primary afferent neurons by thermal stimulation. Influence of ruthenium red. *Naunyn-Schmiedeberg's Arch Pharmacol* 341: 108–113

21 Szallasi A, Blumberg PM (1990) Specific binding of resiniferatoxin, an ultrapotent capsaicin analog, by dorsal root ganglion membranes. *Brain Res* 524: 106–111

22 Caterina MJ, Schumacher MA, Tominaga M, Rosen TA, Levine JD, Julius D (1997) The capsaicin receptor: a heat-activated ion channel in the pain pathway. *Nature* 389: 816–824

23 McIntyre P, McLatchie LM, Chambers A, Phillips E, Clarke M, Savidge J, Toms C, Peacock M, Shah K, Winter J et al. (2001) Pharmacological differences between the human and rat vanilloid receptor 1 (VR1). *Br J Pharmacol* 132: 1084–1094

24 Ohkubo T, Shibata M, Takahashi H (1993) The analgesia induced by intrathecal injection of ruthenium red. *Pain* 54: 219–221

25 Santos AR, Calixto JB (1997) Ruthenium red and capsazepine antinociceptive effect in formalin and capsaicin models of pain in mice. *Neurosci Lett* 235: 73–76

26 Szallasi A (2001) Vanilloid receptor ligands: hopes and realities for the future. *Drugs Aging* 18: 561–573

27 Smith GD, Gunthorpe MJ, Kelsell RE, Hayes PD, Reilly P, Facer P, Wright JE, Jerman JC, Walhin JP, Ooi L et al. (2002) TRPV3 is a temperature-sensitive vanilloid receptor-like protein. *Nature* 418: 186–190

28 Maggi CA (1991) Capsaicin and primary afferent neurons: from basic science to human therapy? *J Auton Nerv Syst* 33: 1–14

29 Bevan S, Hothi S, Hughes G, James IF, Rang HP, Shah K, Walpole CS, Yeats JC (1992) Capsazepine: a competitive antagonist of the sensory neurone excitant capsaicin. *Br J Pharmacol* 107: 544–552

30 Walpole CS, Bevan S, Bovermann G, Boelsterli JJ, Breckenridge R, Davies JW, Hughes GA, James I, Oberer L, Winter J (1994) The discovery of capsazepine, the first competitive antagonist of the sensory neuron excitants capsaicin and resiniferatoxin. *J Med Chem* 37: 1942–1954

31 Bevan S, Rang H, Shah K (1991) Development of a competitive antagonist for the sensory neurone excitant, capsaicin. *Br J Pharmacol* 102: 77P

32 Cholewinski A, Burgess GM, Bevan S (1993) The role of calcium in capsaicin-induced

desensitization in rat cultured dorsal root ganglion neurons. *Neuroscience* 55: 1015–1023

33 Maggi CA, Bevan S, Walpole CS, Rang HP, Giuliani S (1993) A comparison of capsazepine and ruthenium red as capsaicin antagonists in the rat isolated urinary bladder and vas deferens. *Br J Pharmacol* 108: 801–805

34 Santicioli P, Del Bianco E, Figini M, Bevan S, Maggi CA (1993) Effect of capsazepine on the release of calcitonin gene-related peptide-like immunoreactivity (CGRP-LI) induced by low pH, capsaicin and potassium in rat soleus muscle. *Br J Pharmacol* 110: 609–612

35 Urban L, Dray A (1991) Capsazepine, a novel capsaicin antagonist, selectively antagonises the effects of capsaicin in the mouse spinal cord *in vitro*. *Neurosci Lett* 134: 9–11

36 Ellis JL, Undem BJ (1994) Inhibition by capsazepine of resiniferatoxin- and capsaicin-induced contractions of guinea pig trachea. *J Pharmacol Exp Ther* 268: 85–89

37 Fox AJ, Urban L, Barnes PJ, Dray A (1995) Effects of capsazepine against capsaicin- and proton-evoked excitation of single airway C-fibres and vagus nerve from the guinea-pig. *Neuroscience* 67: 741–752

38 Auberson S, Lacroix JS, Lundberg JM (2000) Modulation of capsaicin-sensitive nerve activation by low pH solutions in guinea-pig lung. *Pharmacol Toxicol* 86: 16–23

39 Lou YP, Lundberg JM (1992) Inhibition of low pH evoked activation of airway sensory nerves by capsazepine, a novel capsaicin-receptor antagonist. *Biochem Biophys Res Commun* 189: 537–544

40 Satoh H, Lou YP, Lundberg JM (1993) Inhibitory effects of capsazepine and SR 48968 on citric acid-induced bronchoconstriction in guinea-pigs. *Eur J Pharmacol* 236: 367–372

41 Savidge J, Davis C, Shah K, Colley S, Phillips E, Ranasinghe S, Winter J, Kotsonis P, Rang H, McIntyre P (2002) Cloning and functional characterization of the guinea pig vanilloid receptor 1. *Neuropharmacology* 43: 450–456

42 Vyklicky L, Knotkova-Urbancova H, Vitaskova Z, Vlachova V, Kress M, Reeh PW (1998) Inflammatory mediators at acidic pH activate capsaicin receptors in cultured sensory neurons from newborn rats. *J Neurophysiol* 79: 670–676

43 Habelt C, Kessler F, Distler C, Kress M, Reeh PW (2000) Interactions of inflammatory mediators and low pH not influenced by capsazepine in rat cutaneous nociceptors. *Neuroreport* 11: 973–976

44 Valenzano KJ, Grant ER, Wu G, Hachicha M, Schmid L, Tafesse L, Sun Q, Rotshteyn Y, Francis J, Limberis JT et al. (2003) BCTC (N-(4-tertiarybutylphenyl)-4-(3-chloropyridin-2-yl) tetrahydropyrazine-1(2H)-carbox-amide), a novel, orally-effective vanilloid receptor 1 antagonist with analgesic properties: I. *In vitro* characterization and pharmacokinetic properties. *J Pharmacol Exp Ther* 306: 377–386

45 Liu L, Simon SA (1997) Capsazepine, a vanilloid receptor antagonist, inhibits nicotinic acetylcholine receptors in rat trigeminal ganglia. *Neurosci Lett* 228: 29–32

46 Docherty RJ, Yeats JC, Piper AS (1997) Capsazepine block of voltage-activated calcium channels in adult rat dorsal root ganglion neurones in culture. *Br J Pharmacol* 121: 1461–1467

47 Gill CH, Strijbos PJ, Bates SA, Cairns OD, Davies CH (2001) Capsazepine inhibits a recombinant human HCN1-mediated current. *Br J Pharmacol* 135: 251P

48 Walker KM, Urban L, Medhurst SJ, Patel S, Panesar M, Fox AJ, McIntyre P (2003) The VR1 antagonist capsazepine reverses mechanical hyperalgesia in models of inflammatory and neuropathic pain. *J Pharmacol Exp Ther* 304: 56–62

49 Caterina MJ, Rosen TA, Tominaga M, Brake AJ, Julius D (1999) A capsaicin-receptor homologue with a high threshold for noxious heat. *Nature* 398: 436–441

50 Peier AM, Reeve AJ, Andersson DA, Moqrich A, Earley TJ, Hergarden AC, Story GM, Colley S, Hogenesch JB, McIntyre P et al (2002) A heat-sensitive TRP channel expressed in keratinocytes. *Science* 296: 2046–2049

51 Davis AJ, Perkins MN (1996) Substance P and capsaicin-induced mechanical hyperalgesia in the rat knee joint; the involvement of bradykinin B1 and B2 receptors. *Br J Pharmacol* 118: 2206–2212

52 Kwak JY, Jung JY, Hwang SW, Lee WT, Oh U (1998) A capsaicin-receptor antagonist, capsazepine, reduces inflammation-induced hyperalgesic responses in the rat: evidence for an endogenous capsaicin-like substance. *Neuroscience* 86: 619–626

53 Sakurada T, Matsumura T, Moriyama T, Sakurada C, Ueno S, Sakurada S (2003) Differential effects of intraplantar capsazepine and ruthenium red on capsaicin-induced desensitization in mice. *Pharmacol Biochem Behav* 75: 115–121

54 Planells-Cases R, Aracil A, Merino JM, Gallar J, Perez-Paya E, Belmonte C, Gonzalez-Ros JM, Ferrer-Montiel AV (2000) Arginine-rich peptides are blockers of VR-1 channels with analgesic activity. *FEBS Lett* 481: 131–136

55 Himmel HM, Kiss T, Borvendeg SJ, Gillen C, Illes P (2002) The arginine-rich hexapeptide R4W2 is a stereoselective antagonist at the vanilloid receptor 1: a Ca^{2+} imaging study in adult rat dorsal root ganglion neurons. *J Pharmacol Exp Ther* 301: 981–986

56 Garcia-Martinez C, Humet M, Planells-Cases R, Gomis A, Caprini M, Viana F, De La PE, Sanchez-Baeza F, Carbonell T, De Felipe C et al. (2002) Attenuation of thermal nociception and hyperalgesia by VR1 blockers. *Proc Natl Acad Sci USA* 99: 2374–2379

57 Wahl P, Foged C, Tullin S, Thomsen C (2001) Iodo-resiniferatoxin, a new potent vanilloid receptor antagonist. *Mol Pharmacol* 59: 9–15

58 Seabrook GR, Sutton KG, Jarolimek W, Hollingworth GJ, Teague S, Webb J, Clark N, Boyce S, Kerby J, Ali Z et al. (2002) Functional properties of the high-affinity TRPV1 (VR1) vanilloid receptor antagonist (4-hydroxy-5-iodo-3-methoxyphenylacetate ester) iodo-resiniferatoxin. *J Pharmacol Exp Ther* 303: 1052–1060

59 Rigoni M, Trevisani M, Gazzieri D, Nadaletto R, Tognetto M, Creminon C, Davis JB, Campi B, Amadesi S, Geppetti P, Harrison S (2003) Neurogenic responses mediated by vanilloid receptor-1 (TRPV1) are blocked by the high affinity antagonist, iodo-resiniferatoxin. *Br J Pharmacol* 138: 977–985

60 Appendino G, Harrison S, De Petrocellis L, Daddario N, Bianchi F, Schiano MA, Trevisani M, Benvenuti F, Geppetti P, Di Marzo V (2003) Halogenation of a capsaicin analogue leads to novel vanilloid TRPV1 receptor antagonists. *Br J Pharmacol* 139: 1417–1424

61 Wang Y, Szabo T, Welter JD, Toth A, Tran R, Lee J, Kang SU, Suh YG, Blumberg PM, Lee J (2002) High affinity antagonists of the vanilloid receptor. *Mol Pharmacol* 62: 947–956

62 Park HG, Park MK, Choi JY, Choi SH, Lee J, Park BS, Kim MG, Suh YG, Cho H, Oh U et al. (2003) Synthesis of N,N',N"-trisubstituted thiourea derivatives and their antagonist effect on the vanilloid receptor. *Bioorg Med Chem Lett* 13: 601–604

63 Suh YG, Lee YS, Min KH, Park OH, Seung HS, Kim HD, Park HG, Choi J, Lee J, Kang SW et al (2003) Novel non-vanilloid VR1 antagonist of high analgesic effects and its structural requirement for VR1 antagonistic effects. *Bioorg Med Chem Lett* 13: 4389–4393

64 Lee J, Lee J, Kang M, Shin M, Kim JM, Kang SU, Lim JO, Choi HK, Suh YG, Park HG et al (2003) N-(3-acyloxy-2-benzylpropyl)-N'-[4-(methylsulfonylamino)benzyl]thiourea analogues: novel potent and high affinity antagonists and partial antagonists of the vanilloid receptor. *J Med Chem* 46: 3116–3126

65 Hutchinson A, Desimone RW, Hodgetts KJ, Krause JE, White GG (2002) International Publication Number WO02/08221, World Intellectual Property Organization

66 Swanson DM, Dubin AE, Shah C, Nasser N, Chang L, Dax SL, Jetter M, Breitenbucher JG, Liu C, Mazur C et al (2005) Identification and biological evaluation of 4-(3-trifluoromethylpyridin-2-yl)piperazine-1-carboxylic acid (5-trifluoromethylpyridin-2-yl) amide, a high affinity TRPV1 (VR1) vanilloid receptor antagonist. *J Med Chem* 48: 1857–1872

67 Sun Q, Tafesse L, Islam K, Zhou X, Victory SF, Zhang C, Hachicha M, Schmid LA, Patel A, Rotshteyn Y et al. (2003) 4-(2-pyridyl)piperazine-1-carboxamides: potent vanilloid receptor 1 antagonists. *Bioorg Med Chem Lett* 13: 3611–3616

68 Pomonis JD, Harrison JE, Mark L, Bristol DR, Valenzano KJ, Walker K (2003) BCTC (N-(4-tertiarybutylphenyl)-4-(3-cholorphyridin-2-yl) tetrahydropryazine-1(2H)-carboxamide), a novel, orally-effective vanilloid receptor 1 antagonist with analgesic properties: II. *In vivo* characterization in rat models of inflammatory and neuropathic pain. *J Pharmacol Exp Ther* 306: 387–393

69 Gunthorpe MJ, Rami HK, Jerman JC, Smart D, Gill CH, Soffin EM, Luis HS, Lappin SC, Egerton J, Smith GD et al (2004) Identification and characterisation of SB-366791, a potent and selective vanilloid receptor (VR1/TRPV1) antagonist. *Neuropharmacology* 46: 133–149

70 Culshaw AJ, Gull P, Hallett A, Kim H-Y, Seiler MP, Zimmermann K, Liu Y, Prashad M (2002) International Publication Number WO02/076946, World Intellectual Property Organization

71 Meyer RA, Davis KD, Cohen RH, Treede RD, Campbell JN (1991) Mechanically insensitive afferents (MIAs) in cutaneous nerves of monkey. *Brain Res* 561: 252–261

72 Handwerker HO, Kilo S, Reeh PW (1991) Unresponsive afferent nerve fibres in the sural nerve of the rat. *J Physiol* 435: 229–242

73 Kress M, Koltzenburg M, Reeh PW, Handwerker HO (1992) Responsiveness and func-

tional attributes of electrically localized terminals of cutaneous C-fibers *in vivo* and *in vitro*. *J Neurophysiol* 68: 581–595

74 Amaya F, Oh-hashi K, Naruse Y, Iijima N, Ueda M, Shimosato G, Tominaga M, Tanaka Y, Tanaka M (2003) Local inflammation increases vanilloid receptor 1 expression within distinct subgroups of DRG neurons. *Brain Res* 963: 190–196

75 Simone DA, Baumann TK, Lamotte RH (1989) Dose-dependent pain and mechanical hyperalgesia in humans after intradermal injection of capsaicin. *Pain* 38: 99–107

76 Lamotte RH, Shain CN, Simone DA, Tsai EF (1991) Neurogenic hyperalgesia: psychophysical studies of underlying mechanisms. *J Neurophysiol* 66: 190–211

77 Simone DA, Sorkin LS, Oh U, Chung JM, Owens C, Lamotte RH, Willis WD (1991) Neurogenic hyperalgesia: central neural correlates in responses of spinothalamic tract neurons. *J Neurophysiol* 66: 228–246

78 Guo A, Vulchanova L, Wang J, Li X, Elde R (1999) Immunocytochemical localization of the vanilloid receptor 1 (VR1): relationship to neuropeptides, the P2X3 purinoceptor and IB4 binding sites. *Eur J Neurosci* 11: 946–958

79 Hudson LJ, Bevan S, Wotherspoon G, Gentry C, Fox A, Winter J (2001) VR1 protein expression increases in undamaged DRG neurons after partial nerve injury. *Eur J Neurosci* 13: 2105–2114

80 Fukuoka T, Tokunaga A, Tachibana T, Dai Y, Yamanaka H, Noguchi K (2002) VR1, but not P2X(3), increases in the spared L4 DRG in rats with L5 spinal nerve ligation. *Pain* 99: 111–120

81 Rashid MH, Inoue M, Bakoshi S, Ueda H (2003) Increased expression of vanilloid receptor 1 (VR1) on myelinated primary afferent neurons contributes to the antihyperalgesic effect of capsaicin cream in diabetic neuropathic pain in mice. *J Pharmacol Exp Ther* 304: 940–948

82 Jacobus WE, Taylor GJ, Hollis DP, Nunnally RL (1977) Phosphorus nuclear magnetic resonance of perfused working rat hearts. *Nature* 265: 756–758

83 Tracey DJ, Walker JS (1995) Pain due to nerve damage: are inflammatory mediators involved? *Inflamm Res* 44: 407–411

Index

The PIR-Series
Progress in Inflammation Research

Homepage: http://www.birkhauser.ch

Up-to-date information on the latest developments in the pathology, mechanisms and therapy of inflammatory disease are provided in this monograph series. Areas covered include vascular responses, skin inflammation, pain, neuroinflammation, arthritis cartilage and bone, airways inflammation and asthma, allergy, cytokines and inflammatory mediators, cell signalling, and recent advances in drug therapy. Each volume is edited by acknowledged experts providing succinct overviews on specific topics intended to inform and explain. The series is of interest to academic and industrial biomedical researchers, drug development personnel and rheumatologists, allergists, pathologists, dermatologists and other clinicians requiring regular scientific updates.

Available volumes:

T Cells in Arthritis, P. Miossec, W. van den Berg, G. Firestein (Editors), 1998
Chemokines and Skin, E. Kownatzki, J. Norgauer (Editors), 1998
Medicinal Fatty Acids, J. Kremer (Editor), 1998
Inducible Enzymes in the Inflammatory Response,
D.A. Willoughby, A. Tomlinson (Editors), 1999
Cytokines in Severe Sepsis and Septic Shock, H. Redl, G. Schlag (Editors), 1999
Fatty Acids and Inflammatory Skin Diseases, J.-M. Schröder (Editor), 1999
Immunomodulatory Agents from Plants, H. Wagner (Editor), 1999
Cytokines and Pain, L. Watkins, S. Maier (Editors), 1999
In Vivo *Models of Inflammation*, D. Morgan, L. Marshall (Editors), 1999
Pain and Neurogenic Inflammation, S.D. Brain, P. Moore (Editors), 1999
Anti-Inflammatory Drugs in Asthma, A.P. Sampson, M.K. Church (Editors), 1999
Novel Inhibitors of Leukotrienes, G. Folco, B. Samuelsson, R.C. Murphy (Editors), 1999
Vascular Adhesion Molecules and Inflammation, J.D. Pearson (Editor), 1999
Metalloproteinases as Targets for Anti-Inflammatory Drugs,
K.M.K. Bottomley, D. Bradshaw, J.S. Nixon (Editors), 1999
Free Radicals and Inflammation, P.G. Winyard, D.R. Blake, C.H. Evans (Editors), 1999
Gene Therapy in Inflammatory Diseases, C.H. Evans, P. Robbins (Editors), 2000
New Cytokines as Potential Drugs, S. K. Narula, R. Coffmann (Editors), 2000
High Throughput Screening for Novel Anti-inflammatories, M. Kahn (Editor), 2000
Immunology and Drug Therapy of Atopic Skin Diseases,
C.A.F. Bruijnzeel-Komen, E.F. Knol (Editors), 2000
Novel Cytokine Inhibitors, G.A. Higgs, B. Henderson (Editors), 2000
Inflammatory Processes. Molecular Mechanisms and Therapeutic Opportunities,
L.G. Letts, D.W. Morgan (Editors), 2000

Cellular Mechanisms in Airways Inflammation, C. Page, K. Banner, D. Spina (Editors), 2000
Inflammatory and Infectious Basis of Atherosclerosis, J.L. Mehta (Editor), 2001
Muscarinic Receptors in Airways Diseases, J. Zaagsma, H. Meurs, A.F. Roffel (Editors), 2001
TGF-β and Related Cytokines in Inflammation, S.N. Breit, S. Wahl (Editors), 2001
Nitric Oxide and Inflammation, D. Salvemini, T.R. Billiar, Y. Vodovotz (Editors), 2001
Neuroinflammatory Mechanisms in Alzheimer's Disease. Basic and Clinical Research, J. Rogers (Editor), 2001
Disease-modifying Therapy in Vasculitides, C.G.M. Kallenberg, J.W. Cohen Tervaert (Editors), 2001
Inflammation and Stroke, G.Z. Feuerstein (Editor), 2001
NMDA Antagonists as Potential Analgesic Drugs, D.J.S. Sirinathsinghji, R.G. Hill (Editors), 2002
Migraine: A Neuroinflammatory Disease? E.L.H. Spierings, M. Sanchez del Rio (Editors), 2002
Mechanisms and Mediators of Neuropathic pain, A.B. Malmberg, S.R. Chaplan (Editors), 2002
Bone Morphogenetic Proteins. From Laboratory to Clinical Practice, S. Vukicevic, K.T. Sampath (Editors), 2002
The Hereditary Basis of Allergic Diseases, J. Holloway, S. Holgate (Editors), 2002
Inflammation and Cardiac Diseases, G.Z. Feuerstein, P. Libby, D.L. Mann (Editors), 2003
Mind over Matter – Regulation of Peripheral Inflammation by the CNS, M. Schäfer, C. Stein (Editors), 2003
Heat Shock Proteins and Inflammation, W. van Eden (Editor), 2003
Pharmacotherapy of Gastrointestinal Inflammation, A. Guglietta (Editor), 2004
Arachidonate Remodeling and Inflammation, A.N. Fonteh, R.L. Wykle (Editors), 2004
Recent Advances in Pathophysiology of COPD, P.J. Barnes, T.T. Hansel (Editors), 2004
Cytokines and Joint Injury, W.B. van den Berg, P. Miossec (Editors), 2004
Cancer and Inflammation, D.W. Morgan, U. Forssmann, M.T. Nakada (Editors), 2004
Bone Morphogenetic Proteins: Bone Regeneration and Beyond, S. Vukicevic, K.T. Sampath (Editors), 2004
Antibiotics as Anti-Inflammatory and Immunomodulatory Agents, B.K. Rubin, J. Tamaoki (Editors), 2005
Antirheumatic Therapy: Actions and Outcomes, R.O. Day, D.E. Furst, P.L.C.M. van Riel, B. Bresnihan (Editors), 2005
Regulatory T-Cells in Inflammation, L. Taams, A.N. Akbar, M.H.M Wauben (Editors), 2005
Sodium Channels, Pain, and Analgesia, K. Coward, M. Baker (Editors), 2005